Structural and Comparative Inorganic Chemistry

also by Peter R. S. Murray
Principles of Organic Chemistry
Pan Study Aid: Advanced Chemistry

Structural and Comparative Inorganic Chemistry

A Modern Approach for Schools and Colleges

Peter R. S. Murray, B.Sc., C.Chem., M.R.S.C.
Peel College, Bury

P. R. Dawson, B.Sc.
Head of Science, Margaret Ashton College, Manchester

Heinemann Educational Books
London

Heinemann Educational Books Ltd
22 Bedford Square, London WC1B 3HH
LONDON EDINBURGH MELBOURNE AUCKLAND
HONG KONG SINGAPORE KUALA LUMPUR NEW DELHI
IBADAN NAIROBI JOHANNESBURG
EXETER (NH) KINGSTON PORT OF SPAIN

ISBN 0 435 65644 9

First published 1976
Reprinted with corrections 1978, 1980, 1982, 1984

Filmset by Keyspools Limited, Golborne, Lancashire.
Printed and bound in Great Britain by
Richard Clay (The Chaucer Press) Ltd,
Bungay, Suffolk

Preface

In writing this book, our primary objective has been to produce a textbook of inorganic chemistry which meets the requirements of the Advanced and Scholarship level examinations. The contents more than adequately cover the theoretical aspects of the O.N.C. and O.N.D. syllabuses, and the considerably extended treatment of some of the more important topics should be useful introductory reading for the first-year undergraduate.

The book provides a comparative treatment of modern inorganic chemistry based upon the periodic classification of the elements, drawing attention particularly to the trends and patterns which emerge with regard to both physical and chemical characteristics. In the opening chapters, the emphasis is on the theory of structure and simple thermodynamics, an understanding of which is essential if inorganic chemistry is to have any real meaning. The principles of oxidation-reduction and electrode potentials are introduced and used to explain the reasons for the different techniques adopted for the extraction of metals from their ores.

It is, of course, not possible in a book of this size to include the whole gamut of factual material that one associates with inorganic chemistry. The facts have, therefore, been carefully selected and blended with the fundamental theoretical concepts so as to encourage an understanding of the subject rather than the acquisition of what sometimes appears to be an agglomeration of unrelated facts. It is hoped that this has been achieved without detracting too much from the reference value of the book. Although a number of industrial processes are mentioned, and others briefly explained, no attempt has been made to give a detailed discussion of the technical aspects.

The book is equipped with a number of questions which it is hoped will enable the student to make an objective appraisal of his understanding of the subject.

Nomenclature is in accordance with the recommendations made in the report by the Association for Science Education, *Chemical Nomenclature, Symbols, and Terminology (1972)*.

We would like to convey our gratitude to Hamish MacGibbon of Heinemann Educational Books, and to his advisers, especially Martyn Berry; their many constructive criticisms during the preparation of the manuscript have been invaluable in helping to mould the book into its present form.

P.R.S.M.
1976 P.R.D.

Most of the corrections to this reprint (1980) have been made to bring the book into line with the report by the Association for Science Education, *Chemical Nomenclature Symbols, and Terminology* (1979).

1980 P.R.S.M.
P.R.D.

Contents

1 Electronic Structure and Atomic Orbitals

Basic atomic structure

All atoms consist of a central positively charged nucleus surrounded by negatively charged electrons. With the sole exception of the normal hydrogen atom, whose nucleus comprises a single proton, the nucleus invariably contains neutrons, which together with the protons make up virtually the whole mass of the atom. The number of protons in the nucleus is referred to as the *atomic number* of the element, and the sum of the number of protons and neutrons (these particles being known collectively as *nucleons*) is referred to as the *mass number*. Atoms of the same element differing in mass number, i.e. possessing different numbers of neutrons, are called *isotopes*. For the simpler elements, the number of neutrons is generally about the same as the number of protons; but as the size of the nucleus increases the number of neutrons increasingly exceeds the number of protons and, in certain cases, causes instability of the nucleus, resulting in radioactive emission.

The proton is a particle possessing unit positive electrical charge and is of approximately the same mass as the neutron, which is electrically neutral. Each of the electrons possesses unit negative charge and has a mass about 1/1840 that of either a proton or a neutron. The number of electrons in a neutral atom must be equal to the number of nuclear protons.

The present-day interpretation of electronic theory is fundamentally mathematical. The concepts are derived from wave mechanical calculations, but before a qualitative interpretation of them is considered it is desirable to consider a brief history of the evolution of ideas of electronic structure, starting with the evidence provided by atomic spectra.

Atomic spectra

Each form of electromagnetic radiation is distinguished by its energy, E, which is related to the frequency, ν, by the Einstein-Planck equation:

$$E = h\nu$$

where h is Planck's constant and has a value of $6.626\,196 \times 10^{-34}$ J s.

The frequency is related to the wavelength, λ, and the velocity of propagation of the radiation, c, by the expression:

$$\nu = \frac{c}{\lambda}$$

Hence,
$$E = \frac{hc}{\lambda}$$

Electromagnetic radiation containing all possible wavelengths over a wide range, such as ordinary light, is referred to as a *continuous spectrum* or *continuous radiation* (Figure 1.1).

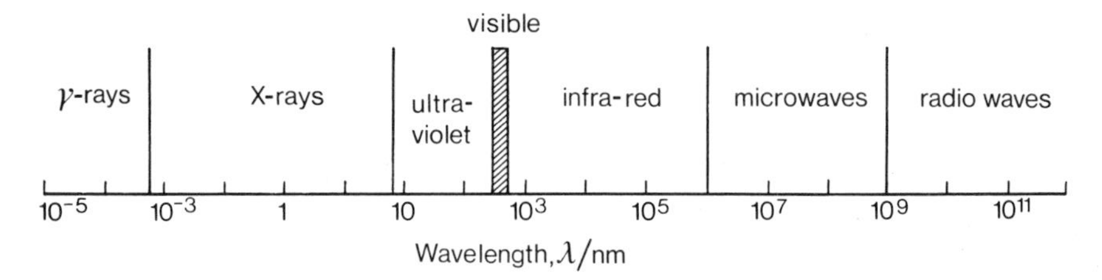

Fig. 1.1 The electromagnetic spectrum

Emission spectra

If the atoms of elements are provided with sufficient energy, they emit light, often of a distinctive and characteristic colour. Gaseous elements produce this effect at low pressures in discharge tubes when subjected to a suitably high potential difference; this is, for example, the basic principle of neon lights. Volatile metallic elements, such as sodium (yellow) and mercury (blue-green), both of which are used in street lighting, can also produce this effect under similar conditions. This phenomenon is not confined solely to discharge tubes; atoms can be excited by heating elements or their compounds in a flame (*flame tests*), by electronic bombardment, or by striking an electric arc with the metal to be investigated forming part of one of the electrodes.

If a discharge is passed through a gas at low pressure in a vacuum tube and the spectrum analysed by a spectrometer, the spectrum is observed not as a continuum but as a series of differently coloured narrow lines against a black background. These may be recorded photographically. In acquiring energy, the atoms become electronically excited (i.e. electrons are promoted to higher energy levels) and then, when they return to lower energy states, they emit radiation. As will be seen, the atoms need not revert in one step to their lower energy state and radiation of several different wavelengths may be emitted. This type of spectrum is referred to as a **line emission spectrum**.

From the equation, $E = hc/\lambda$, it will be seen that each line in a line spectrum corresponds to a definite wavelength. As the lines of the spectrum are a characteristic of the atom, and hence of the element, they may be used for purposes of identification.

Absorption spectra

An **absorption spectrum** can be obtained by passing continuous radiation through an atomic vapour and analysing emerging radiation with a spectrometer. The transmitted light beam is found to be deficient in certain wavelengths as a result of their absorption by the atoms of the atomic vapour as electrons are excited to higher energy levels. The absorptions are observed as narrow black lines against the colours of the visible spectrum. Again, as with emission spectra, these lines may be recorded photographically. The positions of the lines in the absorption spectrum of an atom coincide exactly with those of the emission spectrum of the same element, although they tend to be fewer in number and therefore more easily identified.

An example of an absorption spectrum is provided by the Sun's radiation. When sunlight is observed through a spectrometer, certain wavelengths, corresponding to the so-called Fraunhofer lines, are found to be missing. This is attributable to their absorption by the hydrogen in the Sun's atmosphere.

Atomic spectra are very complex and, under certain conditions, are made more so by the formation of positive ions (see below) of the element, which in turn give rise to their own spectral lines. The simplest of all line spectra is given by hydrogen, and most of the early work in this field was based upon this element.

Similarities between spectra

If increased potentials are used with the discharge tube method of excitation, then new series of lines may be produced with the same gas, these occurring at shorter wavelengths for successively higher potentials. These new spectra result from the loss of one or more electrons from the atoms, forming positive ions, each of which produces a spectrum which is in accordance with the number of electrons that it contains. For example, the normal, or arc, spectrum of the sodium atom, Na, is very similar to the first spark spectrum (singly charged ion) of magnesium, Mg^+, and the second spark (doubly ionized) spectrum of aluminium, Al^{2+}.

Table 1.1

Element or ion	*Atomic number, Z*	*Number of electrons present*
Na	11	11
Mg^+	12	11
Al^{2+}	13	11

The spectrum of hydrogen

Balmer (1885), from observations of monohydrogen (atomic hydrogen) in the visible spectrum of the Sun, showed that the wavelengths of lines in the absorption spectrum of hydrogen are related by the expression:

$$\frac{1}{\lambda} = \text{a constant}\left(\frac{1}{2^2} - \frac{1}{n^2}\right)$$

where n is an integer equal to 3, 4, 5, etc.

This equation was modified by Ritz (1908) who, in his **combination principle**, represented it in the more general form:

$$\frac{1}{\lambda} = R_H \left(\frac{1}{n_1^2} - \frac{1}{n_2^2} \right)$$

The Rydberg constant, R, (after Rydberg, 1890) has units of reciprocal length and is a constant for each element, the latter being denoted by its symbol written as a subscript to the constant. n_1 and n_2 are integers, n_1 being a constant for a given series and n_2 equal to $(n_1 + x)$, where x is also an integer (Table 1.2).

For hydrogen, the Rydberg constant, R_H, is equal to $1.096\,776 \times 10^7\ m^{-1}$ and is one of the most accurately known of all physical constants.

In 1906, Lyman discovered another series of lines for hydrogen in the ultra-violet region. Others were subsequently found in the infra-red spectrum (Paschen, Brackett, and Pfund series), all of which correlated with the Ritz equation (Table 1.2).

Table 1.2

Observer	n_1	n_2	*Spectral region*
Lyman (1906)	1	2, 3, 4, ... , ∞	Ultra-violet
Balmer (1885)	2	3, 4, 5, ... , ∞	Visible
Paschen (1908)	3	4, 5, 6, ... , ∞	Infra-red
Brackett (1922)	4	5, 6, 7, ... , ∞	Infra-red
Pfund (1924)	5	6, 7, 8, ... , ∞	Infra-red

The values of the wavelengths for different values of n_2 given by the Ritz equation for the Balmer series ($n_1 = 2$; see Figure 1.2) are quoted in Table 1.3.

Table 1.3

n_2	*Wavelength, λ/nm*	*Frequency, ν/Hz*
3	656.6	4.568×10^{14}
4	486.4	6.167×10^{14}
5	434.4	6.907×10^{14}
6	410.5	7.309×10^{14}

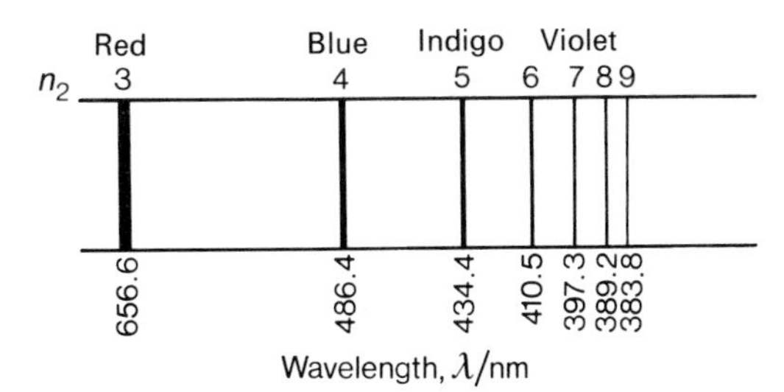

Fig. 1.2 Balmer series, depicting the principal lines in the visible atomic spectrum of hydrogen

As the value of n_2 increases, the lines become closer together, and eventually approach a convergence limit; this corresponds to the point at which the electron leaves the atom and ionization occurs, i.e. n_2 is equal to infinity (see page 80). Beyond this point in the spectrum, *continuous absorption* occurs.

The electronic transitions responsible for the different spectral series are illustrated in Figure 1.3, the direction of the arrows indicating that they correspond to emission spectra; i.e. the electrons are depicted as moving from a higher to a lower energy level.

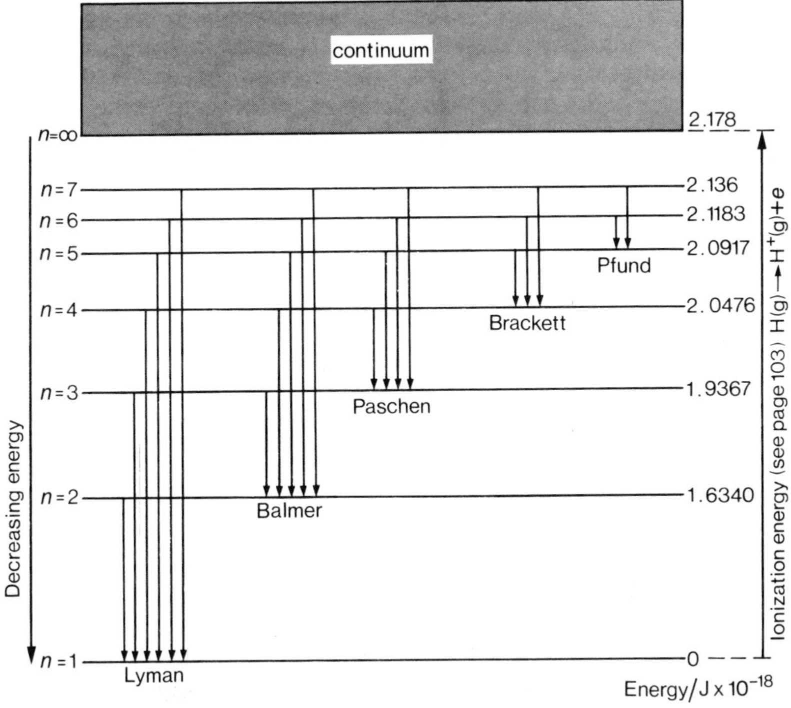

Fig. 1.3 Energy level diagram showing the relationship between the spectral series for the hydrogen atom

Bohr's interpretation of the hydrogen spectrum

Rutherford's planetary model of the atom as a massive nucleus with orbiting electrons moving at various speeds had severe limitations, not least of which was its inability to explain why the electrons did not ultimately fall into the nucleus. However, in 1913, Bohr, who was a colleague of Rutherford at the time, provided an explanation on the basis of the evidence of atomic spectra and the significance of the Ritz equation.

As spectral lines always occur in precisely the same part of the spectrum for each particular element, Bohr proposed in his quantum theory that orbital electrons can emit radiation only in certain definite, or *quantized*, amounts and cannot have all possible energy levels. That is to say, whenever an electron is transferred from one energy level to another, a definite integral amount of energy, a **quantum,** corresponding to a particular frequency of radiation, is absorbed or emitted. The essence of this is that the energy of a body can only change in definite whole-number multiples of this unit of energy, and this energy is emitted or absorbed only when an electron undergoes a transition from one energy level to another.

$$\text{Energy change} = x \times \text{Quantum}$$

where x is a whole number.

The energy emitted by an electron in undergoing a transition between two energy states, E_1 and E_2, appears as a *photon*, and is given by the Einstein-Planck equation:

$$E_1 - E_2 = h\nu$$

For example, if this equation is applied to the red line of the hydrogen spectrum in the Balmer series ($n_1 = 2$ and $n_2 = 3$), which has a wavelength of 656.6 nm and frequency of 4.568×10^{14} Hz (Table 1.3), then the energy of the photon emitted in this case is given by:

$$6.626\,196 \times 10^{-34} \times 4.568 \times 10^{14}\ \text{J}$$
$$= 3.027 \times 10^{-19}\ \text{J}$$

The energies of photons corresponding to the other lines in the spectrum can be calculated in a similar manner.

The state of lowest energy, or ground state, of the hydrogen atom is that for which n_1 of the Ritz equation is equal to unity, and the values denoted by n_2 characterize the higher energy states. The integer assigned to each of these different energy states is called the **principal quantum number**, n, which in addition to determining the energy, defines the velocity and radius of the electron. Bohr also proposed that, as long as an electron remains in a given circular orbit, it neither radiates nor absorbs energy.

As the number of electrons of the atoms of the different elements increases, so also does the number of electron 'shells' or energy levels, the number of electrons possible in each level having a prescribed maximum. The innermost shell is defined by $n = 1$ (or 'K shell') and contains a maximum of two electrons. The second shell is defined by $n = 2$ ('L shell') and contains a maximum of eight electrons. The third shell, $n = 3$ ('*M* shell'), has a maximum of eighteen electrons; the maximum number of electrons in further shells, determined by energy considerations, becomes progressively greater. As a simple aide-memoire, the maximum number of electrons in any shell is given by the formula $2n^2$, where n is the principal quantum number (Table 1.4).

Table 1.4

	Shell			
	K	*L*	*M*	*N*
Principal quantum number, n	1	2	3	4
Maximum number of electrons ($2n^2$)	2	8	18	32

The negatively-charged orbital electrons are attracted towards the positively-charged nuclear protons, the electrons nearest to the nucleus being more firmly held by the strong attractive forces, whereas those further away are less firmly held because of the greater distances over which the forces operate. The more firmly held electrons have a lower potential energy while those less firmly held have a higher potential energy. Therefore, moving further away from the nucleus, each shell represents electrons of progressively higher energy.

Ionization energy from atomic spectra

The amount of energy required to excite the least firmly held electron of a gaseous atom in its lowest energy state (known as its *ground state*) to infinity is referred to as the *ionization energy* (see page 80). In the Lyman series, which is the only series in which the single hydrogen electron is excited from the lowest energy level, i.e. $n = 1$, the wavelength of the convergence limit is therefore equal to the reciprocal of the Rydberg constant:

$$\frac{1}{\lambda} = 1.096\,776 \times 10^7\ \mathrm{m^{-1}}$$

By the Einstein-Planck equation, the energy, ΔE, required for the electronic transition from energy E_1 to E_∞, is given by:

$$E_1 - E_\infty = h\nu = \frac{hc}{\lambda}$$

c being the velocity of light. Therefore:

$$\Delta E = (1.096\,776 \times 10^7\ \mathrm{m^{-1}}) \times (6.626\,196 \times 10^{-34}\ \mathrm{J\,s}) \times (2.997\,925 \times 10^8\ \mathrm{m\,s^{-1}})$$
$$= 2.180 \times 10^{-18}\ \mathrm{J}$$

The molar ionization energy (represented by the enthalpy term, ΔH_{IE}) is then equal to 2.180×10^{-18} J multiplied by the *Avogadro constant*, L (or N_A), this being the number of molecules present in one mole of gas.

$$\begin{aligned} \Delta H_{IE} &= \Delta E \times L \\ &= 2.180 \times 10^{-18} \times 6.022 \times 10^{23}\,\text{J mol}^{-1} \\ &= 1.3 \times 10^{6}\,\text{J mol}^{-1} \\ &= 1300\,\text{kJ mol}^{-1} \end{aligned}$$

The Sommerfeld theory – subsidiary quantum numbers

Bohr's theories adequately explained the existence of the most distinctive lines in the hydrogen spectrum, but Paschen, using a high resolution instrument, discovered that these sharp lines were in fact split into groups of extremely fine lines very closely separated, indicating electronic sub-levels of very small energy difference. Furthermore, careful examination of the atomic spectra of elements possessing more than one extranuclear electron shows them to form distinct series of lines, referred to as the sharp (*S*), primary (*P*), diffused (*D*), and fundamental (*F*) series. Later, the letters *G* and *H* were added to describe two further series.

Sommerfeld (1915) amended the Bohr idea by postulating elliptical as well as circular electronic orbits, explaining the fine structure in terms of electronic transitions between sub-levels. This necessitated introducing a subsidiary quantum number, k, which was later, with the advent of wave mechanics, replaced by the **azimuthal quantum number**, l.

$$l = 0, 1, 2, 3, \ldots, (n-1)$$

The various paths or eccentricities of the electrons, represented by the subsidiary quantum number l, are customarily designated by the letters s, p, d, and f, these being derived from the spectral terms to which they relate.

When $l = 0$ the electrons are called s electrons,
when $l = 1$ the electrons are called p electrons,
when $l = 2$ the electrons are called d electrons,
and when $l = 3$ the electrons are called f electrons.

Even so, not all of the fine lines could be accounted for by the introduction of this single sub-level, and eventually it was found that two more quantum numbers were needed to explain the magnetic and spin properties of the electrons.

The influence of a magnetic field on spectra

If the hydrogen spectrum is studied under the influence of a magnetic field, further splitting of the spectrum occurs (the *Zeeman effect*), leading to the introduction of a **magnetic quantum number**, m_l, where m_l has values of $-l$ to $+l$: e.g. for $l = 2$, m_l has values of $-2, -1, 0, +1$, and $+2$.

Electron spins

If the lines in atomic spectra are examined carefully under high resolution and in the absence of a magnetic field, many of the lines are observed as multiplets of two or more very closely spaced lines. This effect is interpreted in terms of electron spins which give rise to a fourth quantum number, the **spin quantum number**, m_s, which is capable of one of only two values, $+\frac{1}{2}$ or $-\frac{1}{2}$. In the same way that the Earth rotates about its own axis as it orbits the Sun, an electron is thought to spin on its own axis as it orbits the nucleus (Figure 1.4).

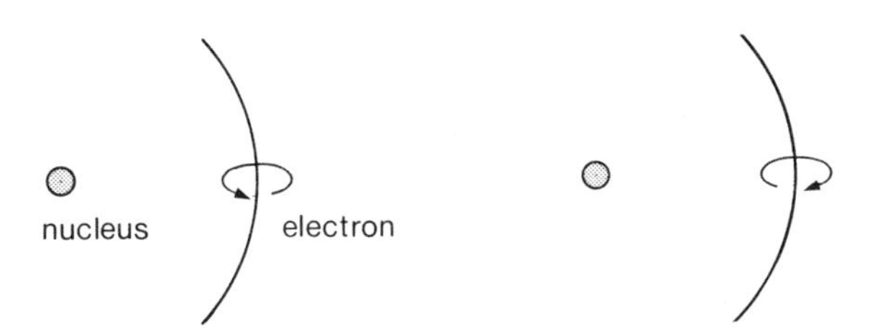

Fig. 1.4 Opposed or paired electron spins

Any two electrons defined by the same n, l, and m_l quantum numbers must have different m_s values (see Pauli exclusion principle below), denoting the clockwise and anti-clockwise spins of the electrons. The spins of two electrons in any one orbital are said to be *opposed* or *paired*.

Molecular spectra

The spectra of molecules are often referred to as *band spectra* (as opposed to line spectra for atoms) as they appear in the form of bands. However, further analysis with a high resolution instrument shows these bands to consist of a large number of closely spaced lines. In order to be able to account for all of these lines, it is necessary to assume that, in addition to the electronic transitions, vibrational and rotational energies of the molecules are quantized.

With the exception of certain diatomic molecules, band spectra are virtually always observed as absorption spectra, as the comparatively high temperatures required for emission spectra would be such as to bring about the dissociation of the molecules.

Electronic structure of atoms

Pauli exclusion principle

Pauli (1925), from his observations on atomic spectra, came to the purely empirical conclusion that *no two electrons in the same atom could be defined by the same four quantum numbers.*

Electronic groupings

The total permissible number of electrons in any one principal quantum level can be determined by considering the four quantum numbers and applying the Pauli principle. For example, the two electrons defined by $n = 1$ and the eight electrons defined by $n = 2$ are distinguished by the quantum numbers l, m_l, and m_s as follows:

For $n = 1$,

l	$= 0$	0
m_l	$= 0$	0
m_s	$= +\frac{1}{2}$	$-\frac{1}{2}$
	s electrons	

For $n = 2$,

l	$= 0$	0	1	1	1	1	1	1
m_l	$= 0$	0	-1	-1	0	0	$+1$	$+1$
m_s	$= +\frac{1}{2}$	$-\frac{1}{2}$	$+\frac{1}{2}$	$-\frac{1}{2}$	$+\frac{1}{2}$	$-\frac{1}{2}$	$+\frac{1}{2}$	$-\frac{1}{2}$
	s electrons				*p* electrons			

A simple way of describing the electrons defined by these first two principal quantum numbers is given by the following notation:

$$1s^2 2s^2 2p^6$$

where the prefix to each letter denotes the principal quantum number, the letters *s* and *p* the type of electron (according to the azimuthal quantum number), and the superscript the number of electrons at each level.

The first four principal quantum levels, when possessing the maximum number of electrons, are described by:

$n = 1$,	(*K* shell)	$1s^2$	(two electrons)
$n = 2$,	(*L* shell)	$2s^2\ 2p^6$	(eight electrons)
$n = 3$,	(*M* shell)	$3s^2\ 3p^6\ 3d^{10}$	(eighteen electrons)
$n = 4$,	(*N* shell)	$4s^2\ 4p^6\ 4d^{10}\ 4f^{14}$	(thirty-two electrons)

Energy levels

Interpretation of the complex spectral patterns observed for different elements indicates that in certain cases the energy levels of electrons defined by a particular principal quantum number overlap with those defined by another. For example, the 4*s* electrons are of lower energy than the 3*d*, and similarly the 5*s* are of lower energy than the 4*d*. Figure 1.5 depicts the approximate relative energy levels of the different electrons, the relative energy of each category of electrons being represented by a horizontal line, and the maximum number of electrons with similar energy being specified in the brackets.

Levels of approximately the same energy content are described as **valence shells** which, unlike the principal quantum shells, need not necessarily be

Potential energy of electrons (increasing upwards)

5p (six e) 4d (ten e) 5s (two e)	Valence shell 5 (eighteen electrons)
4p (six e) 3d (ten e) 4s (two e)	Valence shell 4 (eighteen electrons)
3p (six e) 3s (two e)	Valence shell 3 (eight electrons)
2p (six e) 2s (two e)	Valence shell 2 (eight electrons)
1s (two e)	Valence shell 1 (two electrons)

Fig. 1.5

defined by the same principal quantum number. It is, however, important to appreciate that, despite the terminology, it is only those electrons in the higher energy valence shells which actually determine the valency or oxidation state (see page 64) of an element. In general, electrons may be expected to fill the various levels in the ascending sequence depicted in Figure 1.6, although discrepancies arise for certain elements of high atomic number.

The following scheme, in which the parallel diagonal lines join together electrons of progressively higher energy, provides a useful guide to the way in which the ground state electronic configuration of an atom is attained.

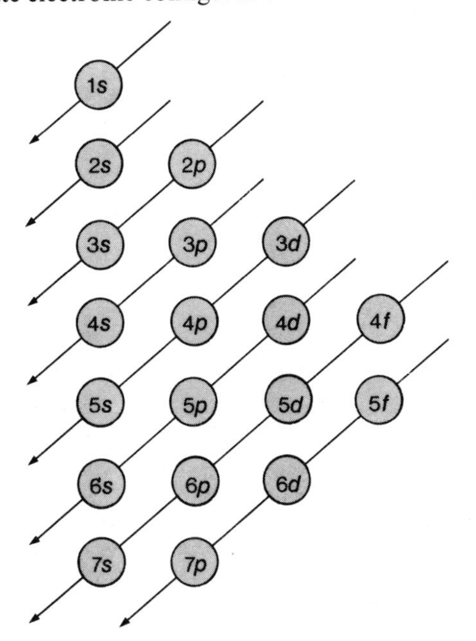

Fig. 1.6

For example, the electronic structure of the bromine atom (atomic number 35) in its lowest energy or ground state, may be written as follows to show the order of the electron energy levels:

Bromine atom, ground state: $1s^2 2s^2 2p^6 3s^2 3p^6 4s^2 3d^{10} 4p^5$

Modern concepts of electronic theory

Probably the greatest limitation of the Rutherford-Bohr theory of electronic structure is its inability to account satisfactorily either for the binding forces between atoms or for the shapes of molecules.

In 1924, de Broglie postulated the *dual nature of electrons.* Based upon the evidence that a beam of electrons can be diffracted by means of a crystal lattice, which serves as a diffraction grating, he suggested that electrons may be thought of as behaving as waves of radiation. Further, Heisenberg (1927), in his **uncertainty principle**, stated that *it is impossible to know both the velocity, or any related property such as energy or momentum, and the position of any particle (such as an electron) at the same time.* This may be interpreted in a more general context as meaning that it is not possible to measure the physical properties of materials without actually interfering with, and therefore modifying, the properties themselves.

Schrödinger (1926) provided a new way of looking at electronic structure, a way which is based on the wave behaviour of the electron and which is much more compatible with the uncertainty principle. Using a branch of mathematics known as wave mechanics, he derived a wave equation which is the basis for present-day interpretations of electronic structure.

The significance of the wave equation is that the solution of it provides *a means of measuring the probability of locating an electron in a particular volume of space.* However, as the number of electrons in an atom increases beyond three or four, the mathematical complexity of the problem requires the introduction of simplifying approximations. The implication of these is of no real significance here, but the solutions of the calculations, treated qualitatively, provide a considerable insight into the study of molecular formulae, the nature and strength of chemical bonds, and also the shapes of molecules.

Atomic orbitals

The volume of space, determined from wave mechanical calculations, *where there is a high probability of finding a particular electron* is called an **orbital**; each orbital can accommodate no more than two electrons.

As has already been seen, each type of electron, whether it be an *s*, *p*, *d*, or *f* electron is defined primarily by the azimuthal quantum number (see page 8). Atomic orbitals containing these different types of electrons can be represented by means of three-dimensional cartesian graphs, with the nucleus as the origin.

When $l = 0$, the orbital is spherical about the nucleus and is called an *s* orbital; the chance of finding a particular electron is the same in all directions (Figure 1.7).

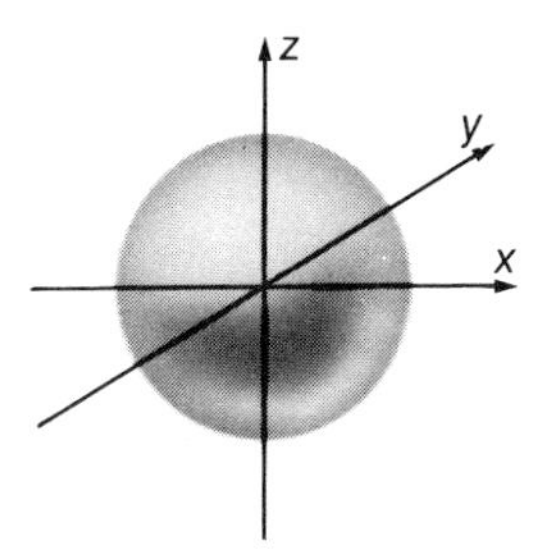

Fig. 1.7

For the hydrogen atom in its ground state (i.e. $n = 1$), a different type of graph showing the relative probabilities of finding the single 1*s* electron at progressively greater distances from the nucleus is illustrated in Figure 1.8. The greatest probability of finding the electron is at a distance of 0.053 nm from the nucleus, which is a value equal to the radius of the first Bohr orbit. The chance of finding any electron in the nucleus is nil.

Solutions of the wave equation enable similar electron probability distribution profiles to be plotted for the hydrogen electron in excited states. Figure 1.9 shows the probability of locating it in the 2*s* level.

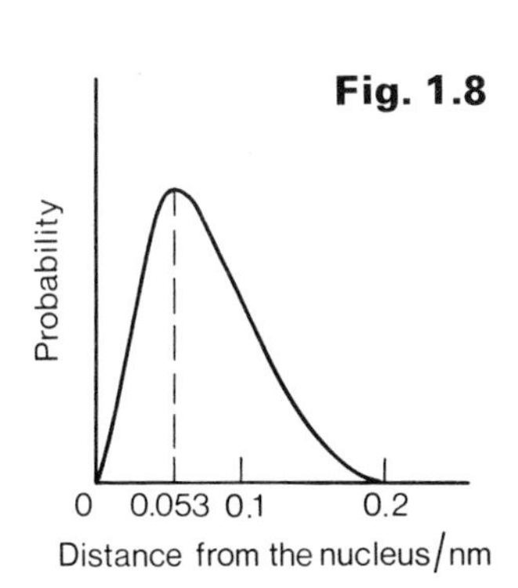

Fig. 1.8

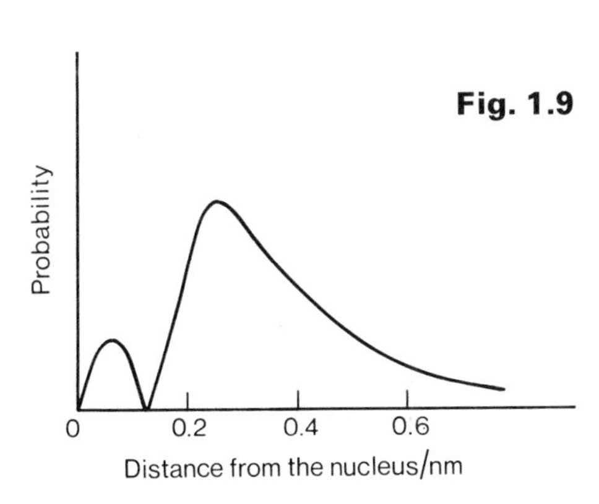

Fig. 1.9

Between spherical *s* orbitals of different energy, there is a region in which the probability of finding an *s* electron becomes negligible and approaches zero. This volume of space is called a *node*.

The boundary surface of an atomic orbital is drawn such that the chance of finding a particular electron within this region is about 95 per cent.

When $l = 1$, the orbital is similar in shape to a dumb-bell and is called a *p* orbital. There are three exactly similar *p* orbitals in any one principal quantum level, and as each orbital can contain a maximum of only two electrons, there are, at the most, six p electrons in any one level. These are divided between three dumb-bell shaped orbitals which are arranged mutually at right angles to each other, i.e. along x, y, and z axes. These orbitals are called p_x, p_y, and p_z respectively (Figure 1.10). All three p orbitals (which are of equivalent energy) are of slightly higher energy than the s orbital governed by the same principal quantum number.

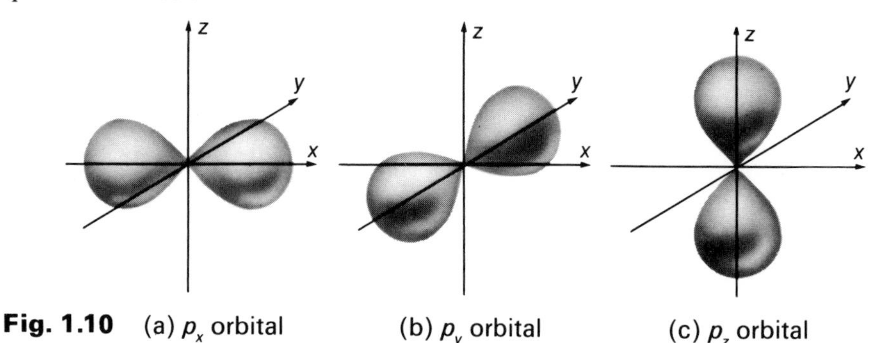

Fig. 1.10 (a) p_x orbital (b) p_y orbital (c) p_z orbital

When $l = 2$, the orbital is called a *d orbital*. In any one level, there are five d orbitals; again each contains a maximum of two electrons, and therefore a maximum of up to ten electrons can be accommodated. The shape of these orbitals is more complex than either an s or a p orbital, but is of little concern here.

The orbitals described by $l = 3$ are referred to as *f orbitals*. There are seven of these, accommodating a maximum of fourteen electrons.

Hund's rule of maximum multiplicity

As electrons are negatively charged, they have a natural tendency to repel each other and distribute themselves between available orbitals of equivalent energy so as to be as far apart as possible. Therefore, when filling p, d, or f orbitals defined by the same principal quantum number, the electrons enter each of the available orbitals before pairing occurs, this process being one

Table 1.5

Element	*Atomic number*	*Electronic structure (ground state)*
Boron	5	$1s^2 2s^2 2p_x^1$
Carbon	6	$1s^2 2s^2 2p_x^1 2p_y^1$
Nitrogen	7	$1s^2 2s^2 2p_x^1 2p_y^1 2p_z^1$
Oxygen	8	$1s^2 2s^2 2p_x^2 2p_y^1 2p_z^1$
Fluorine	9	$1s^2 2s^2 2p_x^2 2p_y^2 2p_z^1$
Neon	10	$1s^2 2s^2 2p_x^2 2p_y^2 2p_z^2$

which requires a certain amount of extra energy. This is in accordance with **Hund's rule of maximum multiplicity**, which is based upon spectroscopic observations and states, that *in the ground state of an atom, the number of unpaired electrons in a given energy level should be a maximum.* Thus the electronic structures of the elements boron to neon in the Periodic Table are represented by the notations shown in Table 1.5.

It is sometimes more convenient to represent the electronic structures diagramatically, using arrows for electrons and boxes for energy levels. For example:

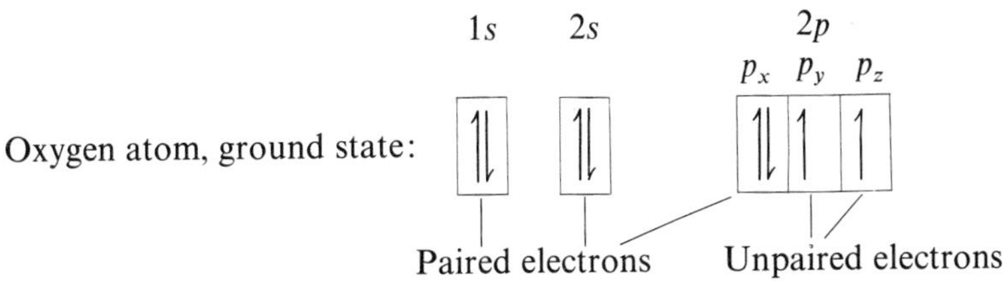

The process of feeding electrons into orbitals of successively higher energy in this way until the electronic structure of an element is complete is in accordance with the *Aufbau* (German for 'building-up') *principle.* The arrows pointing in opposite directions indicate the opposed spins of the two electrons in any one orbital.

It is not until electrons start to fill the orbitals defined by the principal quantum number 3 (*M* shell), which contains a maximum of eighteen electrons, that the filling of *d* orbitals commences.

2 Covalent Bonding and Molecular Structure

This chapter outlines the basic principles involved in the formation of some of the simpler covalent molecules of the *s*- and *p*- block elements (see page 77). Certain covalent structures involving the *d* block elements are discussed later in Chapters 16 and 17. The examples considered here are intended to be purely illustrative and no attempt is made to provide a comprehensive study of molecular structures. It is worth noting, however, that in confining our interest to elements of lower atomic number of the *s*- and *p*-blocks, we are at the same time focusing attention upon many of those elements which have the greatest tendency to form covalent bonds within their respective groups (see page 75).

As the Group I elements (outer electronic structure ns^1) form compounds which are predominantly ionic in character, the compounds of this group are not considered here.

The formation of covalent bonds

Valence bond approach

When a covalent bond is formed, electrons are shared between two atoms. These atoms must approach each other sufficiently closely for an outer orbital of one atom to overlap with an outer orbital of the other. Because each orbital can contain a maximum of two electrons, it follows that in the formation of an ordinary covalent bond an orbital containing a single electron of one atom overlaps with a singly occupied orbital of the other, giving both atoms a share in two bonding electrons. The greater the degree of overlapping between atomic orbitals, the stronger the bond formed. These are the basic concepts of the **valence bond** approach to covalent bond formation.

One of the main disadvantages of this approach, however, is that a single, simple valence bond structure cannot always adequately represent the distribution of electrons within a molecule. Instead it is necessary to represent the molecule by various alternative structures, or *canonical forms*, with the actual structure lying somewhere between the extremes. The actual structure is referred to as a *resonance hybrid* of these canonical forms and is not correctly represented by any one of them. The relative contribution of each canonical form to the resonance hybrid depends upon its energy content, the more stable forms making the greatest contribution. The resonance hybrid is more stable than any one of the various canonical forms, and the difference in energy between the hybrid and the most stable canonical form is known as the *resonance* (or *delocalization*) *energy*. Above all else, it is important to realize that the structure of the resonance hybrid is intermediate between the various canonical forms and is not a mixture of them.

For example, the carboxylate anion, $RCOO^-$, contains two equivalent carbon-oxygen bonds which are intermediate in nature between a single and a double bond, and is therefore not accurately represented by either of the two canonical forms (Figure 2.1):

```
      O⁻            O
     /             //
R—C ←——————→ R—C           (←——→ represents
     \\             \         resonance between
      O              O⁻       structures)
```

Fig. 2.1

Molecular orbital approach

An alternative approach to covalent bond formation is to consider that the overlapping of two atomic orbitals results in the formation of two **molecular orbitals**, one having a lower energy value (i.e. greater stability) than the energies of the constituent atomic orbitals, and the other having a higher energy (i.e. less stability). The molecular orbital of lower energy is referred to as the *bonding orbital*, and the one of higher energy is called the *anti-bonding orbital.* The molecular orbitals are associated with, and envelop, the nuclei of the participating atoms and not the separate atoms, as is the case with atomic orbitals. These combined nuclei are referred to as the *united atom.*

Both the Aufbau principle (see page 15) of feeding electrons into orbitals of successively higher energy, and the Pauli principle (see page 9) for defining electrons, apply to molecular orbitals in the same way as they do to atomic orbitals. The two electrons which are to form the covalent linkage can normally be expected to occupy the more stable bonding molecular orbital and, as a result, the anti-bonding orbital is not considered further in this discussion of stable covalent bond formation.

As both the valence bond and the molecular orbital theories describe limiting conditions, neither method is wholly reliable in providing a full description of all covalent bonds. However, suitable modifications enable both approaches to be improved until they converge and become virtually equivalent.

Sigma and pi bonds

For all single covalent bonds, orbitals overlap so as to form **sigma** (σ) **bonds.** Confining our interest for the moment to s and p orbitals, these bonds result from any of three possibilities (Figure 2.2):

(1) the overlapping of two *s* orbitals,

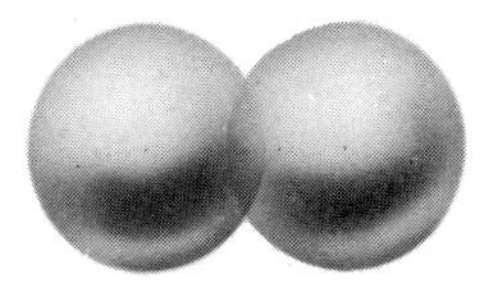

(2) the overlapping of two *p* orbitals linearly opposed to each other,

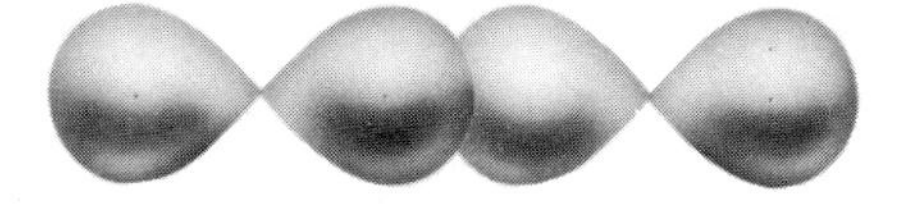

(3) the overlapping of an *s* and a *p* orbital.

Fig. 2.2 σ bonds

The bonding electrons in sigma bonds are most likely to be located in the vicinity of an imaginary line joining the two nuclei of the atoms involved, and are said to be localized.

The lateral overlap of two parallel *p* orbitals of adjacent atoms results in the formation of an entirely different type of covalent bond which is called a **pi** (π) **bond** (Figure 2.3).

Fig. 2.3 π bond

The degree of overlapping is less than in the formation of sigma bonds, and the bonding electrons are located away from the imaginary line joining the nuclei of the atoms. Consequently, they are weaker than the sigma bonds and, being more exposed, are more vulnerable to cleavage by a suitable attacking reagent during a chemical reaction.

The hydrogen molecule, H_2

The hydrogen molecule may be considered to be formed from the overlapping of two singly occupied 1*s* atomic orbitals to form a molecular orbital of the sigma (σ) type in which the electron spins are paired (Figure 2.4).

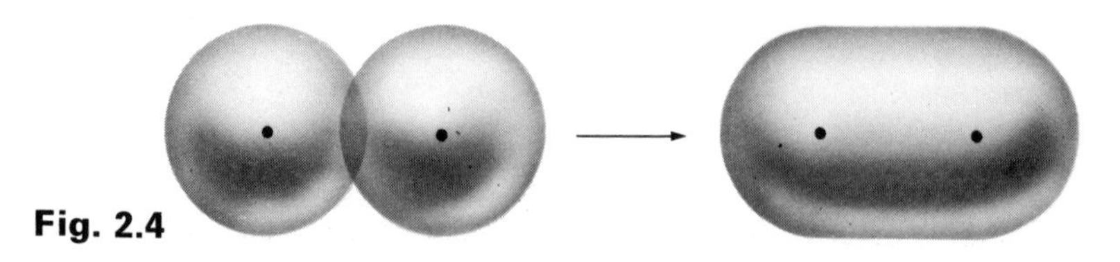

Fig. 2.4

Group II

The concept of hybridization

To account for the fact that beryllium ($1s^2 2s^2$ in the ground state) has a covalency of two, it must be assumed that the 2*s* electrons are unpaired and one of them promoted to the *p* level.

	1*s*	2*s*	2*p*		
Beryllium atom, excited state:	↑↓	↑	↑		

These two electrons should be distinguishable as the 2*s* electron is of marginally lower energy than the 2*p* electron. In fact, *these electrons cannot be distinguished from each other* and are said to be **hybridized**. The two 'new' hybrid orbitals, being a combination of a single *s* and a single *p* orbital, are referred to as *sp hybrid orbitals* and possess a shape different from either the *s* or the *p* orbitals from which they were derived. The spatial orientation of *sp* hybrid orbitals is such that they are linearly opposed to each other (Figure 2.5).

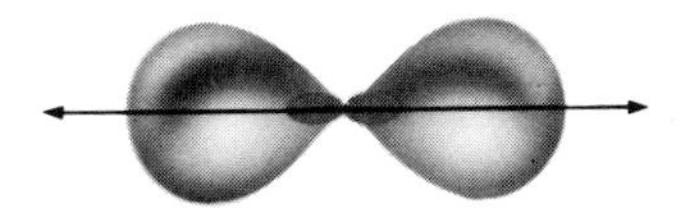

Fig. 2.5 *sp* hybrid orbitals

The structure of BeF_2

The BeF_2 molecule is considered therefore to be formed by the overlapping of each of these two hybrid orbitals with the singly occupied orbital of each of two fluorine atoms ($1s^2 2s^2 2p^5$) giving rise to the formation of a linear structure, F—Be—F, with the fluorine atoms on either side of the central beryllium. This is in accordance with the experimentally observed evidence.

	1*s*	2*s*	2*p*		
Beryllium atom, excited state in BeF_2 molecule:	↑↓	↑⇣	↑⇣		

(2*s* and first 2*p*: *sp* hybrid orbitals)

⇣ represents the electron contributed by each of the two fluorine atoms.

An outer quartet of electrons, as exhibited here by the beryllium atom, is a fairly uncommon occurrence even among the Group II elements. Beryllium is the only element in this group to form predominantly covalent compounds.

Electron-deficient compounds

Covalent compounds in which the central atom does not utilize all potentially available bonding orbitals because of insufficient valency electrons are referred to as *electron-deficient compounds*. Monomeric structures of this type are most prevalent in Groups II and III, and examples include BeF_2 (a quartet of electrons), the boranes, BF_3, BCl_3, and $AlCl_3$ (a sextet of electrons). The Group III compounds in particular have a tendency to dimerize in the anhydrous form, e.g. Al_2Cl_6 (see page 142), so enabling the central atoms to acquire an outer octet of electrons.

Hybridization and stereochemistry

The concept of hybridization is not limited in its application simply to the formation of *sp* hybrid orbitals nor, indeed, is it confined to only *s* and *p* orbitals. Furthermore, the type of hybridization of the atomic orbitals of the central atom in a molecule is directly related to the shape of the molecule. For example, sp^2 hybrid orbitals adopt a planar arrangement (Figure 2.6) and sp^3 hybrid orbitals are orientated tetrahedrally (Figure 2.7).

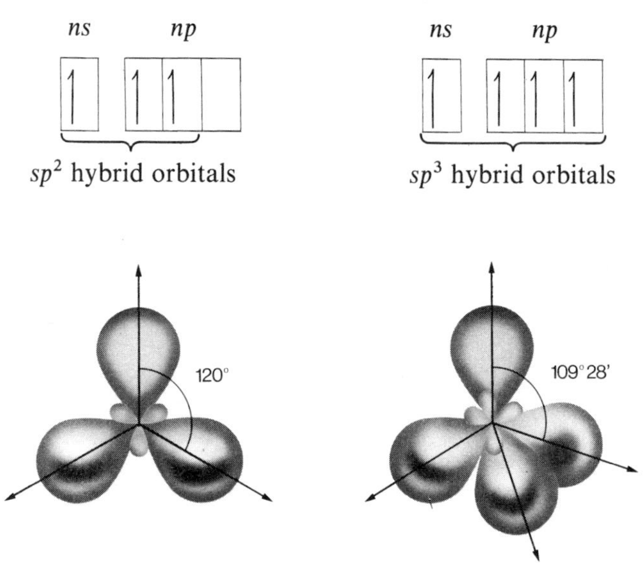

Fig. 2.6 Planar arrangement of sp^2 hybrid orbitals

Fig. 2.7 Tetrahedral arrangement of sp^3 hybrid orbitals

Here the hybridized orbitals of the central atom are depicted as being singly occupied and therefore available to form sigma bonds. However, this is not an essential criterion for the concept, as atomic orbitals containing paired electrons can also be included in the hybridization. A simple example of this is provided by the ammonia molecule (see page 29). Furthermore, it is not necessary for all the outer electrons to be involved in the hybridization. This is exemplified by the carbon atom in its formation of multiple bonds (see page 24).

A summary of some of the spatial orientations of different, acceptable permutations of orbitals for hybridization is given in Table 2.1.

Table 2.1 The fundamental relationship between molecular shape, the type of hybridization, and the number of outer electron pairs

Type of hybridization	*Number of outer electron pairs (bond pairs and lone pairs)*	*Shape of molecule*	*Bond angle(s)*
sp	2	Linear	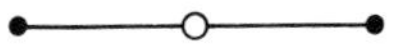 180°
sp^2	3	Planar trigonal	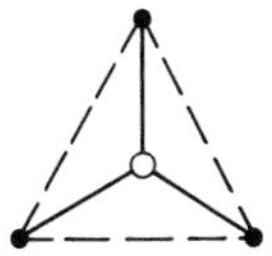 120°
sp^3	4	Tetrahedral	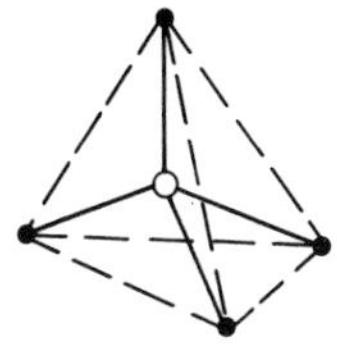 109° 28′
dsp^3 (or sp^3d)	5	Trigonal bipyramidal	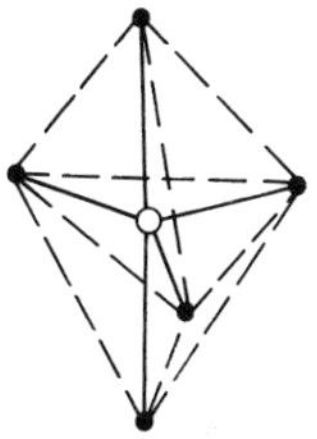 120° and 90°
d^2sp^3 (or sp^3d^2)	6	Octahedral	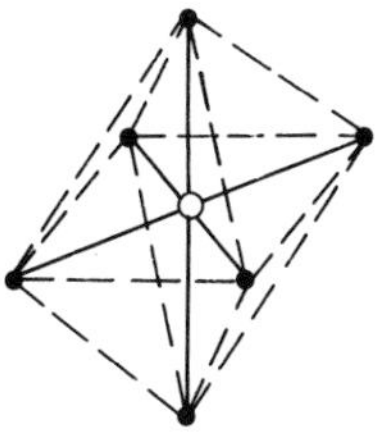 90°

The above generalizations predicting regular structures are valid only when all electron pairs are exactly equivalent. Where they are not, careful consideration should always be given to distortion created by lone pairs of electrons (see page 29), and also to the comparatively minor effect caused by the various polarities of different bond pairs. For example, unlike the methane molecule from which it is derived (see page 24), chloromethane (methyl chloride), CH_3Cl, will not have an exactly regular tetrahedral arrangement as the four bonds are not identical. The repulsive force between each of the C—H bonds differs from that between the C—Cl bond (which is more polar with the electrons displaced towards the more electronegative chlorine atom) and the C—H bonds.

The Sidgwick–Powell theory

Sidgwick and Powell (1940) recognized that atoms having the same outer electronic structure form molecules with a similar spatial arrangement of bonds. Such stereochemical features can often be deduced by making the assumption that electron pairs, irrespective of whether they are non-bonding lone pairs or covalent bond pairs, repel each other and orientate themselves in space in such a way as to be as far apart as possible. As they are associated solely with the central atom, the distance between two sets of lone pairs is shorter than that between other outer electrons, and the repulsive force between them is therefore greater than that between a lone pair and a bond pair, which in turn is greater than that between two bond pairs.

The Sidgwick-Powell electron-repulsion theory of electron pairs, although satisfactorily accounting for the shapes of simple molecules, implies that, in a case like beryllium fluoride, the two bonds in the molecule should be distinguishable in terms of their nature and relative strengths. This, of course, is not the case, and illustrates one of the principal limitations of the theory. The concept of hybridization, on the other hand, does not suffer from this limitation, and successfully accounts for the equivalence of bonds in simple molecules.

In implementing theories such as those of atomic structure, molecular structure, and hybridization, it must always be remembered that, like all other scientific models, they are continually under review and do not pretend to describe the absolute truth about the nature of matter. What these theories do provide is an extremely useful set of tools and models which enable us to describe observed phenomena at each particular stage in the continuous evolution of science. Inevitably, as knowledge increases, all concepts have to be amended and/or superseded by more refined ideas; we must always be prepared for this.

Group III

The boron trifluoride molecule, BF_3

In boron trifluoride, the boron atom ($1s^2 2s^2 2p^1$ in the ground state) is covalently bonded by means of σ bonds to three fluorine atoms (ground state

$1s^2 2s^2 2p^5$) which are arranged at the corners of an equilateral triangle. This is in accordance with the structure predicted by sp^2 hybridization of the boron atom and with the experimentally observed bond angle of 120° (Figure 2.8).

	$1s$	$2s$	$2p$		
Boron atom, ground state:	↑↓	↑↓	↑		
Boron atom, excited state in BF_3 molecule:	↑↓	↑⇣	↑⇣	↑⇣	

sp^2 hybrid orbitals

⇣ represents the electrons contributed by each of the three fluorine atoms.

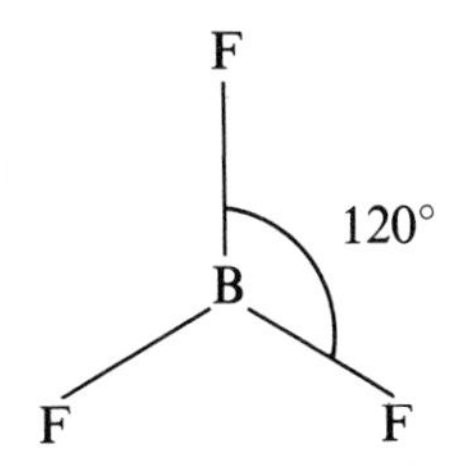

Fig. 2.8 Boron triflouride molecule

Boron trichloride possesses a similar structure.

Group IV

Some structures involving the carbon atom

The methane molecule, CH_4

In methane, CH_4, the carbon is sp^3 hybridized and each hybrid orbital overlaps with the s orbital of each of the four hydrogen atoms to form a σ bond (Figure 2.9).

	$1s$	$2s$	$2p$		
Carbon atom, ground state:	↑↓	↑↓	↑	↑	
Carbon atom, excited state in CH_4 molecule:	↑↓	↑⇣	↑⇣	↑⇣	↑⇣

sp^3 hybrid orbitals

⇣ represents the electrons contributed by each of the four hydrogen atoms.

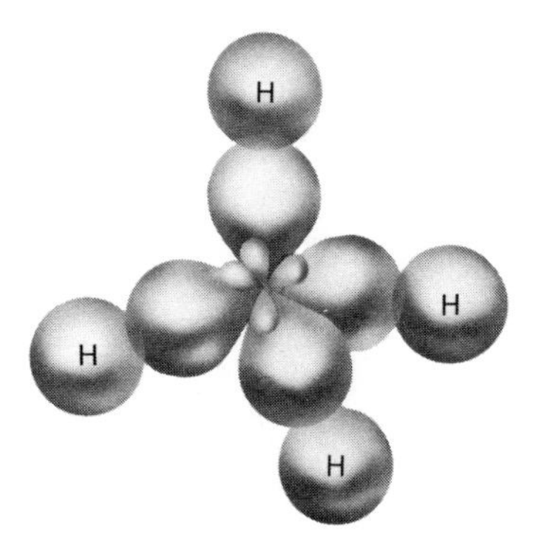

Fig. 2.9 Methane molecule

The methane molecule is completely symmetrical, with the hydrogen atoms located at the four corners of a regular tetrahedron.

Carbon–carbon multiple bonds

The ethene (ethylene) molecule, $CH_2{=}CH_2$

In the formation of carbon-carbon double bonds, $\gt C{=}C\lt$, only three of the four valence electrons are hybridized to form sp^2 orbitals. This leaves the one remaining unpaired $2p$ electron unaffected by the hybridization.

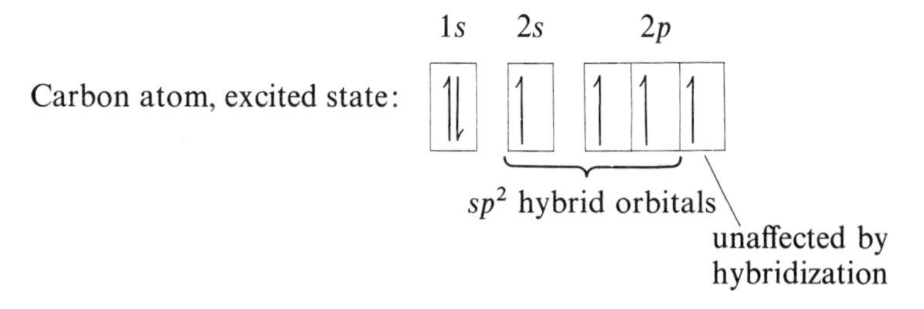

The sp^2 hybrid orbitals are planar and inclined at angles of 120° to each other with the unhybridized p orbital set at right angles to this plane. The simplest and most easily considered unsaturated molecule which contains a carbon-carbon double bond is the alkene, ethene (ethylene), C_2H_4. All the carbon-hydrogen bonds are σ bonds formed by the overlap of the carbon sp^2 hybrid orbitals with the s orbitals of the hydrogen atoms. One of the carbon-carbon bonds is formed by the overlap of two sp^2 orbitals from each of the adjacent atoms, resulting in a σ bond, and the other obtained by the overlapping of the adjacent p orbitals, forming a π bond (Figure 2.10).

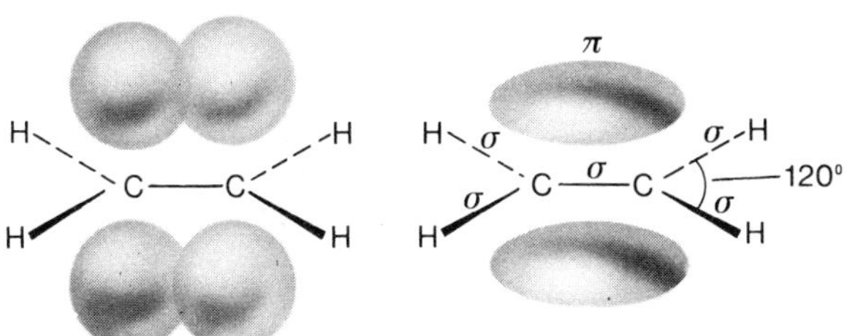

Fig. 2.10 Ethene molecule

The resulting molecular orbital containing the π electrons is spread over both carbon atoms and is located above and below the plane of the rest of the molecule. It must be remembered that although this π bond appears as two electron clouds, it still constitutes only one bond.

The ethyne (acetylene) molecule, CH≡CH

In the formation of carbon-carbon triple bonds, —C≡C—, only two of the valence electrons are hybridized, forming *sp* orbitals, leaving two p orbitals unaffected.

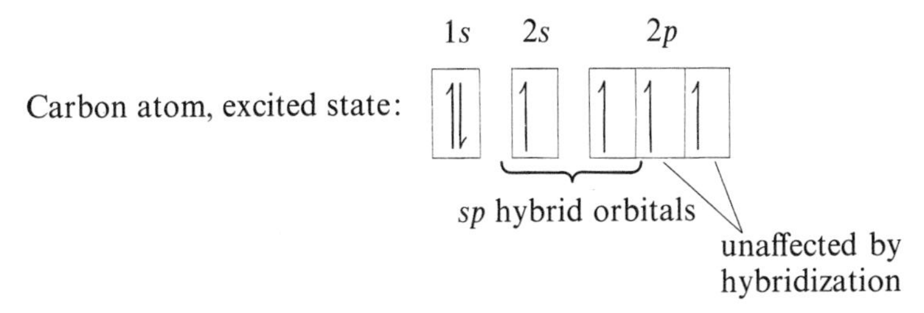

The simplest hydrocarbon containing a triple bond, ethyne (acetylene), is a linear molecule. The carbon and hydrogen atoms are attached by σ bonds, which are formed by the overlapping of the *sp* orbitals of the carbon with the *s* orbitals of the hydrogen atoms. One of the carbon-carbon bonds is a σ bond which results from the overlapping of the other *sp* orbitals of each carbon. The remaining two carbon-carbon bonds are both π bonds which are formed as a result of the overlapping of adjacent p orbitals (Figure 2.11). Since these

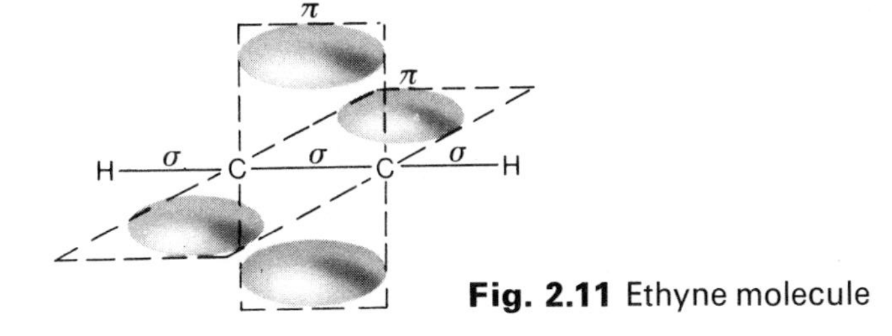

Fig. 2.11 Ethyne molecule

two bonds lie at right angles to each other (the dotted lines clearly illustrate the planes in which these two bonds lie), the ethyne molecule is virtually encased by a cylindrical cloud of negatively charged electrons.

Oxides of carbon

The carbon dioxide molecule, CO_2

Carbon atom, excited state in CO_2 molecule:

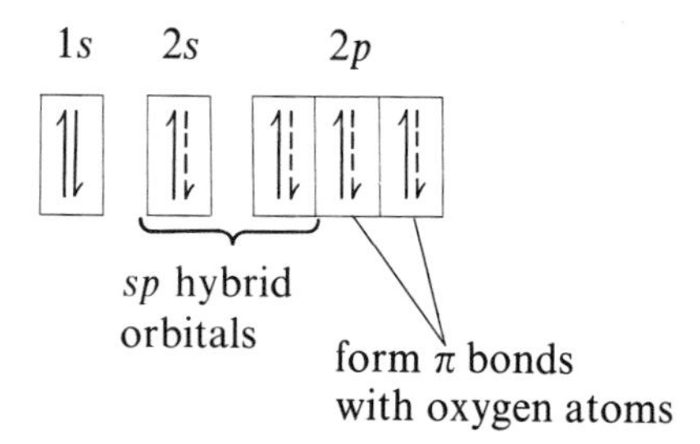

⇂ represents electrons contributed by oxygen atoms.
(Oxygen atom, ground state: $1s^2 2s^2 2p^4$)

The overlapping of the two *sp* hybrid orbitals of the carbon with a singly occupied orbital of each of the two oxygens results in the formation of two σ bonds (Figure 2.12).

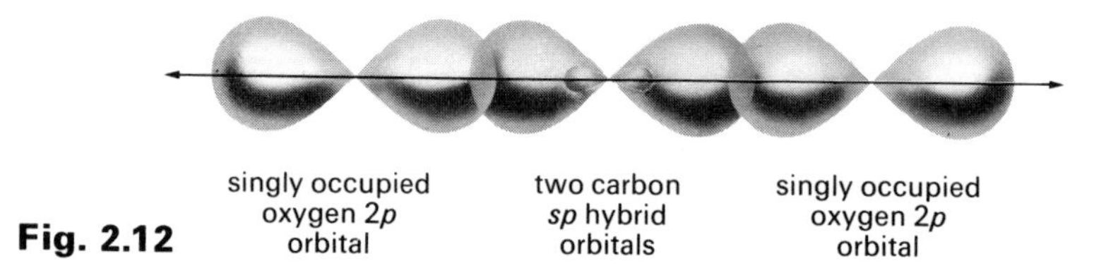

Fig. 2.12

π bonds are then formed by the lateral overlapping of the remaining singly occupied and unhybridized orbitals, leaving each of the oxygen atoms with two lone pairs of electrons (Figure 2.13).

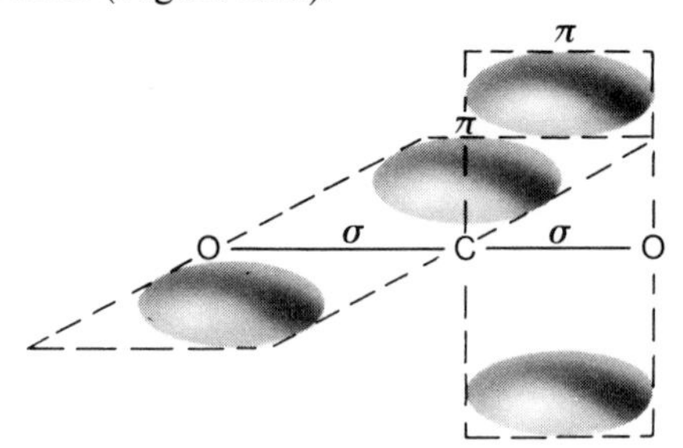

Fig. 2.13

The carbon monoxide molecule, CO

The structure of this molecule may be represented in terms of molecular orbitals only by means of something of an oversimplification. Bond measurements indicate that the carbon monoxide molecule probably contains a carbon-oxygen triple bond which is fundamentally similar to the triple bond of the nitrogen molecule (see page 28).

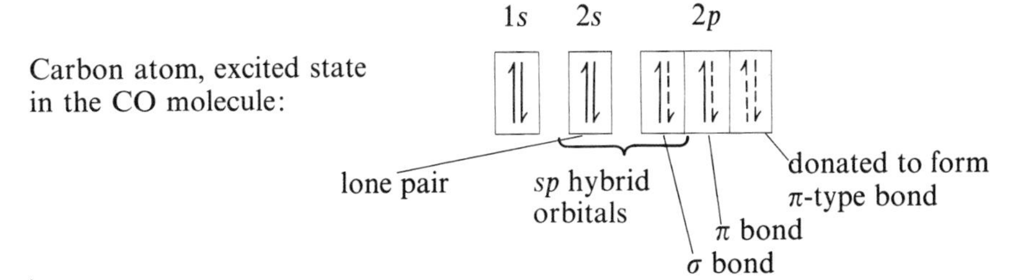

represents electrons contributed by the oxygen atom.

It is likely that one carbon *sp* hybrid orbital contains a lone pair and that the other forms a σ bond with a singly occupied *p* orbital of the oxygen. An ordinary π bond results from the lateral overlapping of the other singly occupied orbitals of each atom, with the two remaining oxygen *p* electrons forming a π-type bond with the vacant *p* orbital of the carbon atom. It has been suggested that this orbital is largely atomic in character, and is localized almost entirely about the oxygen atom.

The two 2*s* electrons of the oxygen atom exist in an atomic orbital as an **inert pair** (as distinct from a lone pair), implying that they are incapable of forming a covalent bond. This is supported by the behaviour of carbon monoxide as a ligand, where it is the lone pair of electrons on the carbon atom which participate in bond formation and not the two electrons on the more electronegative oxygen atom.

The carbonate anion, CO_3^{2-}

The regular trigonal planar geometry of the carbonate anion is compatible with sp^2 hybridization of the central carbon atom.

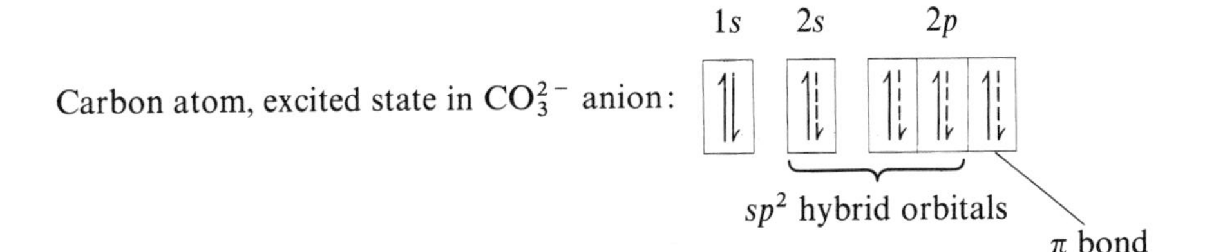

represents the electrons contributed by each of the oxygen atoms.

Overlapping of the sp^2 hybrid orbitals of the carbon with oxygen *p* orbitals results in the formation of three σ bonds, while sideways overlapping of *p* orbitals produces an additional π bond which is *delocalized over all four atoms of the ion.*

Representation of this anion in terms of molecular orbitals tends to be somewhat clumsy, and it is probably more easily represented as a resonance hybrid of the following canonical forms:

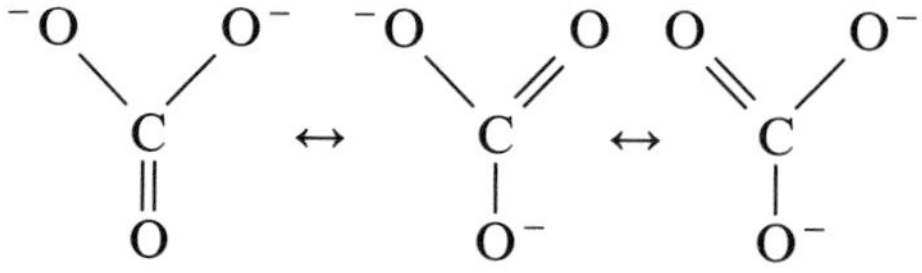

$$\mathrm{{}^{-}O{-}C({=}O){-}O^{-}} \leftrightarrow \mathrm{{}^{-}O{-}C({-}O^{-}){=}O} \leftrightarrow \mathrm{O{=}C({-}O^{-}){-}O^{-}}$$

or by the single structure:

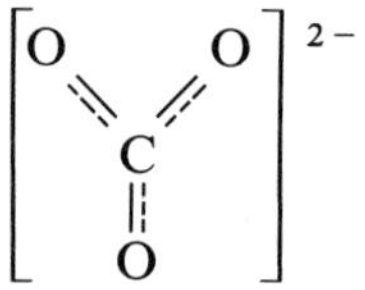

$$\left[\mathrm{C}(\text{:::}\mathrm{O})_3\right]^{2-}$$

which serves to illustrate the equal distribution of the negative charges between all three oxygen atoms.

Group V

Some structures involving the nitrogen atom

The nitrogen molecule, N_2

The nitrogen atom ($1s^22s^22p^3$ in the ground state) possesses three unpaired $2p$ orbitals, which, in the nitrogen molecule, overlap to form a triple bond, N≡N.

	$1s$	$2s$	$2p$		
Nitrogen atom, ground state:	↑↓	↑↓	↑	↑	↑
Nitrogen atom, ground state, in N_2 molecule:	↑↓	↑↓	↑⇣	↑⇣	↑⇣

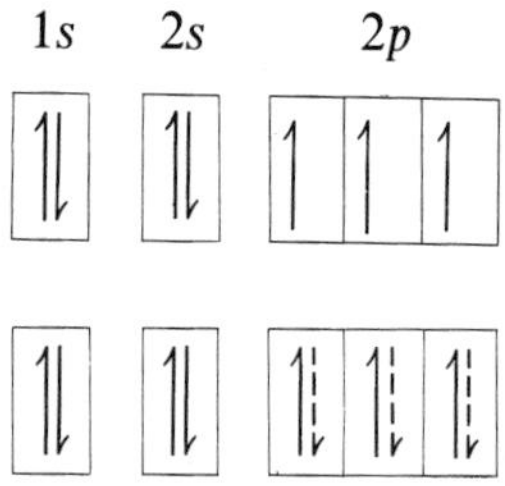

⇣ represents the electrons contributed by the other nitrogen atom.

Linear overlapping of a $2p$ orbital from each atom results in the formation of a σ bond, and lateral overlapping of the other two $2p$ orbitals from each atom produces two π bonds. The two π bonds interact to form what is, in effect, a cylinder of π electrons which encase the σ bond which is formed along the axis between the two nuclei, (cf. ethyne).

The ammonia molecule, NH_3

In the ammonia molecule, the nitrogen atom is sp^3 hybridized with one orbital containing a lone pair of electrons. The remaining three singly-occupied orbitals form σ bonds by overlapping with the singly occupied 1*s* orbital of three hydrogen atoms.

Nitrogen atom, ground state, in NH_3 molecule:

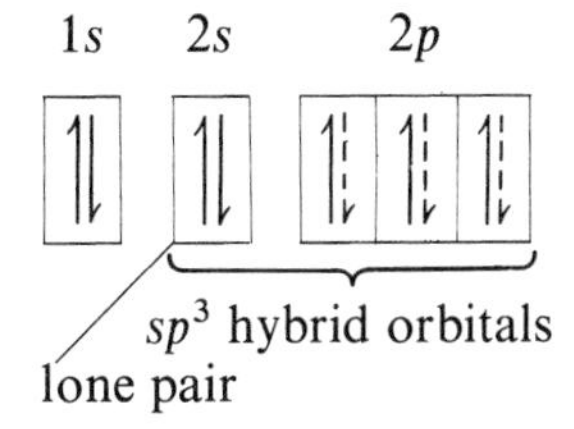

┆ represents the electron contributed by each of three hydrogen atoms.

As the repulsive force between a lone pair and a bond pair exceeds that between bond pairs themselves, the basic tetrahedral arrangement of electrons, predicted by sp^3 hybridization, is distorted. The H—N—H bond angles are 106°45′, compared with 109°28′ for a regular tetrahedron. The arrangement of atoms in the molecule is that of a trigonal pyramid (Figure 2.14).

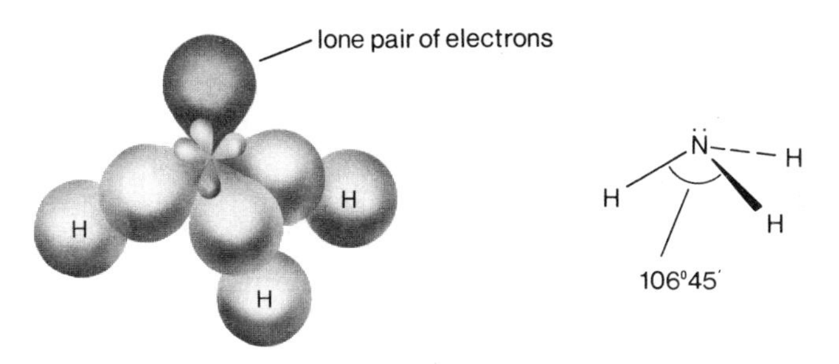

Fig. 2.14 Ammonia molecule

The ammonium ion, NH_4^+

Despite its stable electronic configuration, the ammonia molecule is capable of further combination by allowing the lone pair of electrons to be shared with other molecules and ions possessing vacant orbitals capable of accepting them. In the ammonium ion, for example, the lone pair is *donated* to a proton. The ammonia molecule is the *donor*, the proton is the *acceptor*. The resulting bonds are all covalent and equivalent, and being sp^3 hybridized give rise to a regular tetrahedral structure (Figure 2.15).

Fig. 2.15 Ammonium ion

The ammonia–boron trifluoride(1/1) molecule, $NH_3{\cdot}BF_3$

Another species capable of accepting a lone pair of electrons from ammonia is boron trifluoride (see page 22). As the ammonia molecule approaches the boron trifluoride molecule, the latter adopts a type of sp^3 hybridization, with one orbital completely empty. This empty orbital aligns itself with the orbital of the nitrogen containing the lone pair and overlapping of these orbitals results in the formation of a covalent bond.

$$F_3B \leftarrow NH_3$$

(The arrow depicts the direction of donation of the lone pair.)

Structures containing phosphorus

The halides of phosphorus

Like nitrogen, phosphorus ($1s^2 2s^2 2p^6 3s^2 3p^3$ in the ground state), has three singly occupied orbitals enabling it to form trihalides such as PF_3, PCl_3, PBr_3, and PI_3 by means of ordinary covalent bonds.

Phosphorus atom, ground state:

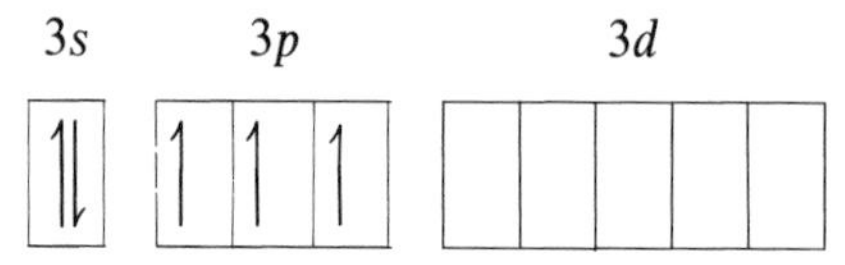

By unpairing the 3*s* electrons and promoting to the 3*d* level, phosphorus can extend its covalency to five, giving molecules such as PF_5 which has a trigonal bipyramidal structure (Figure 2.16).

Phosphorus atom, excited state in PF_5 molecule:

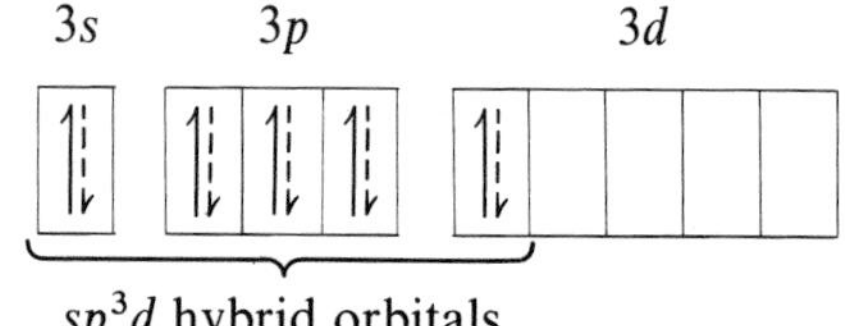

↓ represents electrons contributed by each of the five fluorine atoms.

As PF_5 is not a completely regular structure (bond angles of 120° and 90°), it tends to be somewhat unstable.

PF_5, for example, can readily accept a lone pair of electrons from a fluoride anion to form the more stable octahedral structure of the $[PF_6]^-$ anion (Figure 2.17):

Phosphorus atom, excited state in $[PF_6]^-$ anion:

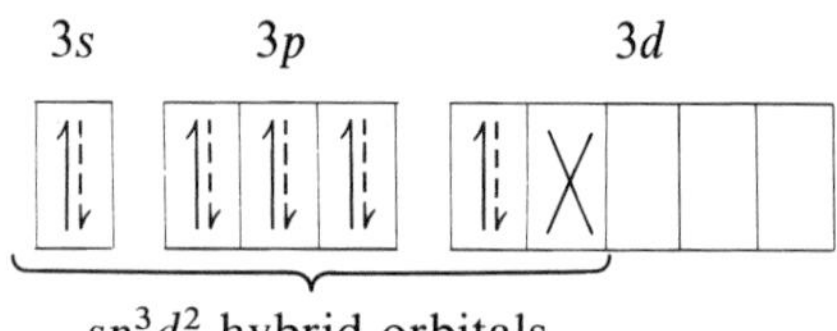

X represents the lone pair of electrons donated by the F^- anion.

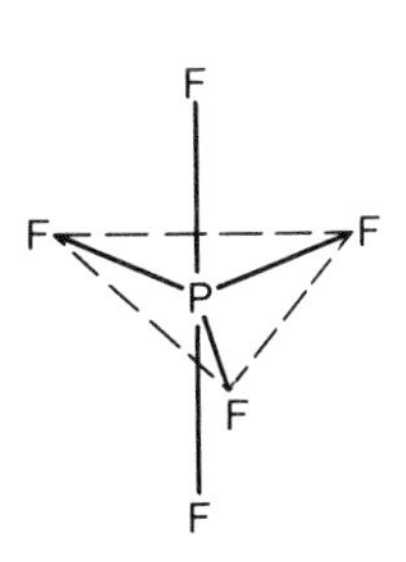

Fig. 2.16

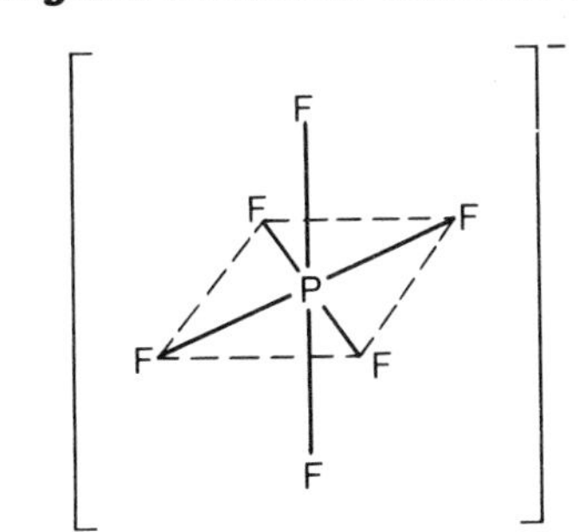

Fig. 2.17

Group VI

The water molecule, H_2O

In exhibiting a covalency of two, the oxygen atom ($1s^22s^22p^4$ in the ground state) in the water molecule may be considered to be sp^3 hybridized with two orbitals containing lone pairs of electrons and the other two, which are both singly occupied, forming σ bond pairs with the 1*s* electron of each of two hydrogen atoms (Figure 2.18).

Oxygen atom, ground state in H_2O molecule:

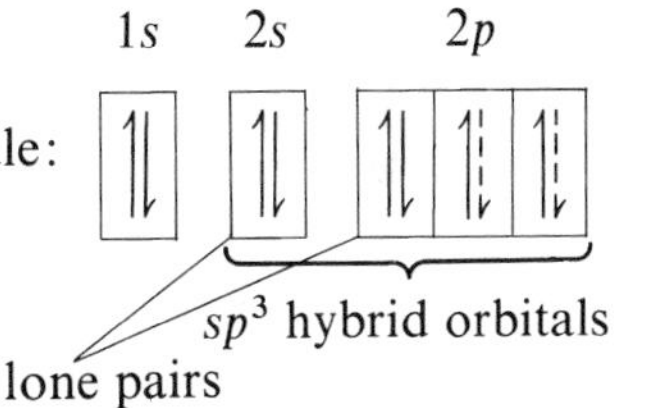

⇂ represents electrons contributed by each of the two hydrogen atoms.

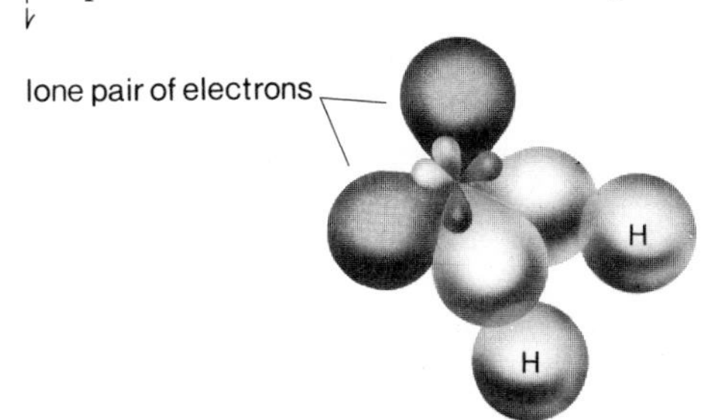

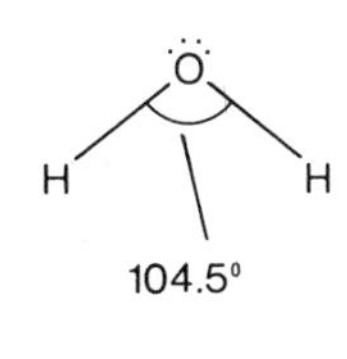

Fig. 2.18 Water molecule

Because of the strong repulsive forces between the two sets of lone pairs of electrons, the H—O—H bond angle is less than the bond angles in the ammonia molecule. The experimentally determined bond angle between the angular arrangement of atoms is 104.5°.

The H_3O^+ ion (see page 105), formed by the donation of a lone pair of electrons of the water molecule to a proton, has a stereochemical structure analogous to that of the ammonia molecule (see page 28).

The sulphur hexafluoride molecule, SF_6

The sulphur atom ($1s^2 2s^2 2p^6 3s^2 3p^4$ in the ground state) exerts a covalency of six by unpairing the paired electrons in the 3*s* and 3*p* levels and promoting one of each to the available 3*d* level.

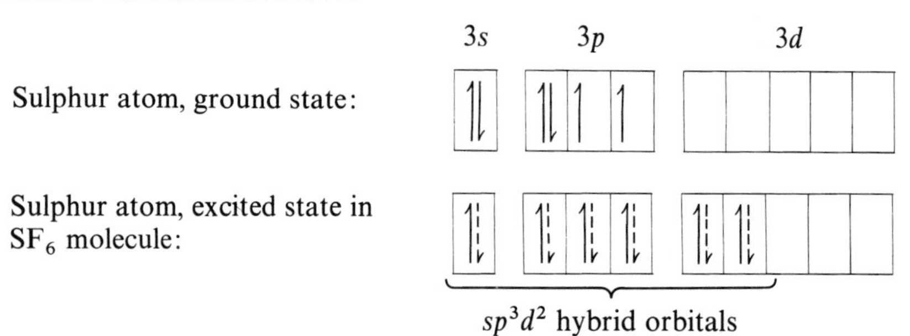

⇂ represents electrons contributed by each of the fluorine atoms.

sp^3d^2 hybridization of the resulting orbitals accounts for the experimentally observed octahedral structure (Figure 2.19).

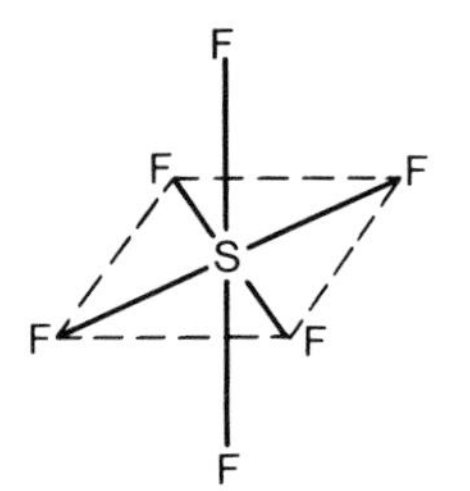

Fig. 2.19

Group VII

The halogen molecules, X_2

The halogen atoms all form diatomic molecules, F_2, Cl_2, Br_2, and I_2, which may be considered to be formed by the overlapping of the singly occupied outer orbitals to give a single σ bond. Iodine is a solid, and the nature of its crystal structure is discussed on page 48.

The hydrogen halide molecules, HX

	1s	2s	2p		
Fluorine atom, ground state:	⇅	⇅	⇅	⇅	↑

A slightly oversimplified approach to the formation of the covalent linkage between hydrogen and fluorine (and other halogen atoms) is to consider the overlapping of the hydrogen 1*s* orbital with the singly occupied 2*p* orbital of fluorine.

A better picture involves the unpaired *p* orbital and the 2*s* orbital of fluorine in the formation of two *sp*-type hybrid orbitals. One of these hybrid orbitals forms a covalent bond with the hydrogen atom, and the other, which is linearly or digonally opposed to this bonding orbital, possesses a lone pair of electrons. On this basis, both of the remaining 2*p* orbitals, each of which contains a lone pair of electrons, overlap to form a ring of delocalized electrons.

Group 0

The xenon tetrafluoride molecule, XeF_4

Owing to their filled outer shells of electrons, the noble gases form relatively few compounds. Formation of the XeF_4 molecule requires the unpairing of two pairs of 5*p* electrons and exciting one of each to the available 5*d* level.

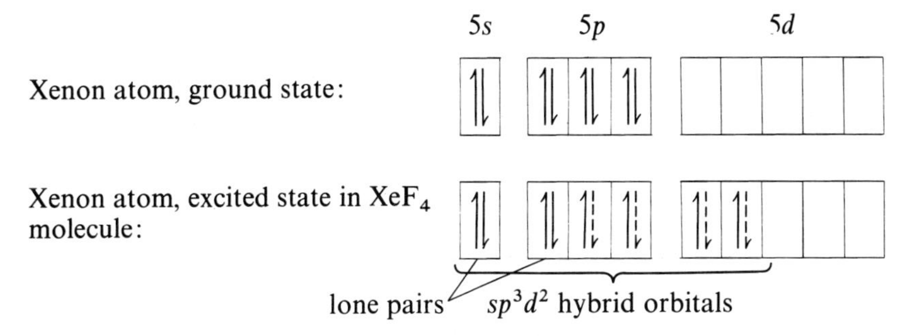

⇂ represents electrons contributed by four fluorine atoms.

The four fluorine atoms can then form covalent bonds with the singly occupied xenon orbitals and adopt a square planar structure about the central xenon atom. The two sets of lone pairs occupy orbitals above and below this plane, giving an octahedral arrangement of electrons as predicted by sp^3d^2 hybridization (Figure 2.20).

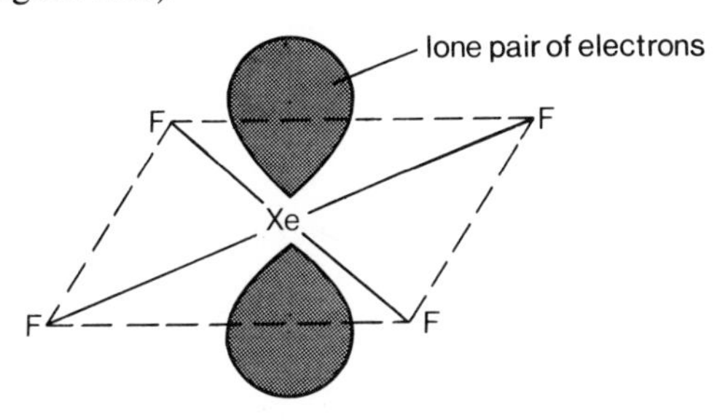

Fig. 2.20 Xenon tetrafluoride molecule

3 *X-Rays and Crystals*

All crystals have a characteristic, regular, geometrical form (a *lattice*) which is a consequence of the systematic arrangement of the ions, atoms, or molecules of which they are built up. The structural pattern is three-dimensional and comprises a certain basic structure which is repeated over and over again. With the development of X-ray crystallography, it has become possible to determine the actual nature of these individual units and the repeating pattern that occurs in a crystal.

The original pioneer of X-ray crystallography was von Laue who, in 1912, suggested that a crystal might be expected to diffract X-rays in a way similar to that in which light is diffracted by a grating. The subsequent implementation of these ideas has become the most useful means for the determination of crystal structures.

The crystal lattice as a diffraction grating

When light falls upon a diffraction grating, i.e. a perforated screen containing holes arranged in a regular pattern with the spacings comparable to the wavelength of the incident light, diffraction occurs, causing some of the light which passes through the grating to deviate from the original direction of propagation. The angle of deviation depends upon the ratio of the wavelength of the light relative to the spacings of the grating.

The pattern of repeating similar units in a crystal occurs several hundred times within a single wavelength of light, and is therefore incapable of diffracting light in the visible spectrum. X-rays, however, have a wavelength about ten thousand times shorter than that of visible light, so enabling them to be diffracted by a crystal.

When a beam of X-rays is passed through a crystal and allowed to fall on a photographic plate, after development a central white spot, due to the direct beam, is observed surrounded by a pattern of smaller spots. This corresponds to the symmetry of the crystal which has, in effect, acted as a three-dimensional diffraction grating. The intensity of the X-rays diffracted by the planes of the crystal is only a small fraction of that of the incident beam, and a long exposure is therefore necessary.

In a typical crystal, such as sodium chloride, the distance between the nuclei is 0.282 nm (2.82×10^{-10} m) whereas a typical X-ray has a wavelength of 0.06 nm (0.6×10^{-10} m), which illustrates the order of comparability of the bond lengths between common atoms or ions and the wavelengths of X-rays.

The Bragg equation

Simple crystal structures were first analysed quantitatively by W.L. Bragg (1913), who used crystals to *reflect* a beam of monochromatic (single wavelength) X-rays. He suggested that the atoms, ions, or molecules of a crystal space-lattice (see page 37) may be considered as lying in a set of similar, equally spaced, parallel planes.

A single crystal is said to be homogeneous, implying that it has identical properties at all points within it. Such crystals reflect X-rays regularly.

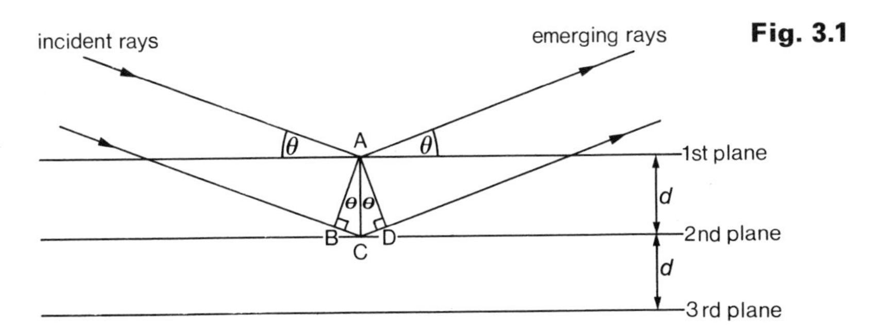

Fig. 3.1

The relationship developed by Bragg is derived from the consideration of two parallel monochromatic X-ray beams of wavelength λ incident on two adjacent crystal planes at a glancing angle, θ. The path difference between the two rays is BC+CD, BC and CD being equal. This represents the distance that the ray incident on the second plane in Figure 3.1 has travelled further than the ray incident on the first plane.

If the spacing between the crystal planes is d, then:

$$\mathrm{BC} + \mathrm{CD} = 2d \sin \theta$$

Maximum reinforcement (waves in phase, Figure 3.2) of the reflected X-rays

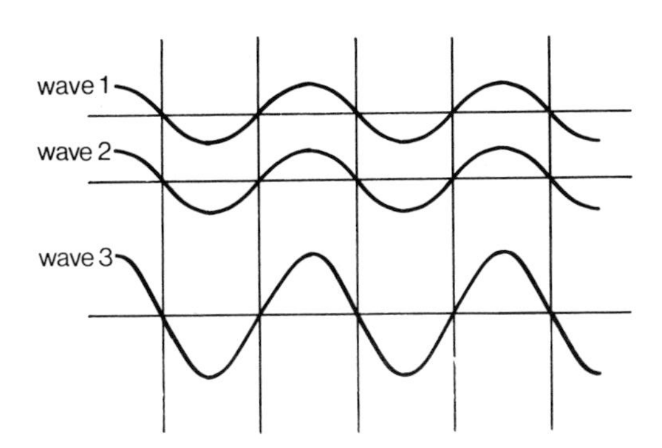

Fig. 3.2 Waves in phase; wave 3 is the reinforced wave of 1 and 2

will therefore only occur for certain values of θ, when the path difference corresponds to a whole number of wavelengths.

i.e. $$BC+CD = n\lambda, \quad n \text{ being an integer}$$

or, $$2d \sin\theta = n\lambda$$

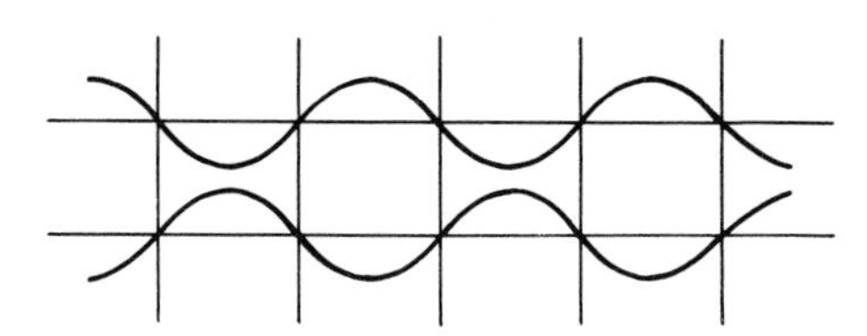

Fig. 3.3 Waves out of phase; in this particular case the waves are completely cancelling

The values of n are referred to as the *order of interference or reinforcement*: when $n = 1$ it is first order, when $n = 2$ it is second order, and so on. In practice, the intensity of the maximum reinforcements increases with increasing values of n.

By making use of the fact that, in being diffracted by the planes, X-rays obey the same laws as when they are reflected, the Bragg equation is usually employed to determine accurate values of d. X-ray analysis is also widely used to determine the angles at which the different planes intersect and hence the arrangement of atoms or ions in the crystal.

Interpretation of the diffraction pattern

The distance between the spots on a photograph can be interpreted to give the separation of the repeating units and determine the dimensions of the unit cell (see page 37), but precise positions of the atoms or ions relative to each other are not so easily ascertained. As X-rays are scattered primarily by electrons, the intensity of diffraction increases with the number of electrons associated with an atom or ion, and as these are arranged regularly or periodically in a crystal, the electron pattern which they produce must also be periodic. (It is assumed that the centre of a high electron density spot represents the centre of the atom or ion.) As this technique is dependent on the crystal containing atoms or ions possessing a high number of electrons, it will obviously prove suitable for those of reasonably high atomic number and is of little use where hydrogen atoms are concerned.

Because diffraction occurs at all planes in a crystal, the X-ray diffraction patterns are extremely complex and calculations are nowadays always done by computer. The pattern is then represented as an electron-density contour map, each contour representing regions of equal density.

The unit cell

The smallest fundamental part of a crystal structure which is a characteristic of the whole crystal is called a **unit cell**. Any unit cell is defined by the dimensions of its *primitive translations* (i.e. the lengths of its sides) denoted by the letters a, b, and c and by the three angles between the pairs of sides denoted by α, β, and γ (see Figure 3.4). These unit cells are, in effect, the *building blocks* of which the crystals are composed, and three-dimensional repetition of these identical units gives the space-lattice, or continuous framework, of the whole crystal (Figure 3.5).

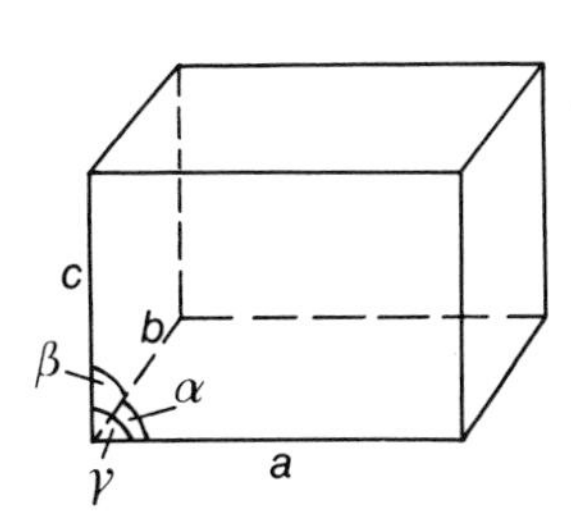

Fig. 3.4 Dimensions of a unit cell

Fig. 3.5 Space lattice comprising an assembly of primitive unit cells

Whatever symmetry a crystal is capable of, this symmetry is also a feature of its unit cell. It is, in fact, possible to have seven different crystal systems (Table 3.1).

Table 3.1 The seven crystal systems

	Dimensions	*Angles*
Cubic	$a = b = c$	$\alpha = \beta = \gamma = 90°$
Tetragonal	$a = b, c$	$\alpha = \beta = \gamma = 90°$
Orthorhombic	a, b, c	$\alpha = \beta = \gamma = 90°$
Rhombohedral	$a = b = c$	$\alpha = \beta = \gamma \neq 90°$
Hexagonal	$a = b, c$	$\alpha = \beta = 90°, \gamma = 120°$
Monoclinic	a, b, c	$\alpha = \gamma = 90°, \beta \neq 90°$
Triclinic	a, b, c	$\alpha \neq \beta \neq \gamma = 90°$

Simple geometry enables the seven basic structures to be subdivided into fourteen types of space-lattice in accordance with their internal arrangements of points. There are three types of cubic structure, two tetragonal, four orthorhombic, two monoclinic, and one each of rhombohedral, hexagonal, and triclinic. This was first recognized by A. Bravais (1848) – long before the advent of X-ray diffraction – and such structures are known as *Bravais lattices.* Consider, for example, the three cubic structures (Figure 3.6).

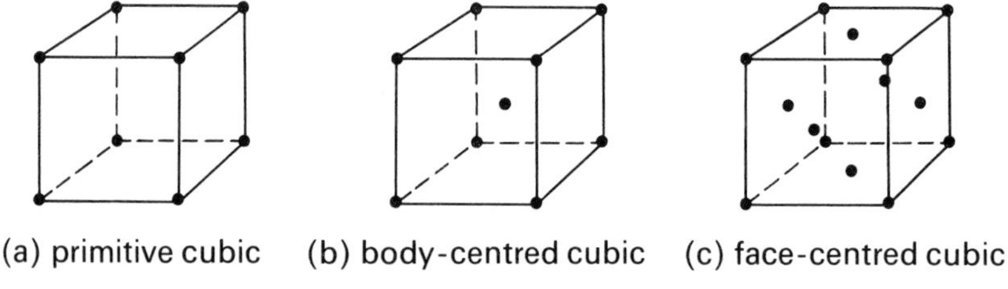

(a) primitive cubic (b) body-centred cubic (c) face-centred cubic

Fig. 3.6

Where the lattice points, which may be either atoms or ions, are at the cell corners, the structure is described as a **primitive** or **simple cube**. In the corresponding space lattice, the number of points is equal to the number of unit cells, as the eight points of each unit cell are shared between eight unit cells (Figure 3.6(a)).

The **body-centred cubic structure** (Figure 3.6(b)) has points at each corner of the cube and one at the centre of the unit, the point in the body-centre being associated with only the one cell. In this space lattice, there are two points per unit cell.

In the **face-centred cubic structure** (Figure 3.6(c)), there are points at the corners of the cube and at the centre of each face, the latter being shared by two unit cells. For the corresponding space-lattice, there are four points per unit cell.

Types of crystal structure

Packing of spheres

Crystal analysis provides important information about the arrangements of atoms and ions in the crystal structure. Each atom or ion may be regarded as a sphere of definite structure, and crystals are being formed by the packing together of such spheres in the most stable and economical manner, thus reducing the matter to purely geometrical considerations.

Metallic bonding

Metallic bonding in the solid state may be conveniently regarded as comprising an *array of cations* (positively charged ions, although not necessarily monopositive as shown in Figure 3.7) *held together in a 'sea' of electrons.*

This is achieved by the metal atoms releasing outer electrons, which then become delocalized and move with a certain degree of freedom among the

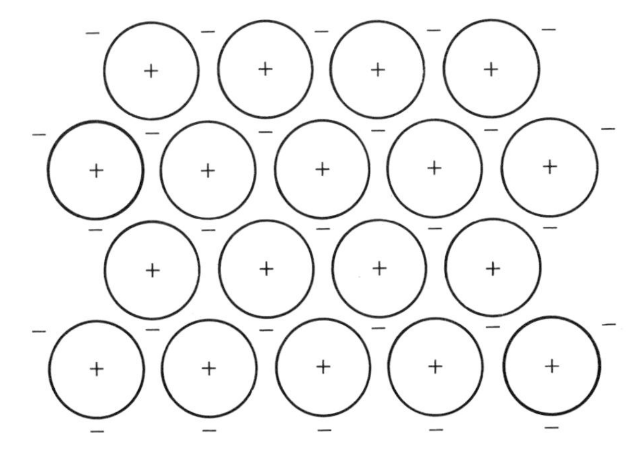

Fig. 3.7 Representative outline of cations and electrons in metallic bonding

metallic ions, providing the strong binding forces operative within the metallic crystal. As the cations and the electrons originate from neutral metal atoms, the total number of positive and negative charges within the structure must balance exactly. The metal cations adopt either a close-packed or body-centred cubic arrangement in which each one has 12 or 8 immediate neighbours (see page 41).

Metallic crystals: close-packing

Metal atoms may be regarded as spheres of equal size packed together in such a way that adjacent atoms touch one another. When, in a particular layer (or plane), one sphere is surrounded by six other equal spheres so that all seven spheres are in contact, as shown in Figure 3.8, the six outer spheres will correspond to the apexes of a regular hexagon.

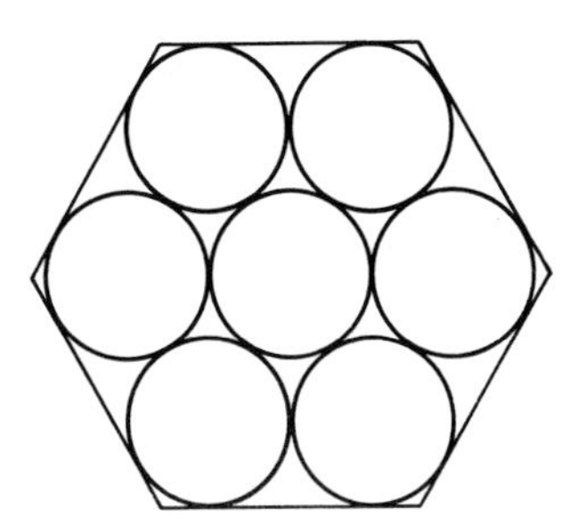

Fig. 3.8

Consider this layer as layer *A* and suppose that a second layer, layer *B*, is now placed on top of it in such a way that every sphere in *B* is touching three spheres in *A* (Figure 3.9).

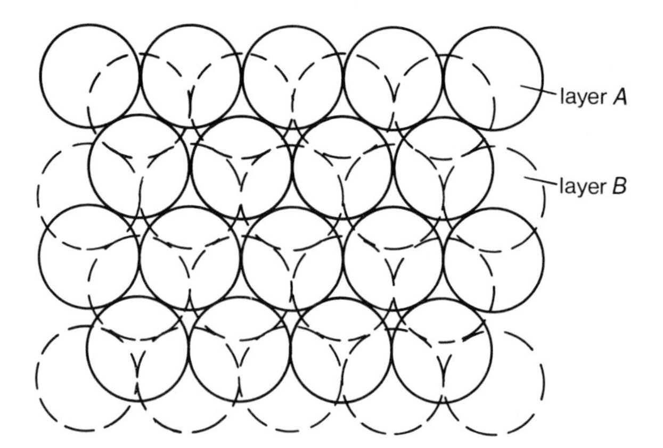

Fig. 3.9

A third layer of spheres can now be placed in one of two ways on top of layer *B*. If the spheres of the third layer are placed directly above those of layer *A*, giving rise to an *ABABAB*... packing arrangement, with each pair of layers being repeated indefinitely, then it is referred to as a **hexagonal close-packed** structure (Figure 3.10).

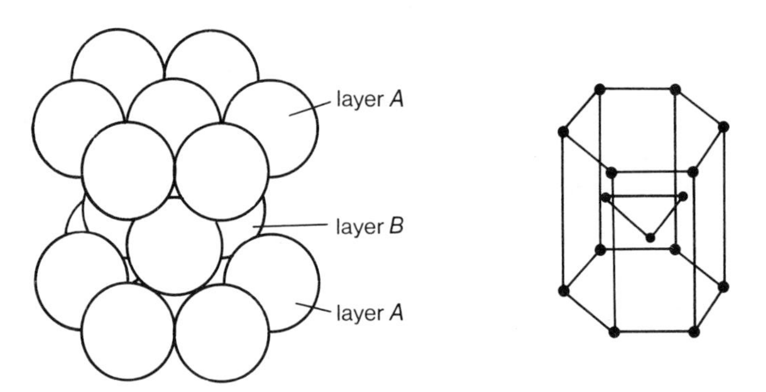

Fig. 3.10 Hexagonal close-packed structure of identical spheres depicting the *ABAB* . . . layer sequence

Alternatively, if the spheres in the third layer are placed neither over the spheres of layer *A* nor over those of layer *B*, but instead cover 'holes' in the structure of layer *A* that are not 'occupied' by spheres of layer *B*, it gives rise to an *ABCABC*... packing arrangement, which is referred to as a **cubic close-packed structure**.

In a close-packed arrangement, 74 per cent of all available space is occupied and each sphere is in contact with twelve other spheres, i.e. has a coordination number (see page 43) of 12. A better appreciation and understanding of close-packing and body-centred cubic arrangements (see next section) can be obtained by constructing models from poly(phenylethene) (polystyrene) spheres.

Metallic crystals: body-centred cubic packings

In this case, the spheres are less efficiently packed (Figure 3.11), occupying only 68 per cent of the available space, with each sphere having a coordination number of 8.

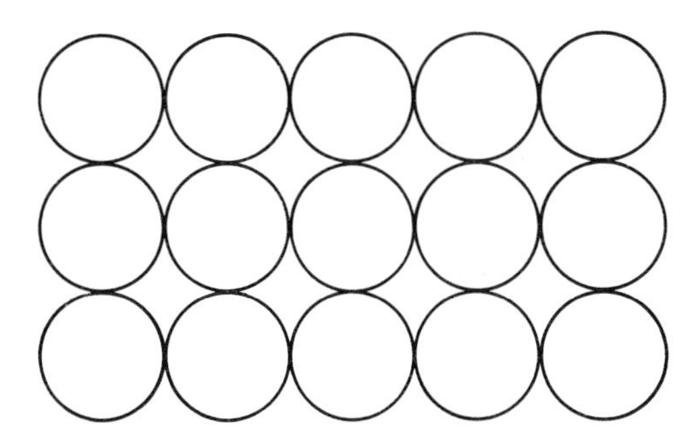

Fig. 3.11 A single plane of more openly packed spheres

A second layer fits on the first layer such that each sphere of the first layer is in contact with four other spheres of the second layer, giving an *AB* arrangement. Repetition of this pattern produces *ABABAB*... packing, leading to a **body-centred cubic** structure (Figure 3.12).

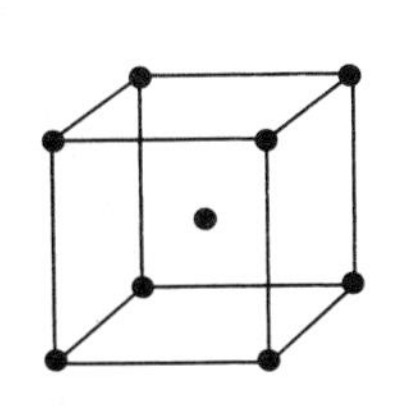

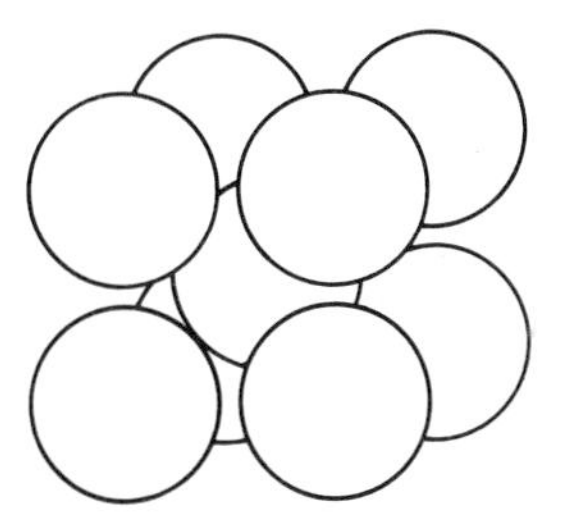

Fig. 3.12 Body-centred cubic packing of identical spheres

Many elements are monatomic, and their atoms are neutral and effectively spherical in shape, with a non-bonded structure. This allows them to be placed near to each other in a manner of least restriction and adopt one (or more) of the types of packing that have just been described. The noble gases are typical non-metallic elements and, with the exception of helium, possess a close-packed cubic structure in the solid state. The majority of the elements are essentially metallic in character, about 50 exhibiting cubic close-packing (e.g. Cu, Ag, Au) or hexagonal close-packing (e.g. Mg, Zn) structures, and about 20 having body-centred cubic packing (e.g. the alkali metals).

Interstitial structures

Tetrahedral and octahedral holes in close-packed structures

The gaps, or interstices, in close-paced structures are of two distinct types:

(1) **tetrahedral holes**, *t* in Figure 3.13, which are formed by four spheres, the centres of which form a tetrahedral arrangement;

(2) **octahedral holes**, o in Figure 3.14, which are formed by six spheres disposed towards the corners of an octahedron.

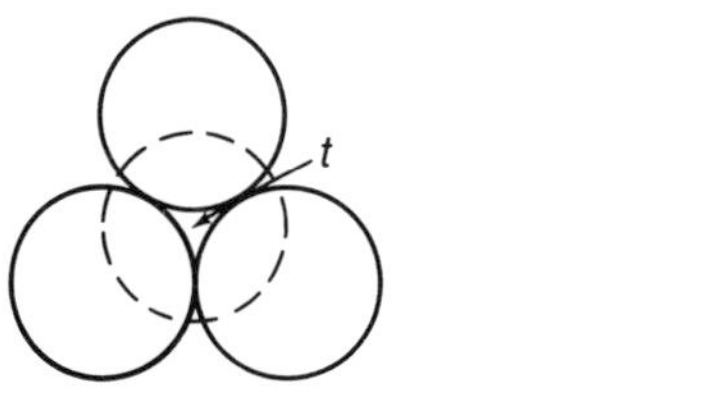

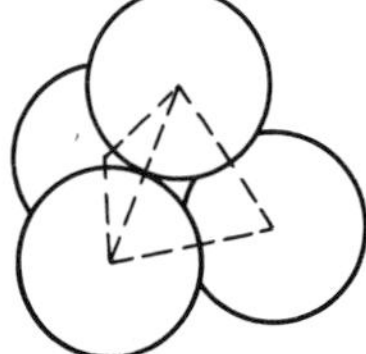

Fig. 3.13 Tetrahedral holes between two layers of close-packed spheres

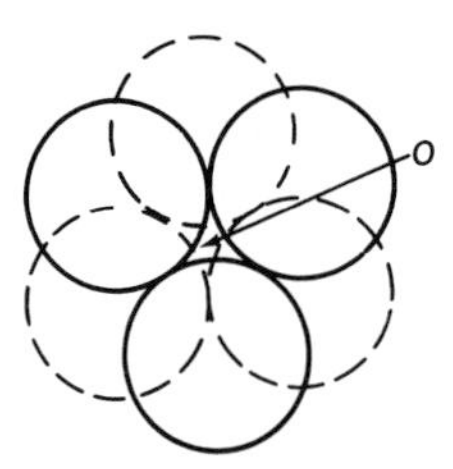

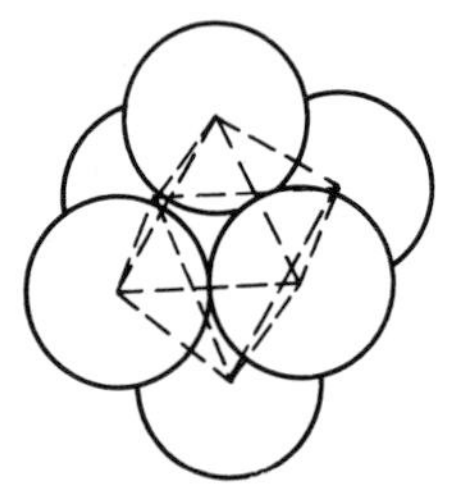

Fig. 3.14 Octahedral holes between two layers of close-packed spheres

The occupation of some of these interstitial sites by small atoms of non-metallic elements such as hydrogen, boron, carbon (as in steel), nitrogen, and oxygen brings about modifications in the physical properties of the base material and diminishes the tendency to undergo structural deformation. On the other hand, if too many of those sites are occupied by foreign atoms, the structure is weakened. The precise nature of the bonding involved is uncertain and is better regarded as being derived from the structural arrangement of the metal atoms rather than as some form of 'chemical bond'. Such structures are formed mainly by the transition metals. Where the interstices are not large enough to admit the atoms, then slight distortion of the close-packing structure occurs in order to accommodate them.

Counting the *number of tetrahedral holes* in a close-packed structure of this type shows there to be twice as many holes as there are spheres. If all the holes are filled by small spheres, Y, and the larger spheres comprising the close-packed arrangement are labelled, M, then a compound of empirical formula MY_2 is obtained, e.g. TiH_2. It is not, however, essential that all interstitial sites should be occupied. For example, in ZnH and TiH, only half of the sites are filled and in Pd_2H, only a quarter. Because of the relatively small size of tetrahedral holes, they tend to accommodate only the very small hydrogen atom.

The *number of octahedral holes* in a close-packed structure is equal to the total number of metal spheres. If, therefore, all the holes are occupied, the empirical formula of the compound corresponds to MY. Interstitial structures containing boron, carbon, and nitrogen have tremendous technical importance as they are extremely hard and resistant to oxidation, and have very high melting points (e.g. HfC, 4160 °C). Furthermore, the appearance of such compounds usually resembles that of the metal; and many of the other metallic properties, such as electrical conductivity and the ability to catalyse reactions, are also retained.

Structures of an interstitial nature are sometimes *non-stoichiometric*, i.e. the atoms of the different elements are not present in simple, whole-number ratios. The palladium-hydrogen system is typical and, under normal conditions, has a limiting composition approximating to $PdH_{0.7}$.

Coordination number

The maximum number of spheres (atoms or ions) that can be packed around a central sphere is referred to as the **coordination number** of the central sphere. In the case of ionic compounds, it relates to the maximum number of ions which can be accommodated around a central, oppositely-charged ion. For those compounds in which there are equal numbers of oppositely charged ions, e.g. Cs^+Cl^-, both ions must have the same coordination number. In this particular case the coordination number is eight.

Hexagonal and cubic close-packed structures having a coordination number of 12, or body-centred cubic with a coordination number of 8, are possible only when all the spheres, or atoms, are identical, and hence arise only for elements in the solid state.

Coordination numbers must, of course, be related to the relative sizes of the participating ions. This physical distribution influences the packing and hence the geometrical arrangement of the crystal. The anion (the ratio, electrons/protons > 1, see page 84) is larger than the cation (the ratio, electrons/protons < 1), and the ratio of the radius of the cation, r_c, to the radius of the anion, r_a, is known as the **radius ratio** (r_c/r_a).

There are therefore certain **limiting ratios**, based on geometrical considerations, which determine the maximum number of ions which can surround other oppositely-charged ions (Table 3.2).

Table 3.2

Radius ratio	*Coordination number*	*Geometrical arrangement*
1.00 to 0.73	8	Primitive cubic
0.73 to 0.41	6	Octahedral
0.41 to 0.22	4	Tetrahedral
0.22 to 0.15	3	Plane triangular

The limiting condition for a small positive ion, A^+, surrounded by a plane triangular arrangement of negative ions, X^-, is illustrated in Figure 3.15.

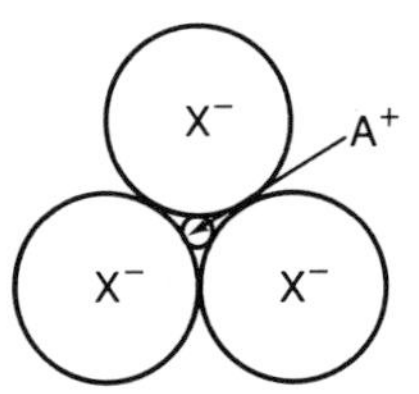

Fig. 3.15

In the majority of cases, there is a close correlation between the geometrical arrangements of ionic crystals observed experimentally, and those predicted by theoretical radius ratios. A knowledge of ionic radii, and hence radius ratio, can therefore be used to predict with reasonable accuracy the coordination number and shape of unknown ionic crystals.

Some ionic crystals

Ionic (electrovalent) bonding

In covalent structures, the constituent atoms acquire a stable electronic configuration (usually that of a noble gas) by sharing electrons. In *ionic compounds*, however, this ideal state is achieved by one type of atom losing electrons to form *positive ions* (*cations*) and another type gaining these electrons to form *negative ions* (*anions*). **Ionic bonds** *involve these oppositely charged species being held together in the form of a crystal lattice by means of strong attractive forces.*

In the formation of, say, sodium chloride from its constituent elements, the outer 3*s* electron of the sodium is lost and acquired by a chlorine atom, with the result that both ions have filled outer shells electrons, thus:

$$\underset{([Ne]3s^1)}{Na} + \underset{([Ne]3s^23p^5)}{Cl} \longrightarrow \underset{([Ne])}{Na^+}\ \underset{([Ar])}{Cl^-}$$

where [Ne] and [Ar] represent the electronic structures corresponding to neon and argon respectively.

Unlike covalent molecules, which are generally capable of existing as discrete entities in their own right, ionic solids comprise positive ions surrounded by a number of negative ions and *vice versa*. These arrangements of ions form a three-dimensional lattice of indefinite size. Consequently, each ion is associated with several oppositely-charged ions and not specifically with any one in particular. As oppositely-charged ions are nearer to each other than ions of like charge, the attractive forces outweigh the repulsive forces. Therefore, bonding within the crystal is strong, generally giving rise to high melting and boiling points for such compounds.

Once the centres of these oppositely-charged ions come within a certain close proximity, this attractive force is counter-balanced by strong repulsive forces between the outer electrons of the different ions. An equilibrium position is therefore reached between oppositely-charged ions, and it is in this sense that they are considered to possess a definite ionic radius (see page 43).

Sodium chloride structure

The experimentally determined radius of the sodium cation, Na^+, is 0.095 nm and that of the chloride anion, Cl^-, is 0.181 nm, giving a radius ratio of 0.55. This value lies within the range 0.73–0.41 nm and corresponds to a coordination number of six. As the stoichiometric ratio (i.e. the ratio between component ions) for sodium chloride is 1:1, then both the cation and the anion must have the same coordination number which, in this case, is referred to as 6:6 *coordination.* Each sodium ion is therefore octahedrally surrounded by six chloride ions and *vice versa* (Figures 3.16 and 3.17).

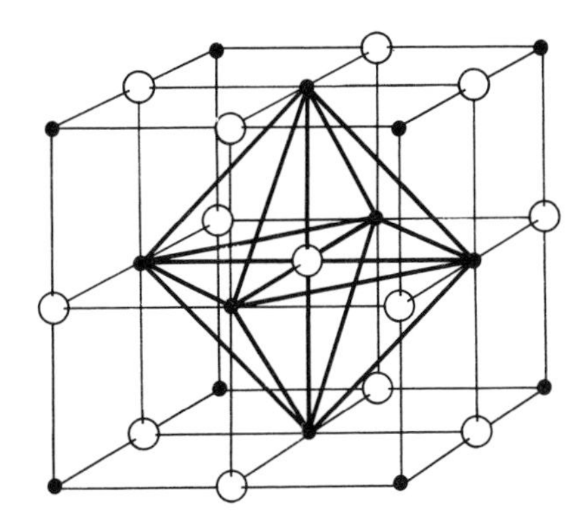

Fig. 3.16 Unit cell of sodium chloride depicting the octahedral coordination

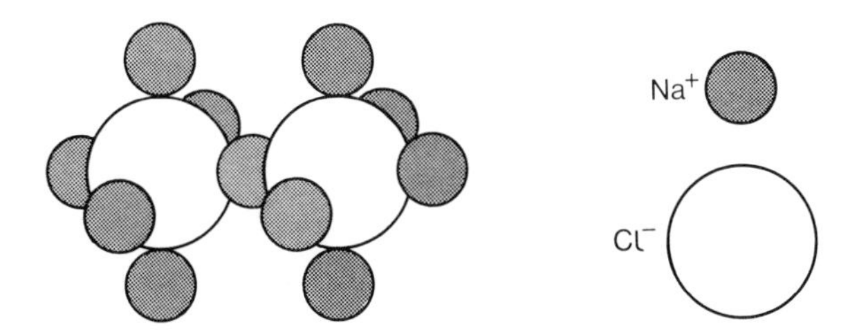

Fig. 3.17 Type of coordination in sodium chloride

One way of visualizing the construction of the crystal is to consider the chloride ions as forming an idealized close-packed cubic lattice with the smaller sodium ions occupying the octahedral holes. However, as the sodium ions are too large to fit directly into these holes, the 'host' chloride ions have to be 'pushed' further apart, therefore creating a more open structure.

From a cursory observation of the layers of close-packed chloride ions (or alternatively sodium ions), it is not at all obvious how the unit cell is constructed, mainly because the faces of the cell are not parallel to the packing layers which are, in fact, perpendicular to the diagonal faces of the unit cell (Figure 3.18).

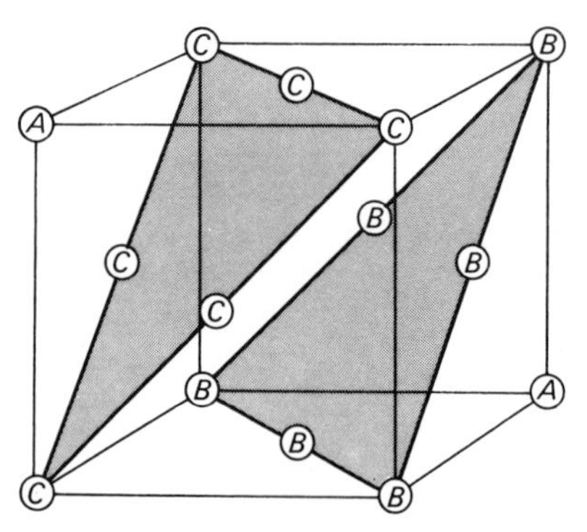

Fig. 3.18 *ABCA* . . . packing layers in a face-centred unit cell

The resultant structure is composed of an interlocking of the face-centred cubic structure of chloride ions with the face-centred cubic structure of sodium ions.

Most of the alkali metal halides and the alkaline earth oxides and sulphides have a sodium chloride-type structure (Table 3.3). In some of these, there is a lack of correlation between the radius ratio value and the expected coordination number (refer to Table 3.2), probably arising from energy and stability factors.

Table 3.3 Further examples of face-centred cubic structures and their corresponding radius ratios

Compound	*Radius ratio*
Potassium chloride	0.73
Potassium bromide	0.68
Potassium iodide	0.62
Calcium oxide	0.71
Calcium sulphide	0.54

A compound which shows the face-centred cubic structure does not have to be ionic; many covalent compounds also crystallize in this way.

Caesium halide structure

All caesium halides (with the exception of CsF, which has a sodium chloride type structure) have a simple cubic lattice structure in which each caesium ion is surrounded by eight halide ions and each halide ion is similarly surrounded by eight caesium ions, i.e. 8:8 *coordination*. A coordination number of 8 is the maximum possible for a crystal containing equal numbers of two kinds of ion (i.e. stoichiometric ratio of 1:1) and is not common.

For caesium chloride, the radius ratio is given by:

$$\frac{r_{Cs+}}{r_{Cl-}} = \frac{0.169}{0.181} = 0.93 \quad \text{(see Table 3.2)}$$

The resultant structure is composed of two interpenetrating primitive cubic lattices, arranged as shown in Figures 3.19 and 3.20.

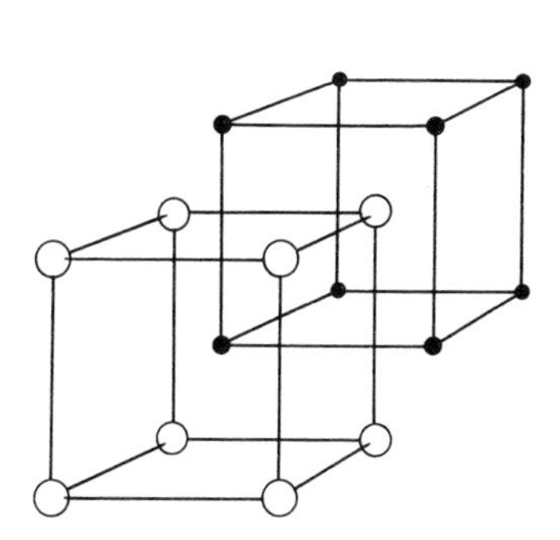

Fig. 3.19 Two interpenetrating cubic lattices in caesium chloride, showing the oppositely charged ion at the centre of each cube

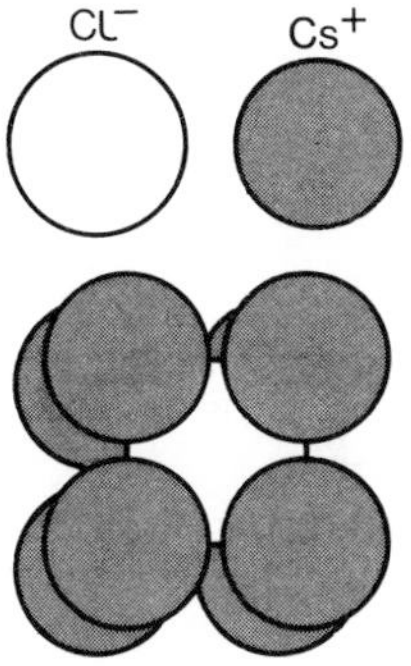

Fig. 3.20 Type of coordination in caesium chloride

Covalent crystals

Simple molecular crystals

Unlike the ionic crystals considered so far, many crystals – particularly those of organic compounds (e.g. benzenecarboxylic (benzoic) acid and naphthalene) – contain well-defined molecules. Such structures are referred to as molecular crystals. The individual molecules are held together by weak van der Waals-type forces and their general properties are characteristic of this type of bonding. In general, they are far less rigid structures, softer, and with notably lower melting and boiling points than typical ionic crystals. As van der Waals forces do not in themselves lead to delocalization of electrons, or to the presence of mobile electrons, molecular crystals are poor conductors of heat and electricity. Even in the gaseous state, when these intermolecular forces are broken down, the covalent nature of the individual molecules still ensures virtually no conduction of electricity. Iodine (Figure 3.21) is a case in point. The individual molecules crystallize out in the orthorhombic form, in which the bond length of the molecules is 0.27 nm and the intermolecular distance is 0.354 nm.

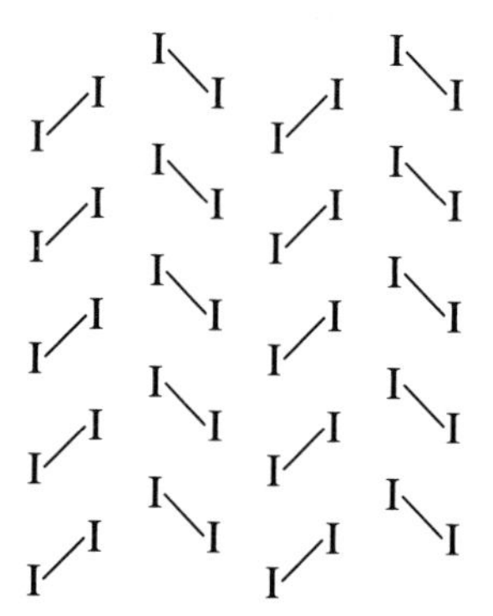

Fig. 3.21 Part of an assembly of iodine molecules in a molecular crystal of the element

In solid phosphorus (white), tetrahedral P_4 units pack together in a cubic system. Its low melting point (44 °C) serves to emphasize its molecular crystallinity.

Naphthalene, $C_{10}H_8$, is another illustration; the flat, discrete organic molecules (which may be likened to two fused benzene rings) crystallize in the monoclinic form.

Giant structures

Crystals of some non-metallic elements possess a network of ordinary covalent bonds throughout the whole of the structure. In the majority of cases, all available electrons are involved in covalent bond formation and hence the crystals generally tend to be poor conductors of heat and electricity. In contrast to simple molecular crystals, certain giant crystal structures are relatively hard and have high melting and boiling points, the latter two properties being largely attributable to the great quantity of energy required to break down the covalent linkages throughout the structure.

Diamond and graphite crystal structures

Carbon exists in two allotropic, crystalline forms: diamond and graphite. The atoms of diamond, which is colourless, are sp^3 hybridized giving a 4:4 co-ordination in which each carbon is tetrahedrally surrounded by four other carbons, the distance between each of the atoms in this saturated system being 0.154 nm. This three-dimensional structure (Figure 3.22) causes diamond to be extremely hard and to have an abnormally high melting point (3550 °C). As all available electrons are involved in bond formation, diamond does not conduct electricity and is a poor conductor of heat.

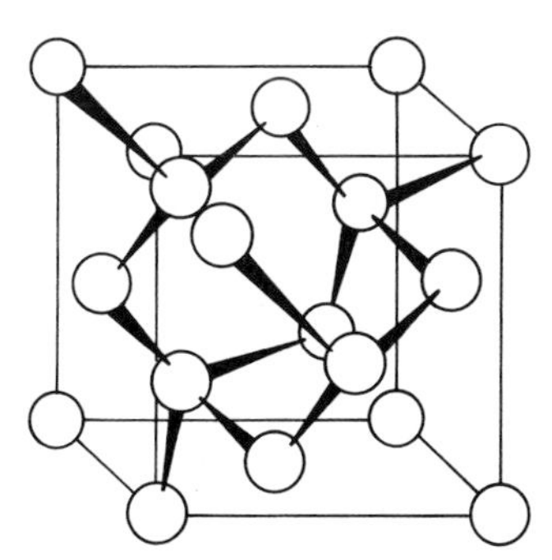

Fig. 3.22 Crystal structure of diamond

Not all giant crystals of this type are three-dimensional. In graphite, Figure 3.23, the carbon atoms are sp^2 hybridized, leading to a planar structure of fused benzene-like hexagonal rings in which the different layers are held together by weak van der Waals forces. The fourth electron is contained in a p orbital at right angles to the plane of the carbon atoms and forms a π bond. These π electrons are delocalized and account for the exceptionally good electrical conductivity of graphite.

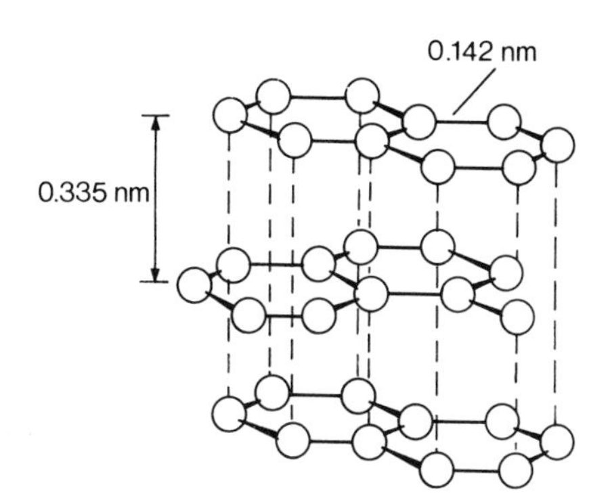

Fig. 3.23 Crystal structure of graphite

Owing to the unsaturated nature of the crystal, the bond lengths in graphite (0.142 nm) are predictably shorter than those in diamond. The distance between the planes is comparatively large (0.335 nm), enabling the crystal to be easily cleaved. It is this wide separation of the layers which gives graphite its lubricating properties, contrasting with the abrasive properties of diamond.

Some other covalent crystals

Two-dimensional giant structures are displayed by the elements arsenic and antimony, and by certain compounds which include $AlCl_3$, $Al(OH)_3$, $FeCl_3$, SiO_2, and the complex micas, e.g. kaolin and talc.

Where a layer structure is formed which has only one or two atoms across its width and covalent bonding prevails throughout, it is termed a one-dimensional chain. Selenium (Figure 3.24), plastic sulphur, and copper(II) chloride are included in this category.

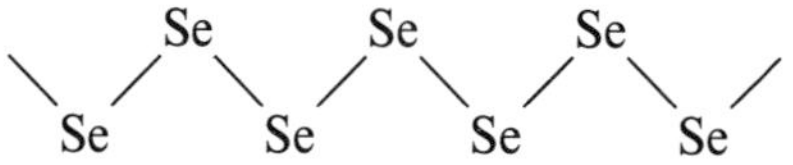

```
       Se       Se       Se
 \    /  \     /  \     /  \    /
   Se      Se       Se       Se
```

Fig. 3.24 One-dimensional chain structure of selenium

4 Energetics

In all chemical reactions, energy, usually in the form of heat, is required to enable the necessary bonds in the reacting substances to be broken. Conversely, the making of new bonds during the formation of the products involves an evolution of energy. For reactions taking place at constant pressure, the net energy change in forming the products from the reactants is referred to as the **enthalpy of reaction**, ΔH. When more energy is evolved in bond-making than is absorbed in bond-breaking, *ΔH is negative*, implying an overall loss of energy by the system*. Such a reaction is said to be **exothermic** (Figure 4.1(a)). Conversely, when less energy is evolved than absorbed, *ΔH is positive*, and the reaction is said to be **endothermic** (Figure 4.1(b)).

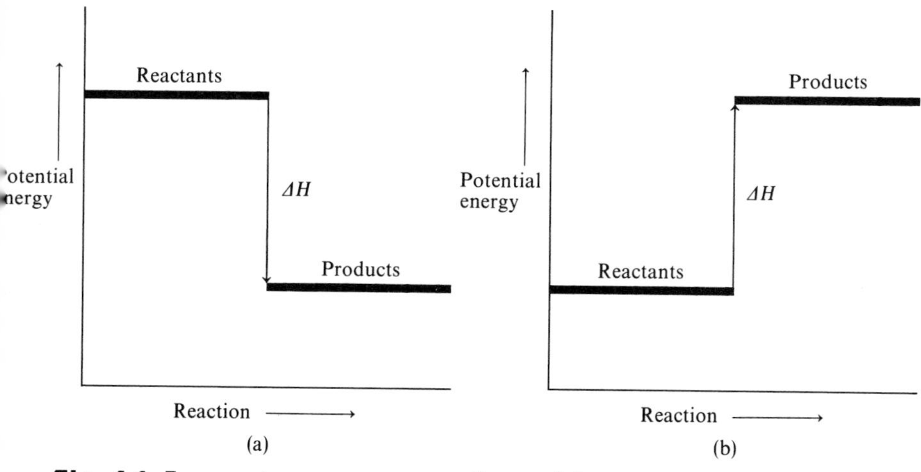

Fig. 4.1 Energy changes corresponding to (a) an exothermic reaction and (b) an endothermic reaction

Products formed via an exothermic process (referred to as *exothermic compounds*) have a lower energy content than the reactants from which they are formed, whereas products formed via an endothermic process (referred to as *endothermic compounds*) have a higher energy content than the parent reactants. The majority of compounds are formed exothermically and are generally more stable than their endothermic counterparts.

---

*In the context of thermodynamics, the particular quantity of material under investigation is referred to as the *system*, while everything else is called the *surroundings*. With regard to a chemical reaction, the system relates to the species taking part in, and formed by, the reaction.

Standard enthalpies

Before any realistic comparison can be made between enthalpy values, it is necessary to standardize the amount of material involved and the conditions under which the measurements are made.

The standard enthalpy of reaction, $\Delta H^{\ominus}$, *refers to the enthalpy change which takes place when molar quantities of the reactants, as specified by the chemical equation, react to form the products at 298 K (25 °C) and a pressure of one atmosphere* (1.013×10^5 Pa). Generally, pressure tends only to be of significance for reactions involving gases.

Convention requires that *elements in their stable state* under standard conditions be assigned *zero enthalpy*. This proves to be convenient, as we are concerned primarily with changes in enthalpy brought about by a chemical change rather than specific values relating to particular elements or compounds.

The precise physical state and allotropic form, where relevant, must always be clearly specified as changes in state and allotropic form, even at a fixed temperature, involve energy changes.

$$H_2(g)+\tfrac{1}{2}O_2(g) \longrightarrow H_2O(g); \quad \Delta H^{\ominus} = -242\,\text{kJ mol}^{-1}$$
$$H_2(g)+\tfrac{1}{2}O_2(g) \longrightarrow H_2O(l); \quad \Delta H^{\ominus} = -286\,\text{kJ mol}^{-1}$$
$$\text{i.e.} \quad H_2O(g) \longrightarrow H_2O(l); \quad \Delta H^{\ominus} = -\ 44\,\text{kJ mol}^{-1}$$

$$C(\text{graphite})+O_2(g) \longrightarrow CO_2(g); \quad \Delta H^{\ominus} = -393.5\,\text{kJ mol}^{-1}$$
$$C(\text{diamond})+O_2(g) \longrightarrow CO_2(g); \quad \Delta H^{\ominus} = -395.4\,\text{kJ mol}^{-1}$$
$$\text{i.e.} \quad C(\text{graphite}) \longrightarrow C(\text{diamond}); \quad \Delta H^{\ominus} = +\ 1.9\,\text{kJ mol}^{-1}$$

There are, however, various ways in which the enthalpy of reaction may be categorized. Some of the important ones are outlined below and during the further course of this chapter.

The standard enthalpy of combustion, $\Delta H_c^{\ominus}$, *refers to the enthalpy change which takes place when one mole of a substance is completely burned in oxygen under standard thermodynamic conditions*, e.g.,

$$CH_3OH(l)+\tfrac{3}{2}O_2(g) \rightarrow CO_2(g)+2H_2O(l); \quad \Delta H_c^{\ominus} = -715\,\text{kJ mol}^{-1}$$

The enthalpy of hydrogenation, ΔH_{H_2}, *refers to the enthalpy change when one mole of an unsaturated gaseous compound is completely converted into a saturated compound by treatment with gaseous hydrogen under a pressure of one atmosphere* (1.013×10^5 Pa). (Values are sometimes quoted in literature at a specified elevated temperature.) For example,

$$CH_2{=}CH_2(g)+H_2(g) \rightarrow CH_3CH_3(g); \quad \Delta H_{H_2}^{\ominus} = -138\,\text{kJ mol}^{-1}$$

The standard enthalpy of formation, $\Delta H_f^{\ominus}$, *refers to the enthalpy change when one mole of a compound is formed from its constituent elements in their standard states*, e.g.

$$C(s)+\tfrac{5}{2}H_2(g)+\tfrac{1}{2}N_2(g) \rightarrow CH_3NH_2(g); \quad \Delta H_f^{\ominus} = -28\,\text{kJ mol}^{-1}$$

For the majority of compounds, including the above example of methylamine, direct synthesis from the constituent elements is not feasible and the enthalpy of formation has to be derived by the application of Hess's law.

Hess's Law of Heat Summation (1840)

This states that *for a given chemical process the enthalpy change is the same regardless of whether the reaction takes place in a single stage or via several intermediate stages.*

This is just one further aspect of the first law of thermodynamics*, enabling energy values which are not measurable directly to be obtained indirectly by adding, subtracting, and multiplying chemical equations. For example, methane cannot be synthesized by direct combination of carbon and hydrogen:

$$C(s) + 2H_2(g) \longrightarrow CH_4(g)$$

but its enthalpy of formation, $\Delta H_f^\ominus$, can be calculated from the enthalpy changes of combustion of methane, carbon, and hydrogen by application of Hess's law:

$$\begin{array}{lll} (1) & CH_4(g) + 2O_2(g) \longrightarrow CO_2(g) + 2H_2O(l); & \Delta H_1^\ominus = -890\ \text{kJ mol}^{-1} \\ (2) & C(s) + O_2(g) \longrightarrow CO_2(g); & \Delta H_2^\ominus = -393.5\ \text{kJ mol}^{-1} \\ (3) & H_2(g) + \tfrac{1}{2}O_2(g) \longrightarrow H_2O(l); & \Delta H_3^\ominus = -286\ \text{kJ mol}^{-1} \end{array}$$

These enthalpy changes, together with the enthalpy of formation of methane, can be illustrated by means of a simple energy diagram (Figure 4.2):

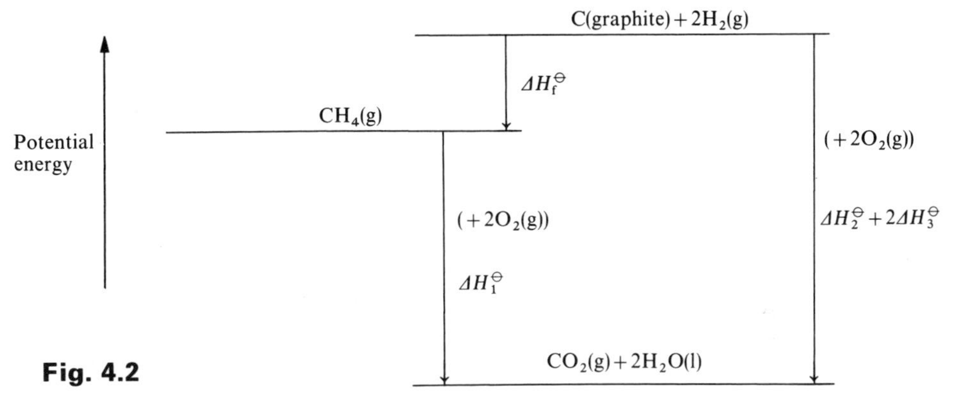

Fig. 4.2

From this,

$$\begin{aligned} \Delta H_f^\ominus &= \Delta H_2^\ominus + 2\Delta H_3^\ominus - \Delta H_1^\ominus \\ &= -393.5 + (-2 \times 286) - (-890)\ \text{kJ·mol}^{-1} \\ &= -75.5\ \text{kJ mol}^{-1} \end{aligned}$$

Because methane is an exothermic compound (like the majority), its constituent elements are represented at a higher energy level than the compound which

---

*Although it may be stated in a variety of ways, the *first law of thermodynamics* is probably best understood as Lavoisier and Laplace's law of *conversation of energy* (1780) which states that *energy can neither be created nor be destroyed.*

they form. Conversely, for endothermic compounds the constituent elements will be represented at a lower level than the compound which they form.

For processes involving several stages, the enthalpy change for the reaction is equal to the algebraic sum of the enthalpy changes for each individual stage.

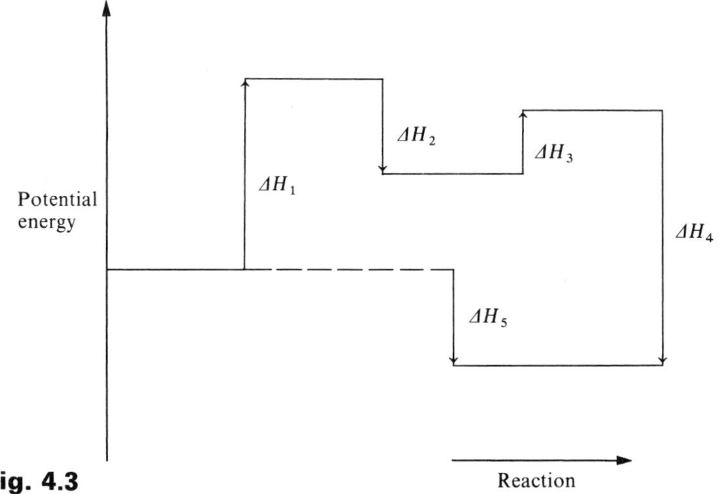

Fig. 4.3

The overall enthalpy change for the exothermic process in Figure 4.3 is given by:

$$\Delta H_5 = \Delta H_1 + \Delta H_2 + \Delta H_3 + \Delta H_4$$

ΔH_1 and ΔH_3 are endothermic quantities and have positive values, whereas ΔH_2, ΔH_4, and, in this particular example, ΔH_5, are exothermic quantities and have negative values.

Bond enthalpy and bond dissociation enthalpies

Standard bond enthalpies and standard bond dissociation enthalpies relate to the separation of the constituent atoms of a molecule in the gaseous state to an infinite distance, under standard conditions.

For a *diatomic molecule, AB*, the **bond enthalpy** (still commonly called bond energy) is defined as *the enthalpy change for the process*:

$$AB(g) \rightarrow A(g) + B(g)$$

For a *molecule of the type*, AB_2, the mean bond enthalpy is defined as *half the enthalpy change for the process*:

$$AB_2(g) \rightarrow A(g) + 2B(g)$$

and so on for polyatomic molecules. As such processes will require energy for breaking the bonds, the enthalpy values will be positive.

Consider methane as a simple example of a polyatomic molecule:

$$CH_4(g) \rightarrow C(g) + 4H(g); \quad \Delta H^{\ominus} = +1648\,kJ\,mol^{-1}$$

$$\text{Therefore, the standard bond enthalpy} = +\frac{1648}{4}\,kJ\,mol^{-1} = +412\,kJ\,mol^{-1}$$

Experimental evidence suggests that there is usually little difference between the strength of a particular type of bond in compounds of similar nature. For example, the strength of a C—H bond in methane (CH_4) is virtually the same as that in ethane (C_2H_6), and that of an O—H bond in water is virtually the same as that in hydrogen peroxide. The values are usually obtained by spectroscopic or thermochemical means.

Bond dissociation enthalpies differ from bond enthalpies in that they *refer specifically to the cleaving of a particular bond in a molecule.* This may be illustrated as follows: the enthalpy change required to remove one hydrogen atom away from water under standard conditions,

$$H_2O(g) \longrightarrow HO(g) + H(g)$$

is equal to $+496\,kJ\,mol^{-1}$, whereas the enthalpy change required to remove the second hydrogen,

$$HO(g) \longrightarrow O(g) + H(g)$$

is appreciably lower, being equal to $+429\,kJ\,mol^{-1}$.

Table 4.1 Some mean standard bond enthalpies

Bond	*$\Delta H^{\ominus}$/kJ mol^{-1}*	*Bond*	*$\Delta H^{\ominus}$/kJ mol^{-1}*
H—H	+436	C—H	+412
C—C	+348	O—H	+463
C⎓C (benzene)	+518	F—H	+562
C=C	+612	Cl—H	+431
C≡C	+837	Br—H	+366
O—O	+146	I—H	+299
O=O	+496	C—O	+360
F—F	+158	C=O	+743
Cl—Cl	+242	C—F	+184
Br—Br	+193	C—Cl	+338
I—I	+151	C—Br	+276
		C—I	+238

A knowledge of bond enthalpies enables the enthalpy of reaction to be determined for a vast number of processes. For example, the standard enthalpy change for the process:

$$H_2(g) + Cl_2(g) \longrightarrow 2HCl(g)$$

may be calculated as follows, the procedure being illustrated by means of the simple energy diagram in Figure 4.4.

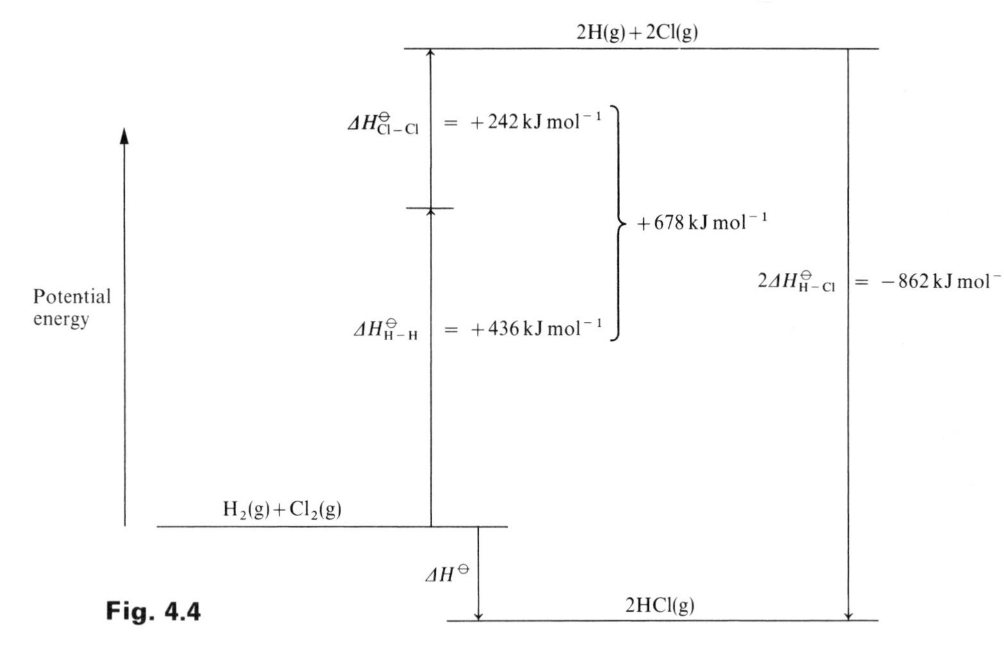

Fig. 4.4

$$\text{Enthalpy change for the reaction, } \Delta H^{\ominus} = +678 - 862 \text{ kJ mol}^{-1}$$
$$= -184 \text{ kJ mol}^{-1}$$

Standard enthalpy of atomization, $\Delta H^{\ominus}_{at}$, is defined as *the enthalpy change needed to form one mole of free gaseous atoms from an element in its normal physical state under standard thermodynamic conditions.* It may equally well be referred to as the standard enthalpy of formation of gaseous atoms.

For elements which exist naturally as gaseous diatomic molecules, the standard enthalpy of atomization will be equal to half the standard bond enthalpy. For example, the standard bond enthalpy of hydrogen corresponds to the change,

$$H_2(g) \rightarrow 2H(g); \quad \Delta H^{\ominus} = +436 \text{ kJ mol}^{-1}$$

whereas the standard enthalpy of atomization corresponds to the change:

$$\tfrac{1}{2}H_2(g) \rightarrow H(g); \quad \Delta H^{\ominus} = +218 \text{ kJ mol}^{-1}.$$

In the case of solid or liquid elements, a further quantity of energy is first of all required to convert the solid or liquid to the gaseous phase. For solid iodine, this involves the enthalpy of sublimation, comprising the enthalpy of fusion and the enthalpy of vaporization:

$$\tfrac{1}{2}I_2(s) \xrightarrow[\text{enthalpy}]{\text{sublimation}} \tfrac{1}{2}I_2(g); \quad \Delta H^\ominus = +31\,\text{kJ mol}^{-1}$$

$$\tfrac{1}{2}I_2(g) \xrightarrow[\text{enthalpy}]{\frac{1}{2}\text{ bond diss.}} I(g); \quad \Delta H^\ominus = \underline{+76\,\text{kJ mol}^{-1}}$$

$$\underline{+107\,\text{kJ mol}^{-1}}$$

The standard enthalpy of atomization is therefore $+107\,\text{kJ mol}^{-1}$.

For liquid bromine, only the enthalpy of vaporization is involved in converting it into the gaseous phase:

$$\tfrac{1}{2}Br_2(l) \xrightarrow[\text{enthalpy}]{\text{vaporization}} \tfrac{1}{2}Br_2(g); \quad \Delta H^\ominus = +15\,\text{kJ mol}^{-1}$$

$$\tfrac{1}{2}Br_2(g) \xrightarrow[\text{enthalpy}]{\frac{1}{2}\text{ bond diss.}} Br(g); \quad \Delta H^\ominus = \underline{+97\,\text{kJ mol}^{-1}}$$

$$\underline{+112\,\text{kJ mol}^{-1}}$$

The standard enthalpy of atomization of bromine is therefore $+112\,\text{kJ mol}^{-1}$.

Such values are not easily measured and, like bond enthalpies, are usually obtained by spectroscopic methods.

Enthalpy changes in solution

Enthalpies of solution and dilution

The standard enthalpy of solution, $\Delta H^\ominus_{sol}$, *relates to the enthalpy change which takes place when one mole of a substance is completely dissolved in a volume of solvent (usually water) which is sufficiently large to prevent any further enthalpy change on the addition of more solvent. At this point the solution is said to be at infinite dilution, under standard thermodynamic conditions.*

The process may be represented by:

$$AB + \text{solvent} \rightarrow AB(\text{solvent})$$

or for a substance dissolving in water,

$$AB + aq \rightarrow AB(aq)$$

where AB(solvent) and AB(aq) represent the solutions at infinite dilution.

In practice, enthalpies of solution are comparatively easily measured and may be used to determine the standard enthalpies of formation of compounds.

(1) $KOH(s) + aq \rightarrow KOH(aq); \quad \Delta H^\ominus = -\ 55\,\text{kJ mol}^{-1}$

(2) $K(s) + H_2O(l) + aq \rightarrow KOH(aq) + \tfrac{1}{2}H_2(g); \quad \Delta H^\ominus = -195\,\text{kJ mol}^{-1}$

(3) $H_2(g) + \tfrac{1}{2}O_2(g) \rightarrow H_2O(l); \quad \Delta H^\ominus = -286\,\text{kJ mol}^{-1}$

Adding equations (2) and (3),

(4) $K(s) + \tfrac{1}{2}H_2(g) + \tfrac{1}{2}O_2(g) + aq \rightarrow KOH(aq); \quad \Delta H^\ominus = -481\,\text{kJ mol}^{-1}$

Subtracting (1) from (4)

(5) $K(s) + \tfrac{1}{2}H_2(g) + \tfrac{1}{2}O_2(g) \rightarrow KOH(s); \quad \Delta H^\ominus = -426\,\text{kJ mol}^{-1}$

i.e. the enthalpy of formation of KOH(s) is $-426\,\text{kJ mol}^{-1}$.

On dilution of a solution which is not infinitely dilute, a further enthalpy change results and is referred to as the **enthalpy of dilution**. For this value to have any real relevance, both initial and final concentrations of the solution must be specified.

e.g. $KOH(2.0M, aq) + aq \rightarrow KOH(\infty, aq); \quad \Delta H = -1.062\,kJ\,mol^{-1}$

Enthalpy of neutralization

The standard enthalpy of neutralization *refers to the enthalpy change which occurs when an acid and a base react to form a salt plus one mole of water, under standard thermodynamic conditions.* This process is always exothermic, and the value for reactions involving strong acids and bases is virtually constant and has a value close to -57.3 kJ mol^{-1}.

$$NaOH(aq) + HCl(aq) \rightarrow NaCl(aq) + H_2O(l); \quad \Delta H^\ominus = -57.3\,kJ\,mol^{-1}$$
$$KOH(aq) + HNO_3(aq) \rightarrow KNO_3(aq) + H_2O(l); \quad \Delta H^\ominus = -57.3\,kJ\,mol^{-1}$$
$$KOH(aq) + \tfrac{1}{2}H_2SO_4(aq) \rightarrow \tfrac{1}{2}K_2SO_4(aq) + H_2O(l); \quad \Delta H^\ominus = -57.3\,kJ\,mol^{-1}$$

All of the above acids and bases are strong, and therefore virtually fully dissociated into ions in aqueous solution. They are more realistically represented as separate hydrated ions, e.g. $Na^+(aq)$, $OH^-(aq)$ and $H^+(aq)$, and $Cl^-(aq)$.

It is apparent in the first reaction that the $Na^+(aq)$ and $Cl^-(aq)$ ions remain unchanged ('spectator' ions) during the chemical process. The heat-producing reaction must therefore be attributable to the process:

$$OH^-(aq) + H^+(aq) \rightarrow H_2O(l)$$

Similar reasoning can be applied to all neutralization reactions involving a strong acid and a strong base.

If either the acid or the base is weak, and therefore not fully dissociated into ions, then the enthalpy of neutralization is smaller in magnitude than -57.3 kJ mol^{-1}, as energy is required to complete the dissociation of the weak acid or base as the reaction proceeds.

The standard enthalpy of neutralization for the reaction between ethanoic (acetic) acid (a weak acid) and sodium hydroxide is equal to -56.1 kJ mol^{-1}.

$$CH_3COOH(aq) + NaOH(aq) \rightarrow CH_3COONa(aq) + H_2O(l); \quad \Delta H^\ominus = -56.1\,kJ\,mol^{-1}$$

Ethanoic acid, being weak, is only partially dissociated into ions:

$$CH_3COOH(aq) \rightleftharpoons CH_3COO^-(aq) + H^+(aq)$$
$$\downarrow\uparrow\; OH^-(aq)$$
$$H_2O(l)$$

and as the $OH^-(aq)$ ions are added, they remove $H^+(aq)$ ions from the equilibrium, forcing the system to produce more $H^+(aq)$ ions (Le Chatelier's principle). This process requires energy and thus reduces the amount of heat evolved during neutralization.

Solubility of ionic compounds

Lattice energy and solvation energy

In order for a solid substance to dissolve in a solvent, energy is first required to break down the crystal structure of the solid. This quantity is referred to as the *lattice energy*. In the case of ionic crystals, this involves breaking them down into discrete ions.

The standard lattice enthalpy, $\Delta H^{\ominus}_{lat}$, *is defined as the enthalpy change for the conversion of one mole of a crystal into its constituent ions in the gaseous state and separated to an infinite distance from each other, under standard thermodynamic conditions.*

e.g. $$MX(s) \rightarrow M^{+}(g) + X^{-}(g)$$

If this energy were to be obtained directly from the solvent, then the amount of heat required would, in many cases, cause the liquid to freeze. Furthermore, as many dissolution processes are exothermic, and even those processes that are endothermic generally involve only a small absorption of heat, then some other factor must clearly be involved. This factor is the **solvation energy**, and as this is a *type of bond-making process* in which positive ions attract and become associated with the negative end of the dipole of the solvent molecules and *vice versa, heat is evolved*:

$$\Delta H_{\substack{\text{solvation}\\ \text{(hydration)}}} = \Delta H_{\text{solution}} - \Delta H_{\text{lattice}}$$

When the solvent molecules are water, which is very often the case, the process is referred to as *hydration* (see Figure 4.5), in which case the solvation energy is usually referred to as the **hydration energy**.

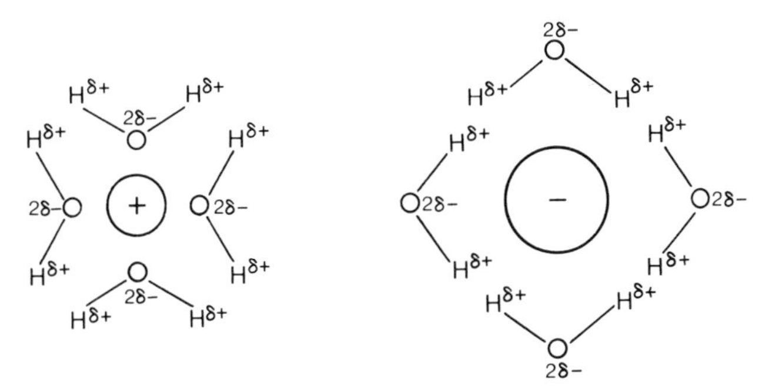

Fig. 4.5 Hydration

All ions, irrespective of their charge, attract the polar water molecules towards themselves and become hydrated. This hydrated ion may be considered as an aqua complex $[M(H_2O)_x]^{n\pm}$, or simply $M^{n\pm}(aq)$. In solution the solvent molecules are generally only loosely attached to the central ion and exist in equilibrium with other solvent molecules. The number of solvent molecules associated with any particular ion at any moment is referred to as the **solvent number** x (cf. coordination number, page 43).

The standard enthalpy of hydration, $\Delta H^{\ominus}_{hyd}$, *relates to the enthalpy change when one mole of an ion in the gaseous state is hydrated, under standard thermodynamic conditions.*

$$M^{n\pm}(g) + aq \rightarrow M^{n\pm}(aq)$$

Small and highly charged ions undergo hydration most easily, evolving a considerable amount of heat, which, in the case of soluble compounds, provides the necessary energy to overcome the lattice energy of the dissolving solid.

Table 4.2

Cation	*Ionic radius/nm*	$\Delta H^{\ominus}_{hyd}$/ *kJ mol*$^{-1}$
H^+		−1075
Li^+	0.060	−519
Na^+	0.095	−406
K^+	0.133	−322
Mg^{2+}	0.065	−1920
Ca^{2+}	0.099	−1650
Al^{3+}	0.050	−4690

Anion	*Ionic radius/nm*	$\Delta H^{\ominus}_{hyd}$/ *kJ mol*$^{-1}$
OH^-		−460
F^-	0.136	−506
Cl^-	0.181	−364
Br^-	0.195	−335
I^-	0.216	−293

Whether or not a compound is soluble in water depends upon the relative values of the lattice enthalpy of the compound and the hydration enthalpies of the ions. If the lattice enthalpy exceeds the sum of the hydration enthalpies of the separate ions to any considerable extent, then there will be insufficient energy to break down the lattice and the substance will remain undissolved.

Table 4.3

Compound	*Lattice enthalpy,* $\Delta H^{\ominus}$/*kJ mol*$^{-1}$	*Enthalpies of hydration of ions,* $\Delta H^{\ominus}$/*kJ mol*$^{-1}$	*Overall enthalpy change,* $\Delta H^{\ominus}$/*kJ mol*$^{-1}$
NaCl	+771	−770	+ 1 (soluble)
NH_4Cl	+640	−665	−25 (soluble)
AgCl	+905	−830	+75 (insoluble)

The overall enthalpy change corresponds to the enthalpy of solution, for those compounds which are soluble.

As the lattice enthalpy of a compound and the hydration enthalpies of its ions are both favoured by small, highly-charged ions, predicting the extent to which the compound will dissolve is a much more difficult process than simply predicting the likelihood of it dissolving. The complexity of the problem lies in the respective relationships between the size of the ions, and the lattice and hydration enthalpies. The *lattice enthalpy is inversely proportional to the sum of the radii of the ions*, whereas *the sum of the hydration enthalpies of the ions is inversely proportional to the individual ionic radii.*

As a general rule, *the most soluble compounds tend to be those in which the disparity in size of the constituent ions is greatest.* For example, caesium fluoride, CsF, which comprises the largest cation, Cs^+, and the smallest anion, F^-, is by far the most soluble of the alkali metal halides. Conversely, the least soluble of these halides are those in which the difference in size between the cation and anion is smallest, e.g. LiF and NaF.

The Born–Haber cycle

This is, in effect, a further application of Hess's law and the first law of thermodynamics. It enables quantities, which are difficult or impossible to measure by direct means, to be acquired indirectly by a step-wise approach. Examples include the oxidizing power of the halogens (see page 199), the acidic strength of the halogen acids (see page 203), enthalpies of hydration, lattice enthalpies, enthalpies of formation of ionic compounds, and electron affinities (see page 83).

The procedure for obtaining the enthalpy of solution, ΔH_{sol}, for an ionic compound, such as an alkali metal halide, $M^+X^-(s)$, may be represented in the form of a simple Born-Haber cycle (Figure 4.6).

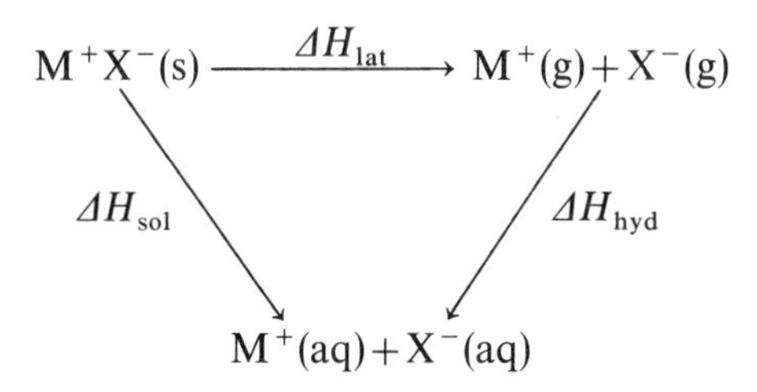

Fig. 4.6 A Born-Haber cycle

i.e. $$\Delta H_{sol} = \Delta H_{lat} + \Delta H_{hyd}$$

remembering that ΔH_{lat} is an endothermic quantity and ΔH_{hyd} is an exothermic quantity. ΔH_{sol} may be positive or negative and its value will be small by comparison with ΔH_{lat} and ΔH_{hyd}, especially if it corresponds to an endothermic quantity.

Consider now a more involved application of this principle to determine the enthalpy of formation of an alkali metal halide, $M^+X^-(s)$, from its constituent elements in their stable state (Figure 4.7).

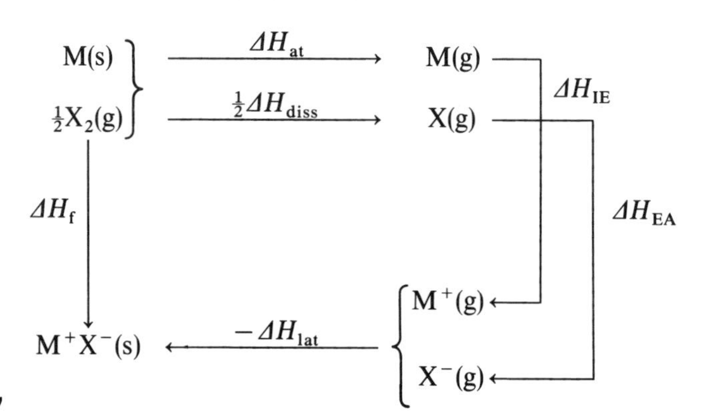

Fig. 4.7

$$\Delta H_f = \Delta H_{at} + \tfrac{1}{2}\Delta H_{diss} + \Delta H_{IE} + \Delta H_{EA} - \Delta H_{lat}$$

The energy involved in the conversion of gaseous atoms into ions is known as the ionization energy, ΔH_{IE}, (for the formation of positive ions) of the electron affinity, ΔH_{EA}, (for the formation of negative ions). These are discussed further on pages 80 and 83.

In order to form a stable crystal lattice, it is essential that the overall process for the enthalpy of formation is exothermic. The magnitude of ΔH_f therefore gives an indication of the stability of a compound.

Although, like the enthalpy of formation, electron affinities are not measurable directly, many were originally obtained by applying the Born-Haber cycle to compounds whose lattice enthalpies were already known. These values were then used to determine the lattice enthalpy of different compounds containing the same negative ions.

The thermodynamic criterion determining the feasibility of a chemical change is that the *change in free energy*, ΔG, for the process must be negative. ΔG is related to ΔH by the expression:

$$\Delta G = \Delta H - T\Delta S$$

T being the temperature in kelvin, and ΔS *the change in entropy.*

Entropy is a fundamental thermodynamic concept which, for simplicity, is probably most conveniently considered as being a measure of *the degree of disorder within a system.* Orderly systems have low entropy and highly disordered systems have high entropy. Increases in entropy occur when a crystalline solid is dissolved in a solvent, when a few gas molecules react to form a larger number of gas molecules, when water diffuses through a semi-permeable membrane into an aqueous solution, when two gases mix spontaneously when placed in contact, and so on.

ΔG represents the maximum quantity of work that a reaction can provide for external work under constant conditions of temperature and pressure, and it is only when the term $T\Delta S$ can be disregarded that ΔH can really satisfactorily replace ΔG.

A simple summary guide as to the spontaneity (or otherwise) of a chemical reaction is outlined below.

(1) If ΔH is positive and ΔS is positive such that $\Delta H < T\Delta S$, then ΔG is negative and the reaction proceeds spontaneously.

(2) If ΔH is negative and ΔS is negative or positive such that $\Delta H > T\Delta S$, then ΔG is negative and the reaction proceeds spontaneously.

(3) If ΔH and ΔS are both either positive or negative such that $\Delta \text{H} = T\Delta S$, then ΔG is equal to zero. This indicates that there is no tendency for the composition of the system to change and the reaction is therefore in equilibrium.

(4) If ΔH is positive and ΔS is positive such that $\Delta H > T\Delta S$, then ΔG is positive and the reaction does not take place.

5 Oxidation and Reduction: Redox Potentials

Oxidation is a process in which electrons are lost and **reduction** is a process in which electrons are acquired. As these processes inevitably occur simultaneously, they are referred to as *redox reactions* (an abbreviated form of *red*uction–*ox*idation).

Oxidation state or oxidation number

During the combination of a distinctly electropositive element with a distinctly electronegative one (see page 86) to form an ionic compound, electrons are lost by the electropositive element, which is oxidized, and gained by the electronegative element, which is reduced. The number of electrons lost by the electropositive element denotes the *positive oxidation state, or number*, of that element, and the number of electrons gained by the electronegative element denotes the *negative oxidation state, or number*, of that element. For example:

$Na^{+}Br^{-}$	$Ca^{2+}O^{2-}$
+1 −1	+2 −2

In *covalent compounds*, the negative oxidation state is assigned to the element acquiring the greater share of the bonding electrons (i.e. the more electronegative element), and the positive oxidation state to the more electropositive element. *The signs designated always depend upon the elements involved.* Atoms in the elementary state have an oxidation state of zero. Oxidation states of zero are also encountered in compounds formed between metals and neutral molecules, e.g. metal carbonyls (see page 224).

CH_4	CH_3Cl	CCl_4	CO_2	NH_3	Cl_2
−4+1	−2+1−1	+4−1	+4−2	−3+1	0

The algebraic sum of the oxidation states of the constituent atoms of an uncharged species is zero, and in the case of an ion, is equal to the charge on the ion:

NO_3^-	SO_4^{2-}	MnO_4^-
+5−2	+6−2	+7−2

The oxidation states of the most electronegative and the most electropositive elements can generally be considered to be fixed. For example, fluorine is the most electronegative element and therefore always exhibits an oxidation state of −1 in its compounds. The other halogens (Cl, Br, and I) are always in the −1 state in the halides. Oxygen is the second most electronegative element and therefore has an oxidation state of −2 in most of its compounds, the exceptions being the fluorides and peroxides. At the other end of the scale, the alkali metals of Group I, being the most electropositive elements in the Periodic Table, invariably exhibit an oxidation state of +1. The Group II elements are also sufficiently electropositive to always adopt the +2 state in their compounds.

Table 5.1

Element	*Oxidation state*	*Exceptions*
Group I	+1	—
Group II metals	+2	—
Hydrogen	+1	Metallic hydrides (−1)
Fluorine	−1	—
Chlorine	−1	Fluorides (+1, +3) Oxides (+1, +4, +6, +7)
Oxygen	−2	Peroxides (−1) Fluorides (+1, +2)

'Fixed' states of this type can be of use in determining the oxidation states of other elements present in compounds or ions. For example, the oxidation state of chromium in the dichromate(VI) ion can be obtained as follows:

$$\underset{x\quad -2}{Cr_2O_7^{2-}} \rightarrow 2x+(-2\times 7) = -2$$
$$x = +6$$

For elements within the same group of the Periodic Table in which more than one oxidation state is possible, the lower state tends to become more stable as the group is descended. This is particularly apparent in those elements in the *p*-block which possess inner *d* electrons and follow immediately after the *d*-block or transition elements. Here, despite the increase in the number of electrons, the shielding effect of those in *d* orbitals is weak, therefore allowing the increased nuclear charge to take a firmer hold on the outer electrons. This leads to an *increase in the tendency for the s electrons to remain paired*, (i.e. the *inert pair effect*), leaving only the *p* electrons available to form bonds. This phenomenon is clearly illustrated by the elements thallium (see page 133), lead (see page 145), and bismuth (see page 161).

Disproportionation

When any species (atomic, ionic, or molecular) undergoes both oxidation and reduction simultaneously, it is said to **disproportionate**. This means that in the oxidized form there will be an increase in the oxidation state and, conversely, in the reduced form there will be a decrease in the oxidation state. For example, nitrous acid disproportionates to nitric acid and nitrogen oxide (nitric oxide).

$$\underset{N(+3)}{3HNO_2(aq)} \rightarrow \underset{N(+5)}{HNO_3(aq)} + \underset{N(+2)}{2NO(g)} + H_2O(l)$$

Other common examples include:

$$2Cu^{+}(aq) \rightarrow Cu^{2+}(aq) + Cu(s)$$
$$+1 \qquad\quad +2 \qquad\quad 0$$

$$Hg_2^{2+}(aq) \rightarrow Hg^{2+}(aq) + Hg(l)$$
$$+1 \qquad\quad +2 \qquad\quad 0$$

and

$$3MnO_4^{2-}(aq) + 4H^{+}(aq) \rightarrow 2MnO_4^{-}(aq) + MnO_2(s) + 2H_2O(l)$$
$$Mn(+6) \qquad\qquad Mn(+7) \qquad Mn(+4)$$

Half reactions

The oxidation and reduction processes for a particular reaction can be considered separately in terms of electron transfer. These are referred to as **half reactions** and written as *partial ionic equations*. Consider, for example, the familiar reaction between zinc metal and an aqueous solution of copper(II) sulphate(VI).

$$Zn(s) + Cu^{2+}(aq) \rightarrow Zn^{2+}(aq) + Cu(s)$$

The oxidation half reaction is that in which the zinc is converted into its ions, i.e.

$$Zn(s) \rightarrow Zn^{2+}(aq) + 2e$$

and the reduction half reaction is that in which the copper ions are converted into the metallic form, i.e.

$$Cu^{2+}(aq) + 2e \rightarrow Cu(s)$$

Some of the common oxidizing agents are the oxoanions of the transition metals, particularly the manganate(VII) (permanganate), MnO_4^-, and dichromate(VI), $Cr_2O_7^{2-}$, anions. (These are discussed further on pages 248 and 250.) In acidic solution the reduction half reaction for the manganate(VII) ion is given by the equation:

$$MnO_4^{-}(aq) + 8H^{+}(aq) + 5e \rightarrow Mn^{2+}(aq) + 4H_2O(l)$$

The five electrons on the left hand side of the equation reduce the manganese from Mn(+7) to Mn(+2), and also balance the charges on each side.

Ethanedioic (oxalic) acid and ethanedioates (oxalates), on the other hand, are reducing agents, being converted into carbon dioxide. The oxidation half reaction for this process is given by the equation:

$$C_2O_4^{2-}(aq) \rightarrow 2CO_2(g) + 2e$$

The overall balanced ionic equation for the manganate(VII) oxidation of an ethanedioate is therefore obtained by multiplying the manganate(VII) half reaction by two and the ethanedioate half reaction by five in order to equate the demand and supply of electrons.

$$2MnO_4^{-}(aq) + 16H^{+}(aq) + 10e \rightarrow 2Mn^{2+}(aq) + 8H_2O(l)$$
$$5C_2O_4^{2-}(aq) \rightarrow 10CO_2(g) + 10e$$

Adding these two equations together gives:

$$2MnO_4^-(aq) + 16H^+(aq) + 5C_2O_4^{2-}(aq) \rightarrow 2Mn^{2+}(aq) + 8H_2O(l) + 10CO_2(g)$$

Electrode potentials

If a zinc plate is placed in an aqueous solution of one of its salts, zinc ions pass from the plate into solution, eventually establishing the following equilibrium:

$$Zn(s) \rightleftharpoons Zn^{2+}(aq) + 2e$$

The electrons released remain on the surface of the zinc plate giving rise to an *electrical double layer* (Figure 5.1(a)). The zinc plate therefore acquires a negative charge and a potential difference, referred to as the **electrode potential**, is established between the metal plate and the solution.

Conversely, if a copper plate is placed in an aqueous solution containing copper(II) ions, some of the latter deposit themselves on the plate, causing it to acquire a positive charge relative to the solution (Figure 5.1(b)).

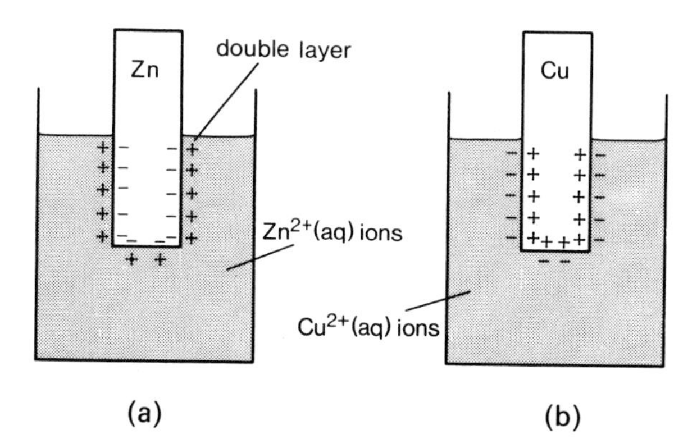

Fig. 5.1

Systems such as $Zn^{2+}(aq)|Zn(s)$ and $Cu^{2+}(aq)|Cu(s)$, in which the reduced state (i.e. the metal) is in equilibrium with the oxidized state (i.e. the cation) are referred to as *redox couples*. As the equilibrium position varies for each couple, so does the magnitude of its electrode potential. The three principal factors influencing the position of the equilibrium are:

(1) the nature of the electrode,
(2) the concentration of metal ions in solution,
(3) the temperature of the system.

The electrode potential of a metal|metal ion system is an absolute quantity and not measurable directly. Instead, it has to be obtained as a relative value by comparison with the standard hydrogen electrode, which is designated an arbitrary zero potential under standard conditions. (The standard hydrogen electrode is a platinum electrode coated with colloidal platinum, saturated with hydrogen at a pressure of one atmosphere and a temperature of 298 K (25 °C), and immersed in a solution of hydrogen ions of unit activity*; this is equivalent to a 1.18M solution of hydrochloric acid.) The hydrogen electrode is represented as:

$$\text{Pt, } H_2(g) \mid 2H^+(aq)$$

where the vertical line represents the boundary between the reacting phases of hydrogen gas and the hydrated hydrogen ions.

$$H_2(g) \rightleftharpoons 2H^+(aq) + 2e$$

The potential difference between the standard hydrogen electrode and any other system in which the concentration of the active ions in solution is of unit activity at a temperature of 298 *K* (25 °*C*) is termed the **standard electrode potential**, or **standard redox potential**, and given the symbol $E^\ominus$.

By convention, the sign attributed to each standard redox potential is that for a complete cell in which the left hand electrode is always the standard hydrogen electrode. For example, the redox couples:

$$Zn^{2+}(aq) \mid Zn(s); \quad E^\ominus = -0.76\ V$$

and

$$Cu^{2+}(aq) \mid Cu(s); \quad E^\ominus = +0.34\ V$$

refer to the cells:

$$\text{Pt, } H_2(g) \mid 2H^+(aq) \;\vdots\; Zn^{2+}(aq) \mid Zn(s); \quad E^\ominus = -0.76\ V$$

and

$$\text{Pt, } H_2(g) \mid 2H^+(aq) \;\vdots\; Cu^{2+}(aq) \mid Cu(s); \quad E^\ominus = +0.34\ V$$

where the broken line represents the porous partition or salt bridge between the two parts of the cell.

This means that species with *negative standard redox potentials* lose electrons more readily than hydrogen and are therefore *better reducing agents*. Conversely, those with *positive values* are weaker reducing agents or, alternatively, *stronger oxidizing agents* (Table 5.2).

Redox couples are not limited to elements and their ions and apply equally well to two ionic species in different oxidation states, e.g. $Fe^{3+}(aq) \mid Fe^{2+}(aq)$, in the presence of an inert electrode, such as platinum.

For practical purposes, it is usually more convenient to use alternative, more manageable, subsidiary standard electrodes such as the calomel or Weston-cadmium electrodes than the somewhat cumbersome hydrogen electrode. Measurements of electrode potential are usually made potentiometrically or by means of a valve voltmeter.

---

*Unit activity with respect to a particular species may be conveniently approximated to its molarity, although this ideal state would only be realized in systems where there is no interference or interaction between the different species present in solution, i.e. where the activity coefficient in the relationship:

$$\text{Activity} = \text{Molarity} \times \text{Activity coefficient}$$

approaches unity.

Table 5.2 A selection of redox half reactions and their standard electrode potentials

Redox couple	*Half reaction* *Oxidized form* ⇌ *Reduced form*	$E^{\ominus}/V$	
Li^+/Li	$Li^+ + e \rightleftharpoons Li$	−3.04	↑
Cs^+/Cs	$Cs^+ + e \rightleftharpoons Cs$	−3.02	
Rb^+/Rb	$Rb^+ + e \rightleftharpoons Rb$	−2.99	
K^+/K	$K^+ + e \rightleftharpoons K$	−2.92	Increasing reducing power of the reduced form
Ba^{2+}/Ba	$Ba^{2+} + 2e \rightleftharpoons Ba$	−2.90	
Sr^{2+}/Sr	$Sr^{2+} + 2e \rightleftharpoons Sr$	−2.89	
Ca^{2+}/Ca	$Ca^{2+} + 2e \rightleftharpoons Ca$	−2.87	
Na^+/Na	$Na^+ + e \rightleftharpoons Na$	−2.71	
Mg^{2+}/Mg	$Mg^{2+} + 2e \rightleftharpoons Mg$	−2.37	
Al^{3+}/Al	$Al^{3+} + 3e \rightleftharpoons Al$	−1.67	
Mn^{2+}/Mn	$Mn^{2+} + 2e \rightleftharpoons Mn$	−1.18	
Zn^{2+}/Zn	$Zn^{2+} + 2e \rightleftharpoons Zn$	−0.76	
Cr^{3+}/Cr	$Cr^{3+} + 3e \rightleftharpoons Cr$	−0.74	
Fe^{2+}/Fe	$Fe^{2+} + 2e \rightleftharpoons Fe$	−0.44	
Ni^{2+}/Ni	$Ni^{2+} + 2e \rightleftharpoons Ni$	−0.25	
Sn^{2+}/Sn	$Sn^{2+} + 2e \rightleftharpoons Sn$	−0.14	
Pb^{2+}/Pb	$Pb^{2+} + 2e \rightleftharpoons Pb$	−0.13	
H^+/H_2	$2H^+ + 2e \rightleftharpoons H_2$	0.00	
Sn^{4+}/Sn^{2+}	$Sn^{4+} + 2e \rightleftharpoons Sn^{2+}$	+0.15	
Cu^{2+}/Cu	$Cu^{2+} + 2e \rightleftharpoons Cu$	+0.34	Increasing oxidizing power of the oxidized form
I_2/I^-	$I_2 + 2e \rightleftharpoons 2I^-$	+0.54	
Fe^{3+}/Fe^{2+}	$Fe^{3+} + e \rightleftharpoons Fe^{2+}$	+0.77	
Ag^+/Ag	$Ag^+ + e \rightleftharpoons Ag$	+0.80	
Br_2/Br^-	$Br_2 + 2e \rightleftharpoons 2Br^-$	+1.07	
MnO_2/Mn^{2+}	$MnO_2 + 4H^+ + 2e \rightleftharpoons Mn^{2+} + 2H_2O$	+1.23	
$Cr_2O_7^{2-}/Cr^{3+}$	$Cr_2O_7^{2-} + 14H^+ + 6e \rightleftharpoons 2Cr^{3+} + 7H_2O$	+1.33	
Cl_2/Cl^-	$Cl_2 + 2e \rightleftharpoons 2Cl^-$	+1.36	
MnO_4^-/Mn^{2+}	$MnO_4^- + 8H^+ + 5e \rightleftharpoons Mn^{2+} + 4H_2O$	+1.52	
Pb^{4+}/Pb^{2+}	$Pb^{4+} + 2e \rightleftharpoons Pb^{2+}$	+1.69	
F_2/F^-	$F_2 + 2e \rightleftharpoons 2F^-$	+2.87	↓

The e.m.f. of electrochemical cells

The Daniell cell comprises a zinc plate (anode) immersed in an aqueous solution of zinc sulphate and a copper plate (cathode) immersed in an aqueous solution of copper(II) sulphate, the two solutions being separated by means of a porous partition which allows the passage of ions but prevents the mixing of the solutions. On connecting the two electrodes by means of a wire, electrons flow from the zinc plate to the copper in the external circuit. (Conventional current flow is in the opposite direction.) This is accompanied by an increase in the concentration of the zinc sulphate solution due to the dissolving of the zinc plate and diffusion of sulphate ions from the copper(II) sulphate solution, the latter becoming correspondingly less concentrated, with copper being deposited on the copper electrode.

The electromotive force, *e.m.f.*, of the cell is denoted by the symbol E and, in accordance with the convention adopted by the International Union of Pure and Applied Chemistry (IUPAC), the system is represented as,

$$\underset{\text{Anode}}{Zn(s)\,|\,Zn^{2+}(aq)} \;\vdots\; \underset{\text{Cathode}}{Cu^{2+}(aq)\,|\,Cu(s)}; \quad E = +1.10\,\text{V}$$

The sign of E is always the same as that of the right hand electrode in the cell diagram.

If the copper is connected to the positive terminal of a battery and the zinc plate to the negative terminal and a voltage in excess of that of the e.m.f. of the cell (1.10 V *for standard electrodes*) is applied, a current is passed through the cell in the opposite direction. Thus, the whole chemical process within the cell is reversed: i.e. zinc is deposited on the zinc plate and the copper plate dissolves. The cell is therefore referred to as a *reversible cell* comprising two **reversible electrodes**.

The **standard e.m.f.** of the cell is conventionally obtained by inspecting the cell diagram and subtracting the standard electrode potential of the negative electrode on the left hand side (i.e. the oxidized state) from the standard electrode potential of the positive electrode on the right (i.e. the reduced state):

$$E = E^{\ominus}_{\substack{\text{reduced}\\\text{state}}} - E^{\ominus}_{\substack{\text{oxidized}\\\text{state}}}$$

or

$$E = E^{\ominus}_{\text{r.h.s.}} - E^{\ominus}_{\text{l.h.s.}}$$

Thus, for the Daniell cell, using standard electrodes,

$$\begin{aligned} E &= (+0.34)-(-0.76)\,\text{V} \\ &= +1.10\,\text{V} \end{aligned}$$

This represents a maximum value for the standard cell which is obtainable only when the cell is not doing any work, i.e. when the voltage is measured on open circuit.

Relative oxidizing and reducing powers

Of the reduced forms of the half reactions listed in Table 5.2, Li metal is the strongest reducing agent and the F^- anion the weakest. Of the oxidized forms, fluorine is the strongest oxidizing agent and the Li^+ cation the weakest. By convention, the oxidized form of the half reaction is written on the left hand side of the equation and the reduced form on the right.

Standard electrode potentials provide a useful guide to the feasibility of a chemical change. In general, provided the difference in the E values of any two couples exceed 0.4 V, the reaction will tend to go to completion. Where the value is less than this, an equilibrium is established. A negative value for this difference implies that the reaction will not take place.

The ability of, say, tin(II) ions to reduce iron(III) ions in aqueous solution,

$$Sn^{2+}(aq) + 2Fe^{3+}(aq) \rightleftharpoons Sn^{4+}(aq) + 2Fe^{2+}(aq)$$

can be predicted in the same way as it can for an electrochemical cell, namely by consideration of the standard electrode potentials of the redox couples $Sn^{4+} \mid Sn^{2+}$ and $Fe^{3+} \mid Fe^{2+}$, and deducting $E^{\ominus}_{oxid}$ from $E^{\ominus}_{red}$:

i.e.
$$E = E^{\ominus}_{\substack{\text{reduced}\\\text{state}}} - E^{\ominus}_{\substack{\text{oxidized}\\\text{state}}}$$

$$= (+0.77) - (+0.15)\text{ V}$$
$$= +0.62\text{ V}$$

As the value of E is positive and greater than 0.4 V, it can be predicted that the reaction will proceed to completion.

The relative oxidizing and reducing powers of metals may be approximated in terms of a Born-Haber cycle and the enthalpy values involved in the process. The electrode potential of a monovalent metal is related to the change,

$$M^{+}(aq) + e \rightarrow M(s)$$

which is, in effect, the reverse of the enthalpy of formation of the hydrated metal ion from the metal and incorporates the enthalpies of atomization, ionization, and hydration (Figure 5.2).

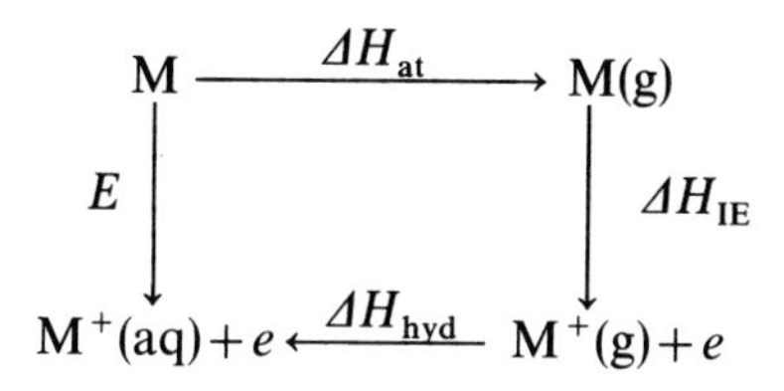

Fig. 5.2

The chemistry of lithium shows it to be less reactive than the other alkali metals and it may at first seem surprising to find it at the top of the redox table (Table 5.2). However, what has to be borne in mind is that the redox potential results from a series of changes as illustrated by a Born-Haber cycle.

A similar cycle involving similar principles can be written for non-metals, such as the halogens, whose relative oxidizing powers are considered on page 199.

Corrosion

Corrosion is familiar to us all; silver spoons become blackened after being used to eat eggs, copper piping and roofing becomes green after exposure to the atmosphere for a period of time, brassware and silverware tarnish, but probably the most familiar of all is the corrosion and rusting of iron and steel. Many millions of tons of steel are used each year in the manufacturing and construction industries, and protection of steel against unwanted corrosion is of considerable economic importance.

Those metals which are most easily obtained from their ores (see page 91), requiring the least energy in smelting, are also those which are most resistant to corrosion. From this we can deduce that the more reactive metals, which have larger negative standard electrode potentials, corrode most easily. The corrosion of a metal is, in effect, the reverse of the extraction process, and the corrosion product is often very similar to the ore from which the metal was extracted.

In dry air, iron is slowly oxidized on the surface to the oxide, but further corrosion does not take place as the oxide film protects the metal from further attack. In moist air, however, the water absorbs carbon dioxide and, particularly in industrial areas, sulphur dioxide to form an acidic solution. Iron objects exposed to such an atmosphere become covered with this acidic solution and are attacked by it, resulting in the gradual corrosion and erosion of the metal.

The electrochemical nature of rusting

The simple metal-acid reaction described above does not provide any real insight into the mechanics of the corrosion and subsequent rusting of iron in neutral or approximately neutral solution. This is more satisfactorily gained by consideration of the theories of electrolysis.

In order to establish a simple cell, it is not, in fact, necessary to have two separate metal electrodes. An iron surface will almost certainly contain some impurities and areas of unequal stress resulting from the mechanical processes to which it has been subjected. Furthermore, the concentration of the solution with which it is in contact is hardly likely to be completely homogeneous. Such heterogeneity in the metal and in the solution results in different electrode potentials being established at different points in the metal surface. The result is the formation of a simple cell, with **anodic points**, at which iron(II) ions pass into solution, and **cathodic points** where hydrogen ions are discharged, leading to an increase in the concentration of hydroxide ions. (In the context of corrosion, it may be convenient to consider an anodic process as being one in which electrons are lost and a cathodic process as being one in which electrons are consumed; i.e. the negative electrode becomes the anode and the positive electrode becomes the cathode. This might seem confusing at first since in the more familiar electrolytic cell the anode is positive and the cathode is negative. However, if the anode is looked upon in this context as referring to the electrode at which anions are liberated (or new cations formed) and where electrons are transferred to the external circuit, then the situation at the *positive* electrode in an **electrolytic cell** and at the negative electrode in a current-producing cell is exactly the same.) Another cathodic reaction involves dissolved oxygen, which is reduced to form more hydroxide ions. Indeed, in neutral or approximately neutral solution, the *oxygen absorption reaction*, as it is called, is primarily responsible for producing the hydroxide ions.

Anodic reaction: $Fe(s) \rightarrow Fe^{2+}(aq) + 2e$

Cathodic reactions: $2H^+(aq) + 2e \rightarrow H_2(g)$

$O_2(g) + 2H_2O(l) + 4e \rightarrow 4OH^-(aq)$

The current flow is in the metal which, in effect, provides the external circuit of the cell.

The combination of the principal products of the anodic and cathodic reactions,

$$Fe^{2+}(aq) + 2OH^-(aq) \rightarrow Fe(OH)_2(s)$$

leads to the formation of iron(II) hydroxide. When in contact with dissolved oxygen or air, this solid undergoes rapid oxidation to hydrated iron(III) oxide or rust. Hence, *rust is a secondary product of corrosion, formed in solution rather than on the surface of the metal.*

If a drop of ordinary water is placed on a clean iron plate, corrosion and pitting of the metal becomes most apparent at the centre of the drop where contact with atmospheric oxygen is least. The iron in the central region of the drop therefore becomes anodic and dissolves while that near the edge of the drop becomes cathodic. Moreover, the greater the oxygen differential between the anodic and the cathodic areas, the more rapidly corrosion occurs. This explains why rusting occurs underneath a layer of paint (which is deprived of oxygen) on a partly painted iron surface. As the pit formed by the rusting becomes progressively deeper, the supply of oxygen at the bottom becomes even less and corrosion is accelerated.

Cathodic protection of iron

One method of protecting iron from corrosion is to make it cathodic with respect to another metal (i.e. one which has a greater negative standard electrode potential). Zinc is widely used for this purpose. If zinc and iron are placed in electrical contact and subjected to a corroding agent, the zinc anode will preferentially corrode, leaving the iron cathode unaffected.

In most cases, iron articles are plated with zinc. This can be done by *galvanizing*, which involves immersing the article in molten zinc, or by spraying. Small articles, however, are usually coated by a method known as *sheradizing*, which involves heating the articles with zinc dust. For objects treated in either of these ways, scratching the zinc layer still leaves the iron underneath with protection; the zinc, being anodic, corrodes preferentially.

Coating iron and steel with tin is also a common method of protection. However, once the tin plating is scratched, the iron is anodic with respect to the tin and therefore the corrosion of the iron is accelerated rather than inhibited. Tin is used as a protective layer for the iron cans used in the storage of food, but here its potential is so greatly lowered, because of the formation of complex ions with the acid juices present, that it becomes the anode and acts as a sacrificial metal. The tin salts which dissolve are not harmful to health; however, zinc forms poisonous soluble salts with the acids present in many fruit juices and therefore cannot be used for this purpose.

In cathodic protection, the anodic metal is referred to as the *sacrificial metal.* As well as zinc- and tin-plating, magnesium and aluminium (and their alloys) are also widely used as sacrificial metals to protect steel.

Steel articles such as pipes are often coated with asphalt, coal-tar enamels, or plastic tapes (e.g. PVC and PTFE). Epoxy resins, vinyl paints, or neoprene-based coatings are used on underwater steel objects, e.g. oil-pipes and metal areas of ships. These coatings used in conjunction with a sacrificial metal give long-lasting protection, the main drawback being a tendency to pitting where-ever corrosion initially starts.

6 The Periodic Classification of the Elements

The format of the modern Periodic Table has been derived from the original classifications of Mendeléeff in Russia and Lothar Meyer in Germany. In 1869, independently and almost simultaneously, they tabulated all the elements that were known at the time in increasing order of relative atomic mass (atomic weight). Each man recorded that the elements could be arranged in families or groups, the members of which exhibited many similar properties, and therefore they concluded that these properties were functions of their relative atomic mass.

At the time of these original classifications, several of the elements with which we are familiar today had not been isolated and identified and a number of gaps had to be left in the table to accommodate them. There were further difficulties in obtaining pure samples and, coupled with the scarcity of a number of the elements, determination of precise values for relative atomic mass was uncertain. Nonetheless, so clear were the patterns relating to both physical and chemical properties that the classifications were not seriously challenged. The profundity of Mendeléeff's insight into his classification was such that he was able to predict the properties of the missing elements with remarkable accuracy.

Shortly after these original compilations, the convention of identifying each element by means of its relative atomic mass was abolished and, instead, each element was assigned an integer in accordance with the order in which it occurred in the table. As a result of work done by van der Broek, Moseley, and Chadwick, we now know this integer, or atomic number as it is known, to be equal to the number of nuclear protons.

Periods and groups

In the modern version of the Periodic Table, elements are arranged in horizontal rows or **periods** and in vertical columns or **groups** in accordance with their increasing atomic number (Table 6.1). Elements possessing similar outer electronic configurations all appear in the same group and exhibit characteristic, and generally predictable, properties.

Period 1 contains only hydrogen and helium. Although helium can be quite reasonably included in Group O (the noble gas elements), hydrogen cannot be satisfactorily included in either Group I (the alkali metals) or Group VII (the halogens) despite exhibiting certain similarities with each of these groups. Instead, these anomalous properties, coupled with its own distinctive characteristics, mean that hydrogen should be classified apart from the other groups (see page 102).

Table 6.1 The electronic build-up (ground states) of the elements in the *s*-, *p*-, and *d*-blocks

1 H $1s^1$																	2 He $1s^2$
3 Li $2s^1$	4 Be $2s^2$											5 B $2s^22p^1$	6 C $2s^22p^2$	7 N $2s^22p^3$	8 O $2s^22p^4$	9 F $2s^22p^5$	10 Ne $2s^22p^6$
11 Na $3s^1$	12 Mg $3s^2$											13 Al $3s^23p^1$	14 Si $3s^23p^2$	15 P $3s^23p^3$	16 S $3s^23p^4$	17 Cl $3s^23p^5$	18 Ar $3s^23p^6$
19 K $4s^1$	20 Ca $4s^2$	21 Sc $3d^14s^2$	22 Ti $3d^24s^2$	23 V $3d^34s^2$	24 Cr $3d^54s^1$	25 Mn $3d^54s^2$	26 Fe $3d^64s^2$	27 Co $3d^74s^2$	28 Ni $3d^84s^2$	29 Cu $3d^{10}4s^1$	30 Zn $3d^{10}4s^2$	31 Ga $4s^24p^1$	32 Ge $4s^24p^2$	33 As $4s^24p^3$	34 Se $4s^24p^4$	35 Br $4s^24p^5$	36 Kr $4s^24p^6$
37 Rb $5s^1$	38 Sr $5s^2$	39 Y $4d^15s^2$	40 Zr $4d^25s^2$	41 Nb $4d^35s^2$	42 Mo $4d^55s^1$	43 Tc $4d^55s^2$	44 Ru $4d^75s^1$	45 Rh $4d^85s^1$	46 Pd $4d^{10}$	47 Ag $4d^{10}5s^1$	48 Cd $4d^{10}5s^2$	49 In $5s^25p^1$	50 Sn $5s^25p^2$	51 Sb $5s^25p^3$	52 Te $5s^25p^4$	53 I $5s^25p^5$	54 Xe $5s^25p^6$
55 Cs $6s^1$	56 Ba $6s^2$	57 La $5d^16s^2$	72 Hf $5d^26s^2$	73 Ta $5d^36s^2$	74 W $5d^46s^2$	75 Re $5d^56s^2$	76 Os $5d^66s^2$	77 Ir $5d^9$	78 Pt $5d^96s^1$	79 Au $5d^{10}6s^1$	80 Hg $5d^{10}6s^2$	81 Tl $6s^26p^1$	82 Pb $6s^26p^2$	83 Bi $6s^26p^3$	84 Po $6s^26p^4$	85 At $6s^26p^5$	86 Rn $6s^26p^6$
87 Fr $7s^1$	88 Ra $7s^2$	89 Ac $6d^17s^2$															
				58 Ce	59 Pr	60 Nd	61 Pm	62 Sm	63 Eu	64 Gd	65 Tb	66 Dy	67 Ho	68 Er	69 Tm	70 Yb	71 Lu
				90 Th $6d^27s^2$	91 Pa $6d^37s^2$	92 U $6d^47s^2$ $6d^57s^1$	93 Np	94 Pu	95 Am	96 Cm	97 Bk	98 Cf	99 Es	100 Fm	101 Md	102 No	103 Lr

Periods 2 and 3 each contain eight elements and are referred to as the *short periods*.

Periods 4 and 5 each contain eighteen elements, and Periods 6 and 7 contain thirty-two. The latter two periods, at a first glance, might appear to contain only eighteen. This is because the elements following lanthanum in Period 6 and actinium in Period 7 are usually represented as two separate rows at the bottom of the table to avoid excessive side-ways expansion.

s, p, d, and f blocks

The elements with their valence electrons only in *s* orbitals occur in Group I, (ns^1), and Group II, (ns^2), which together comprise the *s-block* elements.

Groups III (ns^2np^1), IV (ns^2np^2), V (ns^2np^3), VI (ns^2np^4), VII (ns^2np^5), and 0 (ns^2np^6), whose chemical properties depend upon the behaviour of the outer *p* electrons, are referred to as the *p-block*.

Similarly, those elements in Periods 4, 5, and 6 whose chemistry is governed by valence electrons contained in *d* orbitals which form the penultimate shell of electrons, are known as the *d-block* or *transition elements*, so called because their properties are intermediate or transitional between those of the *s*- and *p*-block elements.

The elements following lanthanum in Period 6 and actinium in Period 7, respectively referred to as the *lanthanoids* (lanthanides before 1965) or *rare earths* and the *actinoids* (actinides), possess partly filled *f* orbitals (which in this case form the antepenultimate shell of electrons) and are referred to as the *f-block* or *inner transition series*.

Uranium is the last of the naturally occurring elements, and the heavy metals which follow it in the table are all synthetically manufactured elements.

Patterns and trends

On traversing a period, the physical and chemical properties are modified by the gradual change in nuclear and outer electronic structure of the elements, and patterns of clearly recognizable general properties emerge.

As chemistry is fundamentally a study of the elements and how they react together, we have in the Periodic Table the 'back-bone' of the whole subject. The main advantage as far as the student is concerned is that the chemistry of the individual elements need not be considered in isolation. Instead, he or she can learn and understand the general properties within a group or period and then predict with considerable accuracy the properties of any particular element, even though this element may be unfamiliar.

The general patterns and trends discussed in the ensuing sections relate primarily to the elements in the *s*- and *p*-blocks. The Group 0 elements, or noble gases, have unique and almost unreactive properties brought about by their stable configuration of electrons, and are not conveniently incorporated into such generalizations. Hydrogen, also, is not suitable for consideration in this particular context owing to its many anomalous properties. The Group 0 elements and hydrogen have therefore been excluded from the summary of general trends discussed in this chapter, although a few specific references are made occasionally in order to facilitate elucidation.

The *d*-block elements are all essentially metals, possessing properties which are intermediate between those of the preceding metals in the *s*-block and those elements that follow in the *p*-block.

It must always be remembered that, when making such generalizations, certain anomalies with respect to particular properties will inevitably occur. These will become more evident in the detailed studies which follow relating to the chemistry of the individual groups.

Size of atoms: atomic radii

On traversing a period from left to right across the Periodic Table, the number of protons in the nuclei and the number of orbital electrons of the respective atoms increases progressively, creating a gradual increase in the electrostatic attractions between these oppositely charged particles. The outer electrons, which, for the elements within any one period, are all defined by the same principle quantum number, are therefore pulled in closer to the nucleus, causing the atoms to decrease in size. Thus, within a period, the alkali metal atom will be the largest and the halogen atom the smallest.

On descending a group, the size of the atoms increases owing to the build-up of extra shells of electrons, the outer of which are defined by progressively higher principle quantum numbers. The size is also influenced by the *screening effect* of the inner electrons which tend to reduce the attractive force between the positive nucleus and the outer electrons. The combined effect of these two factors is sufficient to outweigh that of the increasing population of nuclear protons.

The size of an atom is usually quoted in terms of the *atomic* (*covalent*) *radius, which is defined as half the internuclear distance between two like atoms bound only by a single bond* (see Figure 6.1).

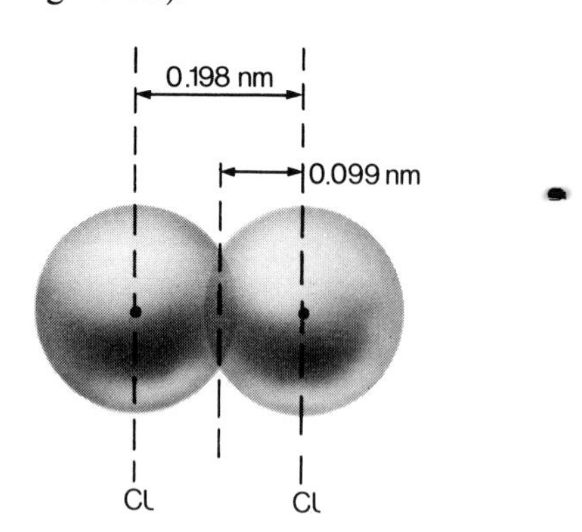

Fig. 6.1 Covalent radius of chlorine

Such values cannot be measured directly for individual atoms and have to be obtained via spectroscopic techniques. The values of the covalent radii of most of the naturally occuring elements are set out in Table 6.2.

Table 6.2 Covalent radii of the elements/nm

1 H 0.037																	2 He 0.120
3 Li 0.123	4 Be 0.089											5 B 0.080	6 C 0.077	7 N 0.074	8 O 0.074	9 F 0.072	10 Ne 0.160
11 Na 0.157	12 Mg 0.136											13 Al 0.125	14 Si 0.117	15 P 0.110	16 S 0.104	17 Cl 0.099	18 Ar 0.191
19 K 0.203	20 Ca 0.174	21 Sc 0.144	22 Ti 0.132	23 V 0.122	24 Cr 0.117	25 Mn 0.117	26 Fe 0.116	27 Co 0.116	28 Ni 0.115	29 Cu 0.117	30 Zn 0.125	31 Ga 0.125	32 Ge 0.122	33 As 0.121	34 Se 0.117	35 Br 0.114	36 Kr 0.200
37 Rb 0.216	38 Sr 0.191	39 Y 0.162	40 Zr 0.145	41 Nb 0.134	42 Mo 0.129	43 Tc —	44 Ru 0.124	45 Rh 0.125	46 Pd 0.128	47 Ag 0.134	48 Cd 0.141	49 In 0.150	50 Sn 0.140	51 Sb 0.141	52 Te 0.137	53 I 0.133	54 Xe 0.220
55 Cs 0.235	56 Ba 0.198	57 La 0.169	72 Hf 0.144	73 Ta 0.134	74 W 0.130	75 Re 0.128	76 Os 0.126	77 Ir 0.126	78 Pt 0.129	79 Au 0.134	80 Hg 0.144	81 Tl 0.155	82 Pb 0.154	83 Bi 0.152	84 Po —	85 At 0.140	86 Rn —
87 Fr —	88 Ra —	89 Ac —															

58 Ce 0.165	59 Pr 0.165	60 Nd 0.164	61 Pm —	62 Sm 0.166	63 Eu 0.185	64 Gd 0.161	65 Tb 0.159	66 Dy 0.159	67 Ho 0.158	68 Er 0.157	69 Tm 0.156	70 Yb 0.170	71 Lu 0.156
90 Th 0.165	91 Pa —	92 U 0.142	93 Np —	94 Pu —	95 Am —	96 Cm —	97 Bk —	98 Cf —	99 Es —	100 Fm —	101 Md —	102 No —	103 Lr —

Ease of formation of positive ions: ionization energy

Chapter 1 discussed how electrons in an atom may be promoted to higher energy levels and, provided sufficient energy is available, may ultimately be removed leaving a positive ion. *The minimum amount of energy required to remove to infinity the least firmly bound electron of an atom in the gaseous state* is referred to as the **first ionization energy** of the element:

i.e. $$A(g) \rightarrow A^{+}(g)+e$$

The **second ionization energy** is the minimum amount of energy required to remove a second electron from a singly charged positive ion,

i.e. $$A^{+}(g) \rightarrow A^{2+}(g)+e$$

Similarly, the **third ionization energy** relates to the process

$$A^{2+}(g) \rightarrow A^{3+}(g)+e$$

and so on.

The values for ionization energies, being fundamentally related to how firmly the outer electrons are bound, are influenced by factors similar to those affecting the size of the atom or ion, namely the nuclear charge, the electronic configuration, and the screening effect of the inner electrons. As a general rule, for elements within a given group or period the first ionization energy decreases with increasing size of atom.

Once an electron has been removed from an atom, forming a singly charged positive ion, there is a dramatic decrease in size (see page 84), causing the remaining electrons to be more firmly held; so the removal of a second electron requires significantly greater energy than the removal of the first. Similarly, the value of the third ionization energy is very much greater than the second. Once the gaseous ion has acquired the electronic configuration of a noble gas, the removal of a further electron is exceedingly difficult and the energy required is considerably greater than for the removal of any previous electrons. Consider, for example, the first three ionization energies of calcium, where the Ca^{2+} ion has the electronic configuration corresponding to that of the noble gas, argon:

$$\begin{aligned} Ca(g) &\rightarrow Ca^{+}(g)+e; & \Delta H^{\ominus} &= +596\,kJ\,mol^{-1} \\ Ca^{+}(g) &\rightarrow Ca^{2+}(g)+e; & \Delta H^{\ominus} &= +1155\,kJ\,mol^{-1} \\ Ca^{2+}(g) &\rightarrow Ca^{3+}(g)+e; & \Delta H^{\ominus} &= +4950\,kJ\,mol^{-1} \end{aligned}$$

As energy has to be provided to remove these electrons, i.e. removal is an endothermic process, ionization energies are assigned positive enthalpy values.

Ionization energies may also be interpreted to provide supporting evidence for the electronic build-up of atoms and their periodic tabulation. Figure 6.2 shows a plot of the first ionization energies of the first nineteen elements and clearly illustrates the way in which the elements form groups of 2, 8, and 8, corresponding to the format of the first three periods of the Periodic Table. On going from the last member of one period to the first of another, i.e. from He to Li, from Ne to Na, and from Ar to K, there is a dramatic change in the first ionization energy values, indicating that the outer electrons of these elements are defined by different principal quantum numbers.

A further study of this plot shows the groups of eight to be apparently sub-divided into three groups of 2, 3, and 3, suggesting that within each particular principal quantum level not all electrons possess the same energy. The sub-group containing the two outer electrons of lowest energy corresponds to those contained in *s* orbitals. The next group of three corresponds to those electrons singly occupying the three separate *p* orbitals, and the other sub-group of three, of higher energy, is accounted for by pairing in the *p* orbitals.

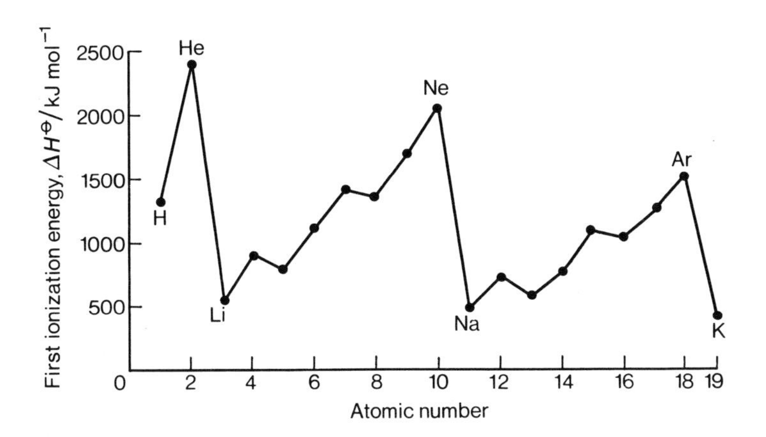

Fig. 6.2 Plot of the first ionization energies of the first nineteen elements in the Periodic Table

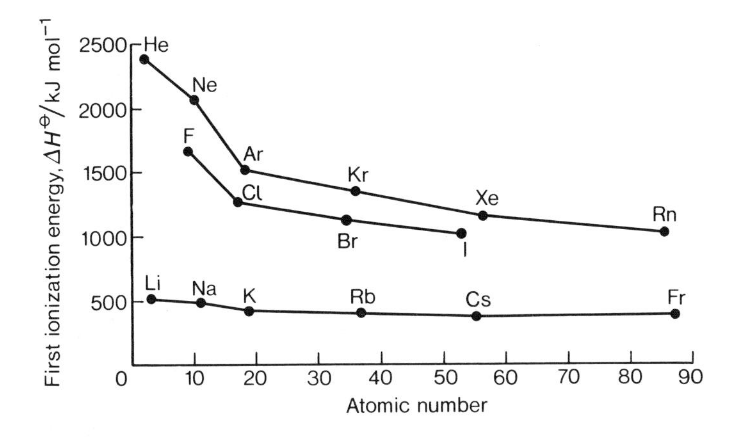

Fig. 6.3 Comparative plot of the first ionization energies of the alkali metals, the halogens, and the noble gases

Table 6.3 First ionization energies $\Delta H^{\ominus}$/kJ mol^{-1}

1 H 1300																	2 He 2370
3 Li 525	4 Be 906											5 B 805	6 C 1090	7 N 1400	8 O 1310	9 F 1680	10 Ne 2080
11 Na 500	12 Mg 742											13 Al 583	14 Si 792	15 P 1060	16 S 1000	17 Cl 1260	18 Ar 1520
19 K 424	20 Ca 596	21 Sc 638	22 Ti 667	23 V 654	24 Cr 659	25 Mn 722	26 Fe 768	27 Co 763	28 Ni 742	29 Cu 751	30 Zn 915	31 Ga 583	32 Ge 768	33 As 972	34 Se 947	35 Br 1140	36 Kr 1350
37 Rb 408	38 Sr 554	39 Y 642	40 Zr 675	41 Nb 659	42 Mo 700	43 Tc 705	44 Ru 730	45 Rh 751	46 Pd 809	47 Ag 738	48 Cd 872	49 In 562	50 Sn 713	51 Sb 839	52 Te 876	53 I 1010	54 Xe 1170
55 Cs 382	56 Ba 508	57 La 546	72 Hf 537	73 Ta 583	74 W 776	75 Re 768	76 Os 847	77 Ir 893	78 Pt 872	79 Au 897	80 Hg 1010	81 Tl 596	82 Pb 722	83 Bi 780	84 Po 818	85 At —	86 Rn 1040
87 Fr 387	88 Ra 516	89 Ac 675															
				58 Ce 671	59 Pr 562	60 Nd 613	61 Pm 562	62 Sm 546	63 Eu 554	64 Gd 600	65 Tb 654	66 Dy 663	67 Ho —	68 Er —	69 Tm —	70 Yb 604	71 Lu 487
				90 Th 680	91 Pa —	92 U 391	93 Np —	94 Pu —	95 Am —	96 Cm —	97 Bk —	98 Cf —	99 Es —	100 Fm —	101 Md —	102 No —	103 Lr —

Ease of formation of negative ions: electron affinities

On traversing the Periodic Table from left to right, it has been shown how the atoms of the elements become progressively smaller as a result of the gradual increase in nuclear charge which exerts a greater attractive pull on the outer electrons. The ability of an atom to acquire a further electron to form a singly charged negative ion is dependent upon the effectiveness of this nuclear charge; that is, the size of the positive charge on the nucleus and the distance over which it operates. The formation of negative ions is therefore favoured by small atoms. Such a process for the formation of monovalent anions is accompanied by the evolution of energy, which is often considerable and is referred to as the **electron affinity** of the element.

The first electron affinity of an element is the energy released when an additional electron is acquired by an isolated neutral atom in the gaseous state:

i.e. $$X(g) + e \rightarrow X^-(g)$$

It follows that within any particular period the halogen atom will possess the highest electron affinity, i.e. it will form a singly charged negative ion (an anion) most easily.

As one might expect, singly charged negative ions strongly repel further electrons within their proximity and therefore a considerable amount of energy is *required* to form a dinegative ion. This is indicated by the endothermic value ($\Delta H^{\ominus} = +850\,kJ\,mol^{-1}$) for the formation of O^{2-} from O^-.

Precise values for the electron affinities of many elements are not known as there is no generally convenient method for obtaining them directly, and indirect methods, utilizing the principles of the Born-Haber cycle (page 61), have to be employed. A selection of values is shown in Table 6.4.

Table 6.4 Electron affinities of a few selected elements

Element	*Electron affinity, $\Delta H^{\ominus}/kJ\,mol^{-1}$*	*Element*	*Electron affinity, $\Delta H^{\ominus}/kJ\,mol^{-1}$*
H	−78	O	−148
		O^-	+850
Li	−58	S	−206
Na	−78	S^-	+538
B	−35	F	−354
		Cl	−370
C	−126	Br	−348
Si	−186	I	−320
N	−9		
P	−76		

Size of ions: ionic radii

The acquisition or loss of electrons in the formation of ions brings about a considerable change in the size of the original atom. If the proton/electron ratio is increased, as in the formation of positive ions, then the outer electrons are pulled in towards the nucleus and the size of the original atom is reduced. This effect is often accentuated by the fact that the formation of the usual positive ion of an element involves the removal of all the outer electrons.

Conversely, if the proton/electron ratio is decreased, as in the formation of negative ions, then the outer electrons become less firmly held and pull away from the nucleus. Another factor is that the incoming electron adds to the existing inter-electron repulsions.

Compare the atomic radii of Na and F with the ionic radii of their commonly encountered ions, Na^+ and F^- (Table 6.5).

Table 6.5

Atom	*Atomic number*	*Number of electrons*	*Atomic (covalent) radius/nm*	*Ion*	*Number of electrons*	*Ionic radius/nm*
F	9	9	0.072	F^-	10	0.136
Na	11	11	0.157	Na^+	10	0.095

From these figures it is apparent that, despite the neutral sodium atom being considerably larger than the fluorine atom, the order is clearly reversed when comparing the sizes of their ions.

The ions, Na^+, Mg^{2+}, and Al^{3+} are all isoelectronic (i.e. have the same electronic configuration), but, as the atomic numbers of these ions increase successively by one, then there is a progressive decrease in their respective ionic radii (Table 6.6).

Table 6.6

Ion	*Atomic number*	*Number of electrons*	*Ionic radius/nm*
Na^+	11	10	0.095
Mg^{2+}	12	10	0.065
Al^{3+}	13	10	0.050

Similar reasoning may also be applied to the negative ions, P^{3-}, S^{2-}, and Cl^- (Table 6.7).

Table 6.7

Ion	*Atomic number*	*Number of electrons*	*Ionic radius/nm*
P^{3-}	15	18	0.212
S^{2-}	16	18	0.184
Cl^-	17	18	0.181

The ionic radii of the elements are listed in Table 6.8.

Table 6.8 Ionic radii of the elements/nm. In certain instances, the existence of the actual ion and the value quoted is purely hypothetical

1 H (−1)0.208																	2 He —
3 Li (+1)0.060	4 Be (+2)0.031											5 B (+3)0.020	6 C (−4)0.260 (+4)0.015	7 N (−3)0.171 (+5)0.011	8 O (−2)0.140 (+6)0.009	9 F (−1)0.136 (+7)0.007	10 Ne —
11 Na (+1)0.095	12 Mg (+2)0.065											13 Al (+3)0.050	14 Si (−4)0.271 (+4)0.041	15 P (−3)0.212 (+5)0.034	16 S (−2)0.184 (+6)0.029	17 Cl (−1)0.181 (+7)0.026	18 Ar —
19 K (+1)0.133	20 Ca (+2)0.099	21 Sc (+3)0.081	22 Ti (+2)0.090 (+4)0.068	23 V (+3)0.074 (+5)0.059	24 Cr (+3)0.069 (+6)0.052	25 Mn (+2)0.080 (+7)0.046	26 Fe (+2)0.076 (+3)0.064	27 Co (+2)0.078 (+3)0.063	28 Ni (+2)0.078 (+3)0.062	29 Cu (+1)0.096 (+2)0.069	30 Zn (+2)0.074	31 Ga (+1)0.148 (+3)0.062	32 Ge (+2)0.093 (+4)0.053	33 As (−3)0.222 (+5)0.047	34 Se (−2)0.198 (+6)0.042	35 Br (−1)0.195 (+7)0.039	36 Kr —
37 Rb (+1)0.148	38 Sr (+2)0.113	39 Y (+3)0.093	40 Zr (+4)0.080	41 Nb (+5)0.070	42 Mo (+4)0.068 (+6)0.062	43 Tc —	44 Ru (+3)0.069 (+4)0.065	45 Rh (+2)0.086	46 Pd (+2)0.050	47 Ag (+1)0.126	48 Cd (+2)0.097	49 In (+1)0.132 (+3)0.081	50 Sn (+2)0.112 (+4)0.071	51 Sb (−3)0.245 (+5)0.062	52 Te (−2)0.221 (+6)0.056	53 I (−1)0.216 (+7)0.050	54 Xe —
55 Cs (+1)0.169	56 Ba (+2)0.135	57 La (+3)0.115	72 Hf (+4)0.081	73 Ta (+5)0.073	74 W (+4)0.064 (+6)0.068	75 Re —	76 Os (+4)0.067	77 Ir (+4)0.066	78 Pt (+2)0.052	79 Au (+1)0.137	80 Hg (+2)0.110	81 Tl (+1)0.140 (+3)0.095	82 Pb (+2)0.120 (+4)0.084	83 Bi (+3)0.120 (+5)0.074	84 Po —	85 At —	86 Rn —
87 Fr (+1)0.176	88 Ra (+2)0.140	89 Ac (+3)0.118															
				58 Ce (+3)0.111 (+4)0.101	59 Pr (+3)0.109 (+4)0.092	60 Nd (+3)0.108	61 Pm (+3)0.106	62 Sm (+3)0.104	63 Eu (+2)0.112	64 Gd (+3)0.102	65 Tb (+3)0.100	66 Dy (+3)0.099	67 Ho (+3)0.097	68 Er (+3)0.096	69 Tm (+3)0.095	70 Yb (+2)0.113	71 Lu (+3)0.093
				90 Th (+3)0.114 (+4)0.095	91 Pa (+3)0.112 (+4)0.091	92 U (+3)0.111 (+4)0.089	93 Np (+3)0.109 (+4)0.088	94 Pu (+3)0.107 (+4)0.086	95 Am (+3)0.106 (+4)0.085	96 Cm —	97 Bk —	98 Cf —	99 Es —	100 Fm —	101 Md —	102 No —	103 Lr —

Electronegativity

The ability of an atom to attract electrons in a chemical bond towards itself when combined with different atoms in a compound is termed the **electronegativity** *of the atom.*

Generally, small atoms, especially those with nearly-filled outer shells of electrons, attract electrons more easily than larger ones and hence tend to have higher electronegativity values. On traversing the Periodic Table, the electronegativity values of the elements generally increase on moving from Group I to Group VII and, within a group, they tend to increase on ascending it.

The alkali metal is therefore the least electronegative (or most electropositive) element in its period and the halogen the most electronegative. Within each group, the first member tends to have the highest electronegativity. The actual values assigned to the elements were derived by Pauling, and are based upon an arbitrary scale ranging from zero to four (Table 6.9).

Electronegativity values are related to the sum of the ionization energy and the electron affinity, although owing to the lack of precise figures for many electron affinities, they are generally obtained from bond enthalpy measurements (see page 54).

Metallic and non-metallic character

Metallic character may be defined as *the tendency of an atom to form positive ions*:

i.e. $$A \rightarrow A^+ + e$$

As large atoms with sparsely filled shells most easily undergo this change, metallic character therefore follows a similar pattern in the Periodic Table as that for ionization energies and atomic size; i.e. it increases towards the left hand side of each period, and on descending a group.

The alkali metal is therefore the most reactive metal in its respective period and, of these, caesium will exhibit the most metallic character. (Francium, being radioactive and unstable, is not generally considered.) Conversely, non-metals are more electronegative in nature and have a greater tendency to acquire electrons and form negative ions. The general pattern for increasing non-metallic character and reactivity is consequently similar to that for increasing electronegativity values.

The halogen atoms are thus the most reactive non-metals in each period and, of these, fluorine is the most reactive.

The periodic and group trends identified so far have been summarized in Figures 6.4(a) and (b) on page 88.

Diagonal relationships of elements

Similarities in properties are apparent between certain pairs of elements which are diagonally related to each other in the Periodic Table. This is particularly significant for the elements *lithium and magnesium* (Table 6.10), *beryllium and*

Table 6.9 Pauling's electronegativity values of the elements

1 H 2.1																	2 He —
3 Li 1.0	4 Be 1.5											5 B 2.0	6 C 2.5	7 N 3.0	8 O 3.5	9 F 4.0	10 Ne —
11 Na 0.9	12 Mg 1.2											13 Al 1.5	14 Si 1.8	15 P 2.1	16 S 2.5	17 Cl 3.0	18 Ar —
19 K 0.8	20 Ca 1.0	21 Sc 1.3	22 Ti 1.5	23 V 1.6	24 Cr 1.6	25 Mn 1.5	26 Fe 1.8	27 Co 1.8	28 Ni 1.8	29 Cu 1.9	30 Zn 1.6	31 Ga 1.6	32 Ge 1.8	33 As 2.0	34 Se 2.4	35 Br 2.8	36 Kr —
37 Rb 0.8	38 Sr 1.0	39 Y 1.2	40 Zr 1.4	41 Nb 1.6	42 Mo 1.8	43 Tc 1.9	44 Ru 2.2	45 Rh 2.2	46 Pd 2.2	47 Ag 1.9	48 Cd 1.7	49 In 1.7	50 Sn 1.8	51 Sb 1.9	52 Te 2.1	53 I 2.5	54 Xe —
55 Cs 0.7	56 Ba 0.9	57 La 1.1	72 Hf 1.3	73 Ta 1.5	74 W 1.7	75 Re 1.9	76 Os 2.2	77 Ir 2.2	78 Pt 2.2	79 Au 2.4	80 Hg 1.9	81 Tl 1.8	82 Pb 1.8	83 Bi 1.9	84 Po 2.0	85 At 2.2	86 Rn —
87 Fr 0.7	88 Ra 0.9	89 Ac 1.1															
				58 Ce 1.1	59 Pr 1.1	60 Nd 1.2	61 Pm 1.2	62 Sm 1.2	63 Eu 1.1	64 Gd 1.1	65 Tb 1.2	66 Dy 1.1	67 Ho 1.2	68 Er 1.2	69 Tm 1.2	70 Yb 1.1	71 Lu 1.2
				90 Th 1.3	91 Pa 1.5	92 U 1.7	93 Np 1.3	94 Pu 1.3	95 Am 1.3	96 Cm 1.3	97 Bk 1.3	98 Cf 1.3	99 Es 1.3	100 Fm 1.3	101 Md 1.3	102 No 1.3	103 Lr —

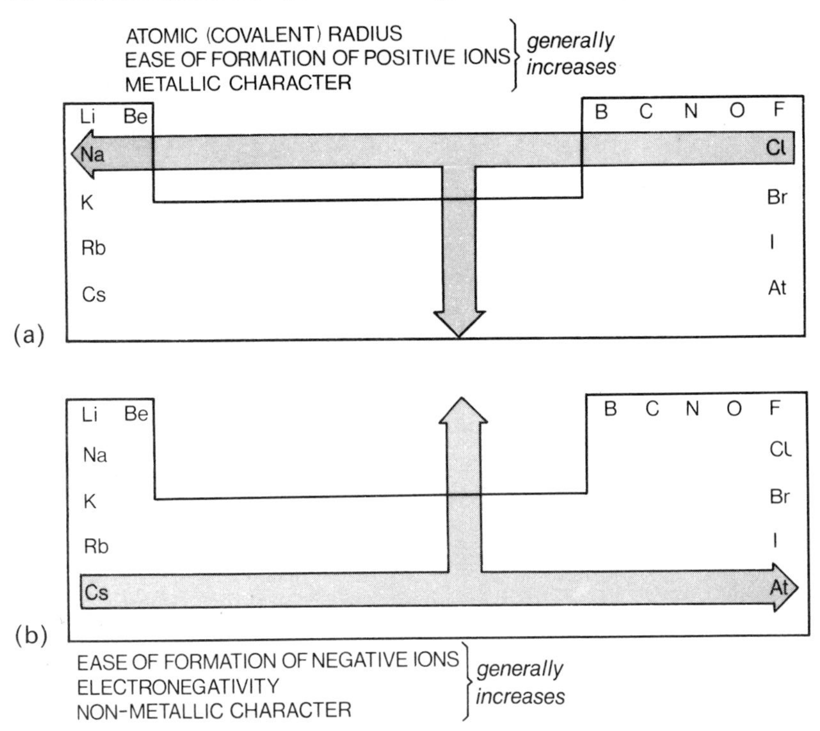

Fig. 6.4 Summary of general periodic and group trends

aluminium (Table 6.11), *and boron and silicon*, although it is important to emphasize that *these similarities are weaker than the group properties of the elements.*

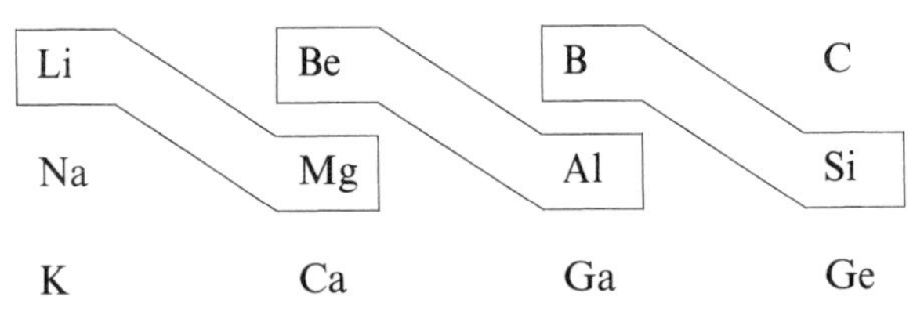

This phenomenon is quite easily explained in terms of the horizontal and vertical relationships already discussed, especially with regard to ionic size and electronegativity values. On traversing a period from left to right, the decrease in the size of positive ions is accompanied by an increase in electronegativity, and on descending a group the increasing size of the ions is accompanied by a decrease in electronegativity. It therefore follows that a number of diagonally related elements will be comparable in terms of ionic size (for the most stable ion) and electronegativity. Tables 6.10 and 6.11 highlight these similarities.

Table 6.10

Property	*Lithium*	*Magnesium*	*Other alkali metals*
Ionic radius/nm	Li^+, 0.060	Mg^{2+}, 0.065	Na^+, 0.095; Cs^+, 0.169
Electronegativity	1.0	1.2	Na, 0.9; Cs, 0.7
Boiling point/°C	1330	1110	Na, 890; Cs, 690
Hardness	Much harder than other alkali metals	Much harder than alkali metals	Soft
Oxides	No peroxide or higher oxide	No peroxide or higher oxide	Form peroxide and higher oxide
Heat on (stability of) hydroxides and carbonates	Decompose to oxide	Decompose to oxide	Stable to heat
Solubility of hydroxides	Sparingly soluble	Sparingly soluble	Very soluble
Solubility of carbonates	Virtually insoluble	Virtually insoluble	Soluble
Heat on (stability of) nitrates	Decomposes to oxide	Decomposes to oxide	Decompose to nitrites
Nitrides	Forms nitride, Li_3N	Forms nitride, Mg_3N_2	Do not form nitride

Table 6.11

Property	*Beryllium*	*Aluminium*	*Other alkaline earth metals*
Ionic radius/nm	Be^{2+}, 0.031	Al^{3+}, 0.050	Mg^{2+}, 0.065; Ba^{2+}, 0.135
Electronegativity	1.5	1.5	Mg, 1.2; Ba^{2+}, 0.9
Boiling point/°C	2477	2470	Mg, 1110; Ba, 1640
Type of oxide	Amphoteric	Amphoteric	Basic
Stability of carbonates	No stable carbonate	No stable carbonate	Form stable, solid carbonates
Nature of chloride	Covalently bridged (see page 126)	Covalently bridged (see page 137)	Ionic crystals
Complex formation	Strong tendency, e.g. $[BeF_4]^{2-}$ (tetrahedral)	Strong tendency e.g. $[AlF_6]^{3-}$ (octahedral)	Weaker tendencies
Formation of hydrated ions	Salts furnish hydrated ions in solution, $[Be(H_2O)_4]^{2+}$	Salts furnish hydrated ions in solution $[Al(H_2O)_6]^{3+}$	Form hydrated ions less easily
Hydrolysis of carbides	Be_2C yields CH_4	Al_4C_3 yields CH_4	Yield $CH{\equiv}CH$ (Mg_2C_3 yields $CH_3C{\equiv}CH$)
Reaction with acids	Fairly resistant unless amalgamated or finely divided	Fairly resistant unless amalgamated or finely divided	React readily to form salts
Reaction with conc. nitric acid	Rendered passive	Rendered passive	React readily to form salts

Occurrence of metals

With the exception of the highly unreactive metals, such as palladium, platinum, silver, and gold, the vast majority of metals exist naturally only in the form of their compounds. Although it is not possible to make any precise correlation between the position of the elements in the Periodic Table and the chemical nature of their ores, certain general patterns emerge on relating these two factors. These patterns fall into four main categories:

1. The highly electropositive metals of Groups I and II are found primarily in the form of their salts, notably the halide, sulphate, nitrate, and carbonate salts. The most abundant ores of the Group II metals are the sulphates and the insoluble carbonates.
2. Naturally occurring oxides occur for aluminium in Group III and for the more electropositive transition elements which are grouped towards the left hand side of the d-block. These include Sc, Ti, V, Cr, Mn, Y, Zr, Nb, and La.
3. The elements grouped towards the right hand side of the *d*-block (e.g. Fe, Co, Ni, Cu, Zn, Cd, and Hg) and the metallic *p*-block elements possessing inner *d*-electrons which succeed the *d*-block (e.g. Ga, In, Tl, Ge, Sn, and Pb) are found largely as their sulphide ores.
4. The chemically unreactive metals, such as Ru, Rh, Pd, Ag, Os, Ir, Pt, and Au, located in the central region of the second and third rows of the transition series, occur naturally in the free state as well as in compound form.

Extraction of metals

Isolating metals from their ores inevitably involves reduction, where metal ions gain electrons, and again certain general relationships can be made between the method of extraction and the position of the metal in the Periodic Table.

Where alternative reduction techniques are available, various factors determining the type of process employed have to be considered; these include the availability and cost of the reducing agent, the magnitude of the operation, the degree of purity required, and the nature and usefulness of the by-products. Two of the most important by-products resulting from the isolation of metals are chlorine from the electrolysis of halides and sulphur dioxide from the roasting of sulphide ores. The latter is of particular importance in the manufacture of sulphuric acid.

Electrolytic techniques

Fused electrolytes

The highly electropositive metals of Groups I and II are among the strongest reducing agents known (Table 5.2), and therefore cannot be obtained from their ores by methods involving other reducing agents. Instead, they are isolated by the electrolysis of the fused halide, which is usually the chloride. In order to economize on electrical power, suitable impurities are added to the electrolyte to lower its melting point. For example, in the Downs' cell used for the extraction of sodium from molten sodium chloride (m.p. 808 °C), calcium chloride is added until the melting point of the electrolyte is approximately 500 °C.

The extraction of the Group I and II metals from aqueous solution is not feasible owing to the preferential discharge of hydrogen. Isolation of these metals from aqueous solution is possible, however, if a mercury cathode is used, enabling an amalgam to be formed with the metal; but recovery of the metal is impracticable on a commercial scale.

Although most processes for the extraction of metals from molten electrolytes involve the chloride, aluminium is obtained from the purified oxide, Al_2O_3 (obtained from hydrated aluminium oxide (*bauxite*), $Al_2O_3{\cdot}2H_2O$), dissolved in molten sodium hexafluoroaluminate(III) (cryolite), Na_3AlF_6.

Aqueous electrolytes

Despite having negative standard electrode potentials, certain metals, such as chromium and zinc, can be isolated by the electrolysis of aqueous solutions. *Zinc blende*, ZnS, one of the principal ores of zinc, can be leached out with sulphuric acid, after first oxidizing the ore by partial roasting. After the removal of impurities, the electrolysis of an acidified solution of zinc sulphate, using a high current density, yields a deposit of zinc on the aluminium cathode. Preferential discharge of hydrogen does not occur owing to a high hydrogen over-potential* at the zinc surface, a factor which is enhanced further by the high current density.

The Group III elements, gallium, indium, and thallium, which only occur in very small quantities, are also usually obtained by electrolysing aqueous solutions of their salts.

Chemical reduction

Of the less highly electropositive metals, the majority are obtained by chemical reduction, usually of the *oxide*. By virtue of its abundance and comparative cheapness, the most common reducing agent is coke. This is particularly important in the iron and steel industry, where production is required on a massive scale to meet the demands of the engineering industries (see page 252).

$$2C(s) + O_2(g) \rightarrow 2CO(g)$$
$$FeO(s) + CO(g) \rightarrow Fe(l) + CO_2(g)$$
$$(Fe_2O_3(s) + 3CO(g) \rightarrow 2Fe(s) + 3CO_2(g))$$

Sulphide ores cannot be reduced directly by coke and are first roasted in air to form the oxide:

e.g.

$$2ZnS(s) + 3O_2(g) \rightarrow 2ZnO(s) + 2SO_2(g)$$
$$ZnO(s) + C(s) \rightarrow Zn(s) + CO(g)$$

Like sulphides, *chlorides* are fairly resistant to reduction. However, certain metal chlorides can be reduced by the highly electropositive metals:

---

*A greater potential is required to discharge hydrogen ions at metal electrodes other than platinum, and this extra potential is called the *hydrogen overpotential* for each particular metal electrode. For example, the overpotential for a zinc electrode is 0.7 volt.

e.g.
$$TiCl_4(s)+4Na(s) \rightarrow Ti(s)+4NaCl(s)$$
$$TiCl_4(s)+2Mg(s) \rightarrow Ti(s)+2MgCl_2(s)$$

Although the technique is rarely used, aluminium chloride can also be reduced to the metal by sodium.

Thermal reduction

Because of the thermal instability of its oxide, mercury can be obtained from the sulphide ore, *cinnabar*, HgS, simply by heating in air.

$$HgS(s)+O_2(g) \rightarrow Hg(l)+SO_2(g)$$

General properties of compounds

Oxides and hydroxides

The ionic and basic character of the oxides follows the same general pattern as that of metallic character. The strongest bases are the oxides of Group I elements (the alkali metals), which dissolve in water to form soluble hydroxides or alkalis, although lithium hydroxide is only sparingly soluble. In fact, the Group I metals form the only genuinely soluble hydroxides.

The gradual change from basic to amphoteric and acidic properties as the elements lose their metallic character is clearly illustrated by some of the oxides of the Period 3 elements:

$$\underbrace{Na_2O \quad MgO}_{\text{Basic}} \qquad \underbrace{Al_2O_3}_{\text{Amphoteric}} \qquad \underbrace{SiO_2 \quad P_4O_6, P_4O_{10} \quad SO_2 \quad Cl_2O}_{\text{Acidic}}$$

Like the oxides, the basic character and thermal stability of the hydroxides increases with increasing metallic or electropositive nature.

Halides

The ionic character of the halides follows the same general pattern as that for increasing metallic or electropositive character.

As the melting point of a compound is related to the amount of energy required to overcome the strong attractive forces in an ionic crystal, or the much weaker intermolecular forces (see page 47) in predominantly covalent compounds, it would appear to be logical for the melting points of halides to generally increase with increasing ionic character. However, although this principle may be taken as a crude general guide, there are many exceptions, notably among the Group I metal halides, as there is *no direct relation* between melting point and the degree of ionic or covalent character in a compound. A closer correlation is obtained with the Group II, or alkaline earth, metal halides. This is shown for the chlorides in Tables 6.12 and 6.13.

Table 6.12

Group I metal chloride	*M.p./°C*
LiCl	614
NaCl	808
KCl	772
RbCl	717
CsCl	645

Table 6.13

Group II metal chloride	*M.p./°C*
$BeCl_2$	410
$MgCl_2$	714
$CaCl_2$	772
$SrCl_2$	875
$BaCl_2$	963

Carbonates and hydrogencarbonates

The thermal stability of the carbonates increases with increasing metallic character. With the exception of lithium carbonate, the Group I metal carbonates are fairly stable to heat whereas those of the Group II metals undergo decomposition to the oxide, the ease of decomposition being related to the electropositive character of the metal (see page 129).

The only stable solid hydrogencarbonates are those formed by the alkali metals (with the exception of lithium), indicating further the high electropositive character of these elements.

The distinctive properties of the first element in each group

The elements heading each group in the *s*- and *p*-block display a number of physical and chemical properties which are sometimes quite distinctive from those of the other elements in the group. These characteristics are largely attributable to the smaller size of the atoms, their higher ionization energies, and their higher electronegativity values, causing their compounds to generally have a noticeably higher degree of covalent nature.

In Groups V and VI the first elements, nitrogen and oxygen respectively, exist under ordinary conditions as diatomic gases. The other elements in both of these groups are solids which can exist in polymorphic (allotropic) forms.

The anomalous properties of these and the other elements which head their groups will be discussed further in the chapters dealing with the group properties.

Table 6.14 Relative atomic masses (atomic weights)

1 H 1.0079																	2 He 4.00260
3 Li 6.941	4 Be 9.01218											5 B 10.81	6 C 12.011	7 N 14.0067	8 O 15.9994	9 F 18.99840	10 Ne 20.179
11 Na 22.98977	12 Mg 24.305											13 Al 26.98154	14 Si 28.086	15 P 30.97376	16 S 32.06	17 Cl 35.453	18 Ar 39.948
19 K 39.098	20 Ca 40.08	21 Sc 44.9559	22 Ti 47.90	23 V 50.9414	24 Cr 51.996	25 Mn 54.9380	26 Fe 55.847	27 Co 58.9332	28 Ni 58.71	29 Cu 63.546	30 Zn 65.38	31 Ga 69.72	32 Ge 72.59	33 As 74.9216	34 Se 78.96	35 Br 79.904	36 Kr 83.80
37 Rb 85.4678	38 Sr 87.62	39 Y 88.9059	40 Zr 91.22	41 Nb 92.9064	42 Mo 95.94	43 Tc (99)	44 Ru 101.07	45 Rh 102.9055	46 Pd 106.4	47 Ag 107.868	48 Cd 112.40	49 In 114.82	50 Sn 118.69	51 Sb 121.75	52 Te 127.60	53 I 126.9045	54 Xe 131.30
55 Cs 132.9054	56 Ba 137.34	57 La 138.9055	72 Hf 178.49	73 Ta 180.9479	74 W 183.85	75 Re 186.2	76 Os 190.2	77 Ir 192.22	78 Pt 195.09	79 Au 196.9665	80 Hg 200.59	81 Tl 204.37	82 Pb 207.2	83 Bi 208.9804	84 Po (210)	85 At (210)	86 Rn (222)
87 Fr (223)	88 Ra 226.0254	89 Ac (227)															
				58 Ce 140.12	59 Pr 140.9077	60 Nd 144.2	61 Pm (147)	62 Sm 150.4	63 Eu 151.96	64 Gd 157.25	65 Tb 158.9254	66 Dy 162.50	67 Ho 164.9304	68 Er 167.26	69 Tm 168.9342	70 Yb 173.04	71 Lu 174.97
				90 Th 232.0381	91 Pa 231.0359	92 U 238.029	93 Np 237.0482	94 Pu (242)	95 Am (243)	96 Cm (247)	97 Bk (247)	98 Cf (251)	99 Es (254)	100 Fm (253)	101 Md (256)	102 No (254)	103 Lr (257)

Table 6.15 Melting points of the elements/ °C

1 H −259																	2 He −270
3 Li 180	4 Be 1280											5 B 2300	6 C (graphite) 3730(subl.)	7 N −210	8 O −218	9 F −220	10 Ne −249
11 Na 97.8	12 Mg 650											13 Al 660	14 Si 1410	15 P 44.2(white) 590(red)	16 S 113(α) 119(β)	17 Cl −34.7	18 Ar −189
19 K 63.7	20 Ca 850	21 Sc 1540	22 Ti 1675	23 V 1900	24 Cr 1890	25 Mn 1240	26 Fe 1535	27 Co 1492	28 Ni 1453	29 Cu 1083	30 Zn 420	31 Ga 29.8	32 Ge 937	33 As subl.	34 Se 217	35 Br −7.2	36 Kr −157
37 Rb 38.9	38 Sr 768	39 Y 1500	40 Zr 1850	41 Nb 2470	42 Mo 2610	43 Tc 2200	44 Ru 2500	45 Rh 1970	46 Pd 1550	47 Ag 961	48 Cd 321	49 In 157	50 Sn 232	51 Sb 630	52 Te 450	53 I 114	54 Xe −112
55 Cs 28.7	56 Ba 714	57 La 920	72 Hf 2220	73 Ta 3000	74 W 3410	75 Re 3180	76 Os 3000	77 Ir 2440	78 Pt 1769	79 Au 1063	80 Hg −38.9	81 Tl 304	82 Pb 327	83 Bi 271	84 Po 254	85 At (302)	86 Rn −71
87 Fr (27)	88 Ra 700	89 Ac 1050															
				58 Ce 795	59 Pr 935	60 Nd 1020	61 Pm 1030	62 Sm 1070	63 Eu 826	64 Gd 1310	65 Tb 1360	66 Dy 1410	67 Ho 1460	68 Er 1500	69 Tm 1540	70 Yb 824	71 Lu 1650
				90 Th 1750	91 Pa 1230	92 U 1130	93 Np 640	94 Pu 640	95 Am (1200)	96 Cm —	97 Bk —	98 Cf —	99 Es —	100 Fm —	101 Md —	102 No —	103 Lr —

Table 6.16 Boiling points of the elements/°C

1 H −252																	2 He −269
3 Li 1330	4 Be 2477											5 B 3930	6 C 4830	7 N −196	8 O −183	9 F −188	10 Ne −246
11 Na 890	12 Mg 1110											13 Al 2470	14 Si 2360	15 P 280 (white)	16 S 445	17 Cl −34.7	18 Ar −186
19 K 774	20 Ca 1487	21 Sc 2730	22 Ti 3260	23 V 3400	24 Cr 2482	25 Mn 2100	26 Fe 3000	27 Co 2900	28 Ni 2730	29 Cu 2595	30 Zn 907	31 Ga 2400	32 Ge 2830	33 As 613 subl.	34 Se 685	35 Br 58.8	36 Kr −152
37 Rb 688	38 Sr 1380	39 Y 2930	40 Zr 3580	41 Nb 3300	42 Mo 5560	43 Tc 3500	44 Ru 4900	45 Rh 4500	46 Pd 3980	47 Ag 2210	48 Cd 765	49 In 2000	50 Sn 2270	51 Sb 1380	52 Te 990	53 I 184	54 Xe −108
55 Cs 690	56 Ba 1640	57 La 3470	72 Hf 5400	73 Ta 5420	74 W 5930	75 Re 5630	76 Os 5000	77 Ir 5300	78 Pt 4530	79 Au 2970	80 Hg 357	81 Tl 1460	82 Pb 1744	83 Bi 1560	84 Po 960	85 At —	86 Rn −61.8
87 Fr —	88 Ra 1140	89 Ac 3200															
				58 Ce 3470	59 Pr 3130	60 Nd 3030	61 Pm 2730	62 Sm 1900	63 Eu 1440	64 Gd 3000	65 Tb 2800	66 Dy 2600	67 Ho 2600	68 Er 2900	69 Tm 1730	70 Yb 1430	71 Lu 3330
				90 Th 3850	91 Pa —	92 U 3820	93 Np —	94 Pu 3240	95 Am (2600)	96 Cm —	97 Bk —	98 Cf —	99 Es —	100 Fm —	101 Md —	102 No —	103 Lr —

Table 6.17 Densities of the solid and liquid elements/g cm^3

1 H —																	2 He —
3 Li 0.53	4 Be 1.85											5 B 2.34	6 C 2.25(graph) 3.51(diam)	7 N —	8 O —	9 F —	10 Ne —
11 Na 0.97	12 Mg 1.74											13 Al 2.70	14 Si 2.33	15 P 1.82(white) 2.34(red)	16 S 2.07(α) 1.96(β)	17 Cl —	18 Ar —
19 K 0.86	20 Ca 1.54	21 Sc 2.99	22 Ti 4.54	23 V 5.96	24 Cr 7.19	25 Mn 7.20	26 Fe 7.86	27 Co 8.90	28 Ni 8.90	29 Cu 8.92	30 Zn 7.14	31 Ga 5.91	32 Ge 5.35	33 As 5.72	34 Se 4.81	35 Br 3.12	36 Kr —
37 Rb 1.53	38 Sr 2.62	39 Y 4.34	40 Zr 6.49	41 Nb 8.57	42 Mo 10.2	43 Tc 11.5	44 Ru 12.3	45 Rh 12.4	46 Pd 12.0	47 Ag 10.5	48 Cd 8.64	49 In 7.30	50 Sn 7.28(white) 5.75(grey)	51 Sb 6.62	52 Te 6.25	53 I 4.93	54 Xe —
55 Cs 1.90	56 Ba 3.51	57 La 6.19	72 Hf 13.3	73 Ta 16.6	74 W 19.4	75 Re 20.5	76 Os 22.5	77 Ir 22.5	78 Pt 21.4	79 Au 19.3	80 Hg 13.6	81 Tl 11.85	82 Pb 11.3	83 Bi 9.80	84 Po 9.4	85 At —	86 Rn —
87 Fr —	88 Ra 5.0	89 Ac 10.1															
				58 Ce 6.78	59 Pr 6.78	60 Nd 7.00	61 Pm —	62 Sm 7.54	63 Eu 5.24	64 Gd 7.95	65 Tb 8.27	66 Dy 8.56	67 Ho 8.80	68 Er 9.16	69 Tm 9.33	70 Yb 6.98	71 Lu 9.84
				90 Th 11.7	91 Pa 15.4	92 U 19.07	93 Np 20.4	94 Pu 19.8	95 Am 11.7	96 Cm —	97 Bk —	98 Cf —	99 Es —	100 Fm —	101 Md —	102 No —	103 Lr —

7 Hydrogen

The structure of the hydrogen atom is the simplest for any element, consisting of a nucleus, comprising one proton, and a single extranuclear electron.

Although the amount of molecular hydrogen in nature is negligible, the presence of its atoms in compounds such as water and the majority of organic compounds makes hydrogen one of the most abundant elements in the Earth's crust. It is the most abundant element in the Universe.

Molecular hydrogen is the least dense of all gases (8.99×10^{-5} g cm^{-3} at s.t.p.). It is colourless, odourless, and virtually insoluble in water. Liquid hydrogen has a boiling point of −252.77 °C (20.38 K) and freezes at −259.23 °C (13.92 K).

Manufacture and preparation

Commercially, hydrogen is obtained from the cracking of hydrocarbons, from natural gas, and from the electrolysis of aqueous solution, e.g. as a by-product in manufacturing sodium hydroxide from sodium chloride solution.

In the steam reformation of natural gas, which consists mainly of methane, the pre-heated gas, together with steam, is passed over a nickel-chromium catalyst at 750 °C and 10 atmospheres pressure:

$$CH_4(g) + H_2O(g) \xrightleftharpoons[]{\text{Ni–Cr catalyst, } 750\,^\circ\text{C, 10 atm.}} CO(g) + 3H_2(g); \quad \Delta H^\ominus = +205\ \text{kJ mol}^{-1}$$

This process can also be applied to other hydrocarbon gases obtained from the refining of petroleum.

In the laboratory, hydrogen is usually prepared by the action of dilute mineral acids (but not dilute nitric acid) on suitably reactive metals.

e.g. $$Zn(s) + 2H^+(aq) \rightarrow Zn^{2+}(aq) + H_2(g)$$

Reactivity

Despite the fact that all elements, with the exception of the noble gases, form compounds with hydrogen, it is not renowned for its reactivity although under certain conditions it does react explosively with oxygen, fluorine and chlorine.

$$2H_2(g) + O_2(g) \rightarrow 2H_2O(l); \quad \Delta H^\ominus = -572\ \text{kJ mol}^{-1}$$
$$H_2(g) + Cl_2(g) \rightarrow 2HCl(g); \quad \Delta H^\ominus = -184.6\ \text{kJ mol}^{-1}$$

The formation of hydrogen chloride proceeds via a free-radical mechanism, initiated by the homolytic fission* of chlorine molecules brought about by the incidence of ultra-violet light, e.g. direct sunlight:

$$Cl_2 \xrightarrow{h\nu} 2Cl\cdot$$
$$H_2 + Cl\cdot \longrightarrow HCl + H\cdot$$
$$Cl_2 + H\cdot \longrightarrow HCl + Cl\cdot$$

The dissociation of the hydrogen molecule is a highly endothermic process and it is primarily this factor which accounts for its comparatively low reactivity under moderate conditions:

$$H_2(g) \rightarrow 2H(g); \quad \Delta H^\ominus = +436\,\text{kJ mol}^{-1}$$

Monohydrogen (atomic hydrogen)

The endothermic dissociation of the hydrogen molecule (dihydrogen) into free atoms can be brought about by means of a tungsten wire heated by an electric current and surrounded by an atmosphere of hydrogen at low pressure, or by electromagnetic radiation of an appropriate wavelength, or by passing hydrogen under low pressure through an electric arc between tungsten electrodes. Permanent recombination of the free atoms does not readily occur as the energy evolved in doing so ($-436\,\text{kJ mol}^{-1}$) causes immediate dissociation of the molecule. However, the presence of a third body such as a metal surface, which absorbs the excess energy, enables recombination to take place.

Even at atmospheric pressure, a considerable proportion of monohydrogen can be formed by passing a stream of hydrogen through an electric arc between water-cooled tungsten electrodes. If the monohydrogen issuing from the arc is directed on to a metal surface, the heat evolved by the combination of the monohydrogen is such that it is capable of melting tungsten (m.p. 3400 °C). This is the principle of a torch used for welding metals. It is particularly suitable for metals which are readily oxidized, such as aluminium, but which have little tendency to absorb the gas.

Monohydrogen is a much more powerful reducing agent than the molecular form:

e.g. $$BaSO_4(s) + 8H(g) \rightarrow BaS(s) + 4H_2O(l)$$

'Nascent' hydrogen

If a small quantity of iron(III) chloride is added to a mixture of zinc and dilute hydrochloric acid, the iron(III) salt (yellow) is rapidly reduced to the iron(II) salt (green). No similar change is produced on bubbling molecular hydrogen

---

**Homolytic fission* of a covalent bond involves each of the two bonded atoms separating with one of the bonding electrons to form two highly reactive species, known as *free radicals*:

$$A{:}B \rightarrow A\cdot + B\cdot$$

through the solution. The peculiar activity of hydrogen in reactions of this type is sometimes incorrectly attributed to the fact that it is *nascent* ('new-born'). In actual fact, the reduction is not brought about by the hydrogen at all but by the zinc metal, being facilitated by the acidic medium, with the metal supplying the electrons necessary for the process:

$$2Fe^{3+}(aq) + Zn(s) \rightarrow 2Fe^{2+}(aq) + Zn^{2+}(aq)$$

Hydrogen is formed by the side-reaction:

$$Zn(s) + 2H^{+}(aq) \rightarrow Zn^{2+}(aq) + H_2(g)$$

Isotopes of hydrogen

Ordinary hydrogen consists of 99.984 per cent ^1H (sometimes called *protium*) and 0.016 per cent ^2H or D (*deuterium*). A third isotope, *tritium*, ^3H or T, whose existence in ordinary hydrogen is doubtful, is formed when lithium-6 is bombarded with neutrons,

$$^6_3Li + ^1_0n \rightarrow ^4_2He + ^3_1H$$

As all three isotopes have the same electronic configuration, they exhibit the same chemical properties although the heavier isotopes tend to undergo changes more slowly. However, because of their considerable difference in mass (their relative values being much greater than for the isotopes of any other element), they possess significantly different physical properties, particularly with regard to those properties related to mass, e.g. rate of diffusion, density, etc.

Table 7.1

	H_2	D_2
Melting point	−259.2 °C (14 K)	−254.5 °C (18.7 K)
Boiling point	−252.6 °C (20.6 K)	−249.4 °C (23.8 K)
Enthalpy of fusion	−117 kJ mol^{-1}	−219 kJ mol^{-1}
Enthalpy of vaporization	+907 kJ mol^{-1}	+1230 kJ mol^{-1}
Bond enthalpy	+436 kJ mol^{-1}	+444 kJ mol^{-1}

Deuterium is particularly useful for isotopic labelling in the study of reactions (especially organic) and their mechanisms.

Tritium is a low energy β-emitter with a half-life of 12.3 years. It is also used as a 'label', often in reactions and processes of biological importance.

Deuterium oxide is obtained by the electrolysis of ordinary water over a considerable period of time. ^1H is liberated with much greater ease than D with the result that the remaining water becomes progressively richer in D_2O. Because deuterium oxide has a lower dielectric constant than ordinary water, ionic compounds tend to be less soluble in it than in water. It has a higher freezing point, boiling point, and density, and also differs with respect to other physical properties (Table 7.2).

Table 7.2

	H_2O	D_2O
Freezing point	0 °C (273.15 K)	3.8 °C (276.95 K)
Boiling point	100 °C (373.15 K)	101.4 °C (3074.59 K)
Density at 20 °C	0.998 g cm^{-3}	1.106 g cm^{-3}

Ortho- and para-hydrogen

In the molecular form, 75 per cent of ordinary hydrogen molecules under normal conditions have the two protons spinning in the same direction. This is referred to as *ortho*-hydrogen. The remaining 25 per cent have the spins of the protons opposed and are referred to as *para*-hydrogen.

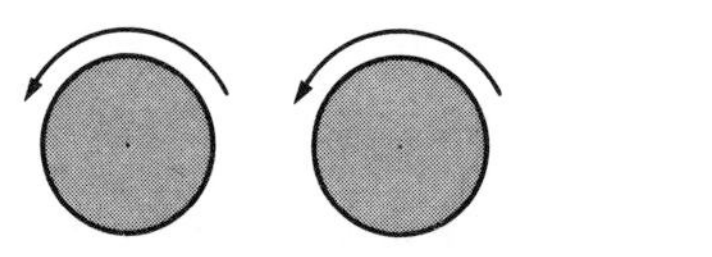

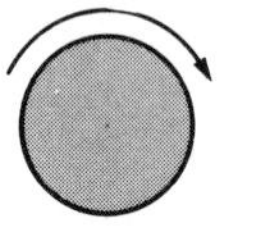

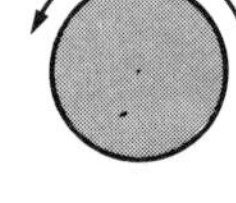

Fig. 7.1 (a) *Ortho*-hydrogen (spins parallel) (b) *Para*-hydrogen (spins opposed)

Of the two forms, *para*-hydrogen is the more stable; and on lowering the temperature, the proportion of this form increases until at absolute zero a sample would exist solely in this form. Ordinary hydrogen gas is a mixture of these two species, the actual composition being determined by temperature.

The chemical properties of the two forms are identical although they differ slightly with respect to certain physical properties (Table 7.3), such as melting point, boiling point, specific heat capacity, and thermal conductivity.

Table 7.3

	Ortho-hydrogen	*Para-hydrogen*	*Ordinary hydrogen*
Melting point	−259.22 °C (13.93 K)	−259.27 °C (13.88 K)	−259.23 °C (13.92 K)
Boiling point	−252.74 °C (20.41 K)	−252.86 °C (20.29 K)	−252.77 °C (20.38 K)

The anomalous position of hydrogen in the Periodic Table

Hydrogen shows certain interesting similarities with the alkali metals of Group I, and others with the halogens of Group VII. If indeed hydrogen were to be included in either of these groups, then it would obviously have to occupy a place at the head of the group, in which case its properties would have to be typical of an element occupying such a position. From the information discussed below, it will become evident that attempts to classify the element rigidly in either of these groups would be totally erroneous, and it is for this reason that it occupies an anomalous and unique position at the top of the Periodic Table.

Its distinctive properties are strongly influenced by its extremely small size and by the fact that its electronegativity value in no way corresponds to the sequence of values for the elements in either Group I or Group VII (Tables 7.4 and 7.5).

Table 7.4

Element	*Atomic radius/nm*	*Electro-negativity*
H	0.037	2.1
Li	0.123	1.0
Na	0.157	0.9
K	0.203	0.8
Rb	0.244	0.8
Cs	0.262	0.7

Table 7.5

Element	*Atomic radius/nm*	*Electro-negativity*
H	0.037	2.1
F	0.072	4.0
Cl	0.099	3.0
Br	0.114	2.8
I	0.133	2.5

Ionization energy of hydrogen

Like the alkali metals, hydrogen has a single outer electron but, whereas the alkali metals form positive ions with comparative ease during the course of a reaction, hydrogen has little tendency to do so. The relative ease of formation of positive ions for hydrogen and the alkali metals in their normal physical state is indicated below.

Hydrogen

$$\tfrac{1}{2}H_2(g) \xrightarrow[\text{energy}]{\frac{1}{2}\text{ bond diss.}} H(g); \qquad \Delta H^{\ominus} = +\ 218\,\text{kJ mol}^{-1}$$

$$H(g) \xrightarrow[\text{energy}]{\text{ionization}} H^+(g) + e; \quad \Delta H^{\ominus} = +1315\,\text{kJ mol}^{-1}$$

$$\text{Overall enthalpy change:} \quad \Delta H^{\ominus} = +1533\,\text{kJ mol}^{-1}$$

Group I elements

Table 7.6

Element	*Enthalpy of atomization, $\Delta H^{\ominus}_{at}$/kJ mol^{-1}*	*First ionization energy, $\Delta H^{\ominus}_{IE}$/kJ mol^{-1}*	*$\Delta H^{\ominus}_{at} + \Delta H^{\ominus}_{IE}$ /kJ mol^{-1}*
Li	+161	+525	+686
Na	+109	+500	+609
K	+90	+424	+514
Rb	+86	+408	+494
Cs	+79	+382	+461

The high ionization energy of hydrogen is attributable to the very small size of the atom and the strong attractive force between the proton and the electron.

Electron affinity of hydrogen

Like the halogens, hydrogen exists as a diatomic molecule and the atom has an electronic structure which is only one electron short of a filled outer shell. However, by comparison with the halogens, it has a very small electron affinity value and has little tendency to form hydride ions, H^-. Because of its low electron affinity, the formation of the hydride ion from molecular hydrogen is an endothermic process whereas the halogens, which have high electron affinities, form halide ions exothermically and with ease (Table 7.7).

Table 7.7

Element	*Enthalpy of sublimation, $\Delta H^\ominus_{sub}$ /kJ mol^{-1}*	*Enthalpy of vaporization, $\Delta H^\ominus_{vap}$ /kJ mol^{-1}*	*$\frac{1}{2}$ bond dissociation energy, $\Delta H^\ominus_{diss}$ /kJ mol^{-1}*	*Electron affinity $\Delta H^\ominus_{EA}$ /kJ mol^{-1}*	*Sum of enthalpy terms*
H_2	—	—	+218	−78	+140
F_2	—	—	+79	−354	−275
Cl_2	—	—	+121	−370	−249
Br_2	—	+15	+96.5	−348	−236.5
I_2	+31	—	+75.5	−320	−213.5

Compounds containing the hydride ion are therefore formed only by the highly electropositive metals in Groups I and II, as these metals require only a relatively small amount of energy to expel electrons and form positive ions.

The instability of the hydride ion as compared to the halide ion is indicated by a comparison between the enthalpies of formation of the hydrides (Table 7.8) and halides (as illustrated by the chlorides in Table 7.9) of the Group I and II metals.

Table 7.8 Enthalpies of formation of some Group I and Group II hydrides

Hydride	*$\Delta H^\ominus_f$/kJ mol^{-1}*
LiH	−91
NaH	−57
KH	−58
RbH	−56
CsH	−84
CaH_2	−189
SrH_2	−177
BaH_2	−171

Table 7.9 Enthalpies of formation of some Group I and Group II chlorides

Chloride	*$\Delta H^\ominus_f$/kJ mol^{-1}*
LiCl	−409
NaCl	−411
KCl	−436
RbCl	−430
CsCl	−433
$CaCl_2$	−795
$SrCl_2$	−828
$BaCl_2$	−860

The hydrogen ion

The ionization energy of hydrogen, $+1315\,kJ\,mol^{-1}$, is very high by comparison with the values of the alkali metals and is in fact higher than the first ionization energy of the noble gas, xenon.

$$H(g) \rightarrow H^+(g)+e; \quad \Delta H^\ominus_{IE} = +1315\,kJ\,mol^{-1}$$

$$Li(g) \rightarrow Li^+(g)+e; \quad \Delta H^\ominus_{IE} = +\ 525\,kJ\,mol^{-1}$$
$$Na(g) \rightarrow Na^+(g)+e; \quad \Delta H^\ominus_{IE} = +\ 500\,kJ\,mol^{-1}$$

$$Xe(g) \rightarrow Xe^+(g)+e; \quad \Delta H^\ominus_{IE} = +1170\,kJ\,mol^{-1}$$

The bonds formed by hydrogen in compounds in which it is the more electropositive element are therefore essentially covalent. The one exception is hydrogen fluoride. Fluorine, being the most electronegative element, reduces the degree of covalent character to about 45 per cent.

The free proton, or hydrogen ion, can be obtained only under extreme physical conditions, e.g. by an electric arc or in discharge tubes, and even here the life of an individual ion is only about half a second. As the ionization energy of hydrogen is so high, the hydrogen ion is encountered only when one of its compounds is dissolved in a solvent which will solvate protons, yielding species such as H_3O^+, ROH_2^+, and NH_4^+, the solvation energy providing the energy necessary to overcome the ionization energy of the hydrogen. Compounds which furnish solvated hydrogen ions or protons in this way are functioning as acids.

The solvation of a proton is a highly exothermic process, especially in water in which it forms the *hydrated proton*, H^+ (aq). This is sometimes depicted as the H_3O^+ ion which is known as the *oxonium ion* (the names hydroxonium or hydronium ion are also commonly used), although it is more accurately represented as $[H(H_2O)_x]^+$

$$H^+(g)+aq \rightarrow H^+(aq); \quad \Delta H^\ominus_{hyd} = -1075\,kJ\,mol^{-1}$$

General classification of hydrogen compounds

Ionic ('salt-like') hydrides

The *salt-like* hydrides formed by the electropositive elements in Groups I and II are generally highly reactive, white crystalline solids with fairly high melting points. The ionic nature of these compounds is further indicated by their ability, when molten or dissolved in alkali metal halides, to undergo electrolysis and liberate hydrogen at the anode.

They are strong reducing agents, LiH and CaH_2 being used as such in preparative work. Their reducing properties are illustrated by their ability to liberate hydrogen from water, indicating that in aqueous solution H^- is a stronger base than OH^-.

$$H^- + H_2O(l) \rightarrow OH^-(aq) + H_2(g)$$

Similarly, in liquid ammonia, H^- behaves as a stronger base than NH_2^-

$$H^- + NH_3(l) \rightarrow NH_2^- + H_2(g)$$

Ionic hydrides also function as powerful reducing agents at elevated temperatures.

$$SiCl_4(l)+4NaH(s) \xrightarrow[\text{temperature}]{\text{high}} SiH_4(g)+4NaCl(s)$$

'Polymer-like' hydrides

Beryllium and magnesium hydrides are essentially covalent with bridged polymer-type structures (see page 126), representing a transition between covalent and ionic hydrides. Because of their non-stoichiometric nature, or variable composition, they are sometimes referred to as *interstitial hydrides.* Similar non-stoichiometric hydrides are also formed by the transition elements (see page 241).

They tend to be less dense than the parent metal and exhibit reducing properties of hydrogen while retaining properties characteristic of the metal component.

Covalent hydrogen compounds

The majority of hydrogen compounds are covalent and possess molecular structures which exert only weak intermolecular attractions.

The simplest hydrogen compounds of the less electropositive and non-metallic elements in Groups IV, V, VI, and VII are generally gaseous under ordinary conditions, the notable exceptions being water (b.p. 100 °C) and hydrogen fluoride (b.p. 19 °C):

Group IV	*Group V*	*Group VI*	*Group VII*
CH_4	NH_3	H_2O	HF
SiH_4	PH_3	H_2S	HCl
GeH_4	AsH_3	H_2Se	HBr
SnH_4	SbH_3	H_2Te	HI

The comparatively low volatility of water and hydrogen fluoride, and the ease with which ammonia can be liquefied, is attributable to the high electronegativities of oxygen (3.5), fluorine (4.0), and nitrogen (3.0) and the availability of lone pairs of electrons on these atoms in their respective compounds, enabling them to form **intermolecular hydrogen bonds**.

Where hydrogen is bonded to oxygen, or to another highly electronegative element, such as fluorine, it may serve as a bridge between two oxygen atoms of different molecules, forming a covalent bond with one and holding the other solely by electrostatic forces, the latter being referred to as an intermolecular hydrogen bond (Figure 7.2). This is, in effect, a very strong dipole-dipole interaction and accounts for the unexpectedly high boiling points of compounds such as water, hydrogen fluoride, alcohols, carboxylic acids, etc.

```
δ+  2δ−   δ+  2δ−               δ+  δ−    δ+  δ−
H—Ö:---H—Ö:                     H—Ö:---H—Ö:
   |       |                       |       |
   H^δ+    H^δ+                    R       R
     (a)                               (b)

              δ−      δ+     δ−
              Ö:---H—Ö
           δ+//          \ δ+
        R—C                C—R
            \ δ−    δ+   δ−//
              Ö—H---:Ö
                 (c)
```

Fig. 7.2 Intermolecular hydrogen bonding in (a) water, (b) alcohols, and (c) monocarboxylic acids

In *ice*, each water molecule is tetrahedrally surrounded by four other water molecules which are bound to the central molecule by means of intermolecular hydrogen bonds (Figure 7.3).

```
            O
            ¦
            H
            |
  O—H---O—H---O
            ¦
            H
            |
            O
```

Fig. 7.3

The crystal lattice of ice has a very open structure, and this accounts for the low density of the solid. Water is unusual in having a solid form which is less dense than the liquid; on melting, the solid structure breaks down enabling the molecules to pack together more closely in the liquid phase.

The most compact arrangement of water molecules occurs at a temperature of 4 °C, at which point the density of water reaches a maximum ($1.0\,g\,cm^{-3}$). Structurally, water is somewhat similar to ice but possesses a greater degree of disorder. It is therefore incorrect to consider water molecules as being arranged randomly.

The strength of a hydrogen bond, with a bond enthalpy of about $+21\,kJ\,mol^{-1}$, is only 10–20 per cent that of most covalent bonds.

Complex hydrides

The only elements, other than the transition metals, to form complex hydrides are the Group III elements, boron, aluminium and gallium; such hydrides include $Li[AlH_4]$, $Na[BH_4]$, $Al[BH_4]_3$, and $Li[GaH_4]$. In each case the $[MH_4]^-$ ion has a tetrahedral structure. All of these compounds are useful reducing agents, especially lithium tetrahydridoaluminate(III) (lithium aluminium hydride) and sodium tetrahydridoborate(III) (sodium borohydride), and are frequently employed as such in organic chemistry. For example, lithium tetrahydridoaluminate(III) will reduce carboxylic acids to the corresponding alcohol.

Sodium tetrahydridoborate(III), $Na^+[BH_4]^-$, is ionic and soluble in water, whereas the majority of the others are predominantly covalent in nature and are generally more soluble in organic solvents, notably ethers. Lithium tetrahydridoaluminate(III) is the only one which is soluble in ethoxyethane (ether).

Group I: The Alkali Metals

Table 8.1

	Lithium	*Sodium*	*Potassium*	*Rubidium*	*Caesium*
Symbol	Li	Na	K	Rb	Cs
Outer electronic structure	$2s^1$	$3s^1$	$4s^1$	$5s^1$	$6s^1$
Principal oxidation state	+1	+1	+1	+1	+1

Electronic structure and size of atoms

The Group I elements have an outer electronic configuration of ns^1, and so have an oxidation state of +1.

With the general exception of the noble gases, the atoms of the Group I elements are the largest (by comparison of covalent radii) in their respective periods and, as their most stable isotopes also possess the lowest number of nuclear particles within each period, they have comparatively low densities (Table 8.2); lithium, sodium, and potassium are all less dense than water. Furthermore, all the elements are unusually soft.

Table 8.2

	Li	*Na*	*K*	*Rb*	*Cs*
Atomic (covalent) radius/nm	0.123	0.157	0.203	0.216	0.235
Ionic radius, M^+/nm	0.060	0.095	0.133	0.148	0.169
Density/g cm^{-3}	0.53	0.97	0.86	1.53	1.90

Ionization energies and electropositive character

The large size of the atoms and the singly occupied outer shell of electrons cause the elements to have low first ionization energies (Figure 8.1 and Table 8.3).

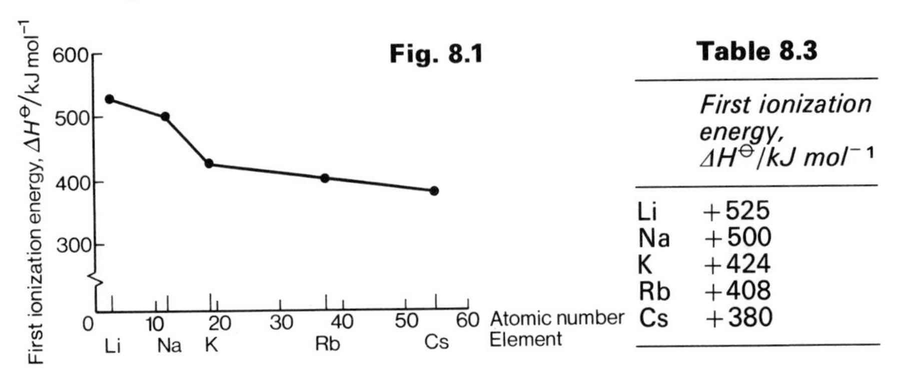

Fig. 8.1

Table 8.3

	First ionization energy, $\Delta H^{\ominus}$/kJ mol^{-1}
Li	+525
Na	+500
K	+424
Rb	+408
Cs	+380

Second ionization energies are extremely high, owing to the stable electronic configuration of the M^+ ions. The chemistry of the Group I elements is, in fact, primarily a study of their M^+ ions.

Hydration energies, which are favoured by small ions, preferably with a high charge, decrease with increasing size of the ions i.e. Li^+ to Cs^+ (Figure 8.2 and Table 8.4).

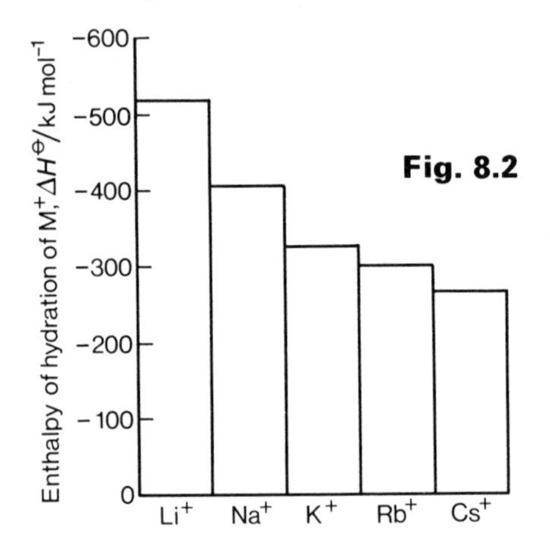

Fig. 8.2

Table 8.4

	Enthalpy of hydration, $\Delta H^{\ominus}$ /kJ mol^{-1}
Li^+	−519
Na^+	−406
K^+	−322
Rb^+	−301
Cs^+	−276

The elements are all highly electropositive metals, i.e. they possess small electronegativity values (see Table 8.5), and are the most reactive metals in their respective periods. Within the group, the reactivity increases on descending from lithium to caesium in accordance with the increasing size of the atoms and the decreasing ionization energy.

On account of their ability to form soluble hydroxides, the Group I metals are collectively referred to as the *alkali metals*.

Table 8.5

	Li	*Na*	*K*	*Rb*	*Cs*
Electronegativity	1.0	0.9	0.8	0.8	0.7

General

Similarities between the elements with regard to general behaviour and nature are particularly strong in this group.

The singly charged positive ions of the alkali metals have all their electrons paired, are diamagnetic (see page 240), and are typically white or colourless. As most of the commonly encountered simple anions, e.g. halide, sulphate, hydrogensulphate, sulphite, hydrogensulphite, carbonate, hydrogencarbonate, nitrate, nitrite, etc., are all similarly without colour, their compounds with the alkali metals are also white or colourless. The chromate(VI) (yellow), dichromate(VI) (orange), and manganate(VII) (permanganate) (purple) anions are themselves coloured and the colour of their compounds with these metals is consequently that of the anion.

The ease of excitation of the outer electron enables the elements to give a sensitive flame test, the colour of the flame being a characteristic of the metal. On heating, the electron is raised to a higher energy level, and when it returns to the lower level the colour of light corresponding to the energy absorbed during the initial excitation is emitted (see page 2). Lithium gives a crimson flame, sodium a golden yellow, potassium a lilac, rubidium a red, and caesium a blue one. An extension of this principle accounts for the use of caesium and potassium in photoelectric cells; in this case the outer electron is excited by irradiation with light.

All the elements have body-centred cubic structures (see page 38) and, by comparison with other metals, have particularly low melting and boiling points, which generally decrease on descending the group (Figure 8.3 and Table 8.6).

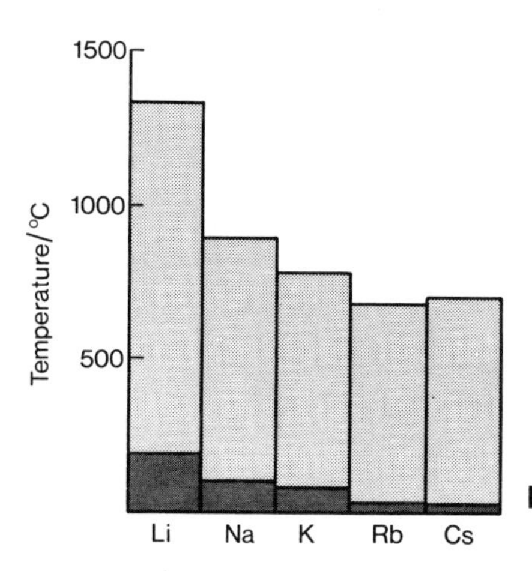

Table 8.6

	M.p./°C	*B.p./°C*
Li	180	1330
Na	98	890
K	64	774
Rb	39	688
Cs	29	690

Fig. 8.3 Melting points and boiling points of the Group I elements

Because of their large size and small charge, the alkali metal ions *have little tendency to form complexes.*

Owing to the great reactivity of the metals, they tarnish rapidly on exposure to air, gradually building up a crust of the oxide:

$$4M(s) + O_2(g) \rightarrow 2M_2O(s)$$

Lithium also forms some nitride (cf. magnesium):

$$6Li(s) + N_2(g) \rightarrow 2Li_3N(s)$$

To prevent such deterioration, the elements are usually stored under a light oil fraction for protection, e.g. naphtha.

The metals react either vigorously or violently with cold water, liberating hydrogen, the reactivity increasing with increasing electropositive character from Li to Cs:

$$2M(s) + 2H_2O(l) \rightarrow 2MOH(aq) + H_2(g)$$

Distinctive properties of lithium

Lithium has many distinctive properties, the majority of which are attributable to its small size and higher electronegativity. For example, unlike the other Group I elements, its carbonate and hydroxide are unstable to heat, its hydrogencarbonate is stable only in aqueous solution, and its nitrate decomposes on heating to the oxide, whereas the nitrate compounds of the other Group I metals form the nitrite. In certain respects it shows similarities to its diagonally related neighbour, magnesium (see page 89).

Compared to the ions of the other elements in the group, lithium has a significantly higher ionization energy and the Li^+ ion has a much higher hydration energy, attracting the polar water molecules with comparative ease. Virtually all lithium salts in the solid state are hydrated. The number of hydrated compounds for the other elements decreases on descending the group, rubidium forming very few hydrated salts.

Hydrides

The alkali metals are among the few elements which are sufficiently electropositive to form genuine ionic hydrides, M^+H^-. These are all white solids and are usually prepared either by direct synthesis,

$$2M(s) + H_2(g) \rightarrow 2MH(s)$$

or by heating the monoxide in a stream of hydrogen:

$$M_2O(s) + H_2(g) \rightarrow MOH(s) + MH(s)$$

The stability of the hydrides decreases on descending the group.

In the molten state, they function as strong electrolytes, H^- ions being discharged at the anode.

They are all hydrolysed by water to the hydroxide, hydrogen being liberated:

$$H^- + H_2O(l) \rightarrow OH^-(aq) + H_2(g)$$

Probably the most important of the alkali metal hydrides is lithium hydride, which on treatment with aluminium chloride in a dry ethereal solution yields one of the most useful reducing agents in organic chemistry, lithium tetrahydridoaluminate(III) (lithium aluminium hydride).

$$4LiH + AlCl_3 \xrightarrow{\text{dry ethoxyethane}} Li^+[AlH_4]^- + 3LiCl$$

Oxides

With the exception of lithium, which forms only a monoxide, all the elements form three types of ionic oxides on controlled air-oxidation:

$$4M(s) + O_2(g) \xrightarrow{180\,^\circ C} 2M_2O(s)$$

$$2M(s) + O_2(g) \xrightarrow{300\,^\circ C} M_2O_2(s)$$

$$M(s) + O_2(g) \xrightarrow{300\,^\circ C} MO_2(s)$$

The monoxides, $(M^+)_2O^{2-}$, are ionic and colourless, the peroxides, $(M^+)_2(O{-}O)^{2-}$, are weakly coloured, e.g. sodium peroxide is usually pale yellow (although it is white when absolutely pure), and the superoxides, $M^+(O{\cdots}O)^-$, which contain an odd number of electrons and a *three-electron bond**, are paramagnetic and coloured (NaO_2 and KO_2 are yellow, RbO_2 is orange, and CsO_2 is red).

All oxides react vigorously with water to form the hydroxides:

$$O^{2-} + H_2O(l) \rightarrow 2OH^-(aq)$$
$$(O{-}O)^{2-} + 2H_2O(l) \rightarrow 2OH^-(aq) + H_2O_2(aq)$$
$$2(O{\cdots}O)^- + 2H_2O(l) \rightarrow 2OH^-(aq) + H_2O_2(aq) + O_2(g)$$

The peroxides and superoxides are powerful oxidizing agents and this is indicated by their reaction with water to give hydrogen peroxide. They are useful in both qualitative and quantitative analysis, for oxidizing green chromium(III) salts to yellow chromate(VI) ions, CrO_4^{2-} (see page 247).

Peroxides react with carbon dioxide to yield carbonates and liberate oxygen:

$$2(O{-}O)^{2-} + 2CO_2(g) \rightarrow 2CO_3^{2-} + O_2(g)$$

Hydroxides

The hydroxides are all pearly white, deliquescent, crystalline solids, which, with the exception of the slightly soluble lithium hydroxide, are all readily soluble in water yielding the strongest known bases – the alkalis. Of these,

---

*A three-electron bond is formed only between atoms of the same or similar electronegativities, and is about half as strong as a covalent bond. It may be considered as a resonance hybrid between the canonical forms A·:B and A:·B.

sodium and potassium hydroxide are the most important, providing a useful source of OH^- ions for qualitative and quantitative analysis, and also for preparative work.

Because of their ability to inflict unpleasant burns upon flesh, they are often referred to as *caustic* alkalis. With the exception of lithium hydroxide, which decomposes to the oxide, they are all stable to heat, indicating the strong electropositive character of the alkali elements.

The majority of metallic ions in aqueous solution react with alkalis, precipitating the insoluble hydroxide. However, with certain metallic elements which possess some degree of amphoteric character, the insoluble hydroxide dissolves in excess alkali to yield salts, further indicating the great basic strength of these compounds.

e.g. $$Pb^{2+}(aq)+2OH^-(aq) \rightarrow Pb(OH)_2(s) \xrightarrow{4OH^-(aq)} \underset{\text{(hexahydroxoplumbate(II) ion)}}{\underset{\text{plumbate(II) ion}}{[Pb(OH)_6]^{4-}(aq)}}$$

$$Zn^{2+}(aq)+2OH^-(aq) \rightarrow Zn(OH)_2(s) \xrightarrow{2OH^-(aq)} \underset{\text{(tetrahydroxozincate(II) ion)}}{\underset{\text{zincate ion}}{[Zn(OH)_4]^{2-}(aq)}}$$

$$Al^{3+}(aq)+3OH^-(aq) \rightarrow Al(OH)_3(s) \xrightarrow{OH^-(aq)} \underset{\text{(tetrahydroxoaluminate(III) ion)}}{\underset{\text{aluminate ion}}{[Al(OH)_4]^-(aq)}}$$

If sodium hydroxide is added to an aqueous solution of a silver salt, the hydrated oxide, instead of the unstable hydroxide, is precipitated as a brown solid:

$$2Ag^+(aq)+2OH^-(aq) \rightarrow Ag_2O(s)+H_2O(l)$$

On isolating and drying at above 80 °C, the oxide is obtained as a black anhydrous solid.

Acidic oxides react with alkalis to form salts:

$$2OH^-(aq)+CO_2(g) \rightarrow CO_3^{2-}(aq)+H_2O(l)$$
$$2OH^-(aq)+SO_2(g) \rightarrow SO_3^{2-}(aq)+H_2O(l)$$

Non-metals react in various ways. Phosphorus yields phosphine:

$$P_4(s)+3OH^-(aq)+3H_2O(l) \rightarrow \underset{\text{phosphine}}{PH_3(g)}+\underset{\text{(hypophosphite) ion}}{\underset{\text{phosphinate}}{3H_2PO_2^-(aq)}}$$

The products obtained from the reaction with chlorine depend upon the reaction conditions (see page 198).

Ammonium salts react to yield ammonia,

$$NH_4^+(aq)+OH^-(aq) \rightleftharpoons NH_3(g)+H_2O(l)$$

providing a simple and convenient qualitative test for the ammonium ion.

The reaction between carbon monoxide and fused or concentrated aqueous sodium hydroxide is used commercially to manufacture methanoic (formic) acid:

$$CO(g) + NaOH(aq) \xrightarrow[\text{high pressure}]{200\,°C} HCOO^{-}Na^{+}(aq) \xrightarrow{\text{dil. HCl}} HCOOH(aq) + NaCl(aq)$$

Sodium hydroxide finds wide application in organic chemistry. Examples include, saponification (hydrolysis of esters) and the extraction of phenols and methylphenols (cresols) from coal tar.

The halides

The halides are all white, ionic, crystalline solids. Their melting points are characteristically high, although there is no direct correlation between their values and the degree of ionic nature of the bonds, which increases on descending the group (Figure 8.4 and Table 8.7).

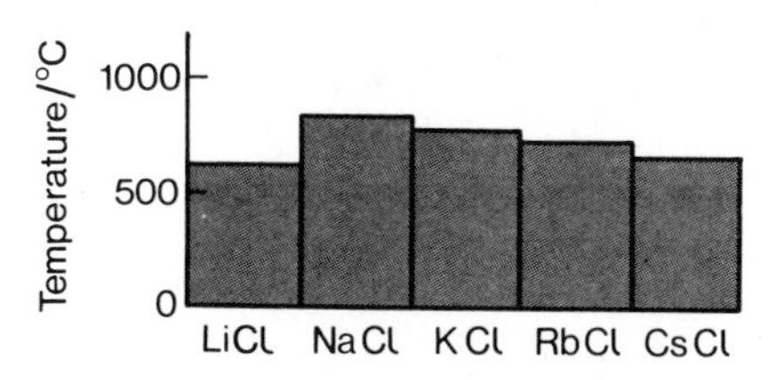

Fig. 8.4 Melting points of the alkali metal chlorides

Table 8.7

	M.p./°C
LiCl	614
NaCl	808
KCl	772
RbCl	717
CsCl	645

They can be prepared by direct halogenation although, owing to the abundance of the more commonly used halides, this process is rarely employed for preparative purposes:

$$2M(s) + X_2(g) \rightarrow 2MX(s)$$

The standard enthalpy of formation of the alkali metal chlorides, obtained on the basis of a Born-Haber cycle, from their elements in the normal state is reasonably constant (Table 8.8). The decrease in sublimation enthalpy and first ionization energy on descending the group is more or less offset by the greater exothermic equivalent of the lattice enthalpy (see page 59).

Table 8.8 The enthalpy of formation of the alkali metal chlorides

Element	*Sublimation enthalpy,* $\Delta H^{\ominus}_{sub}$ / $kJ\,mol^{-1}$	*First ionization energy,* $\Delta H^{\ominus}_{IE}$ / $kJ\,mol^{-1}$	$\frac{1}{2}$ *Dissociation enthalpy of* Cl_2, $\frac{1}{2}\Delta H^{\ominus}_{diss}$ / $kJ\,mol^{-1}$	*Electron affinity of Cl,* $\Delta H^{\ominus}_{EA}$ / $kJ\,mol^{-1}$	*−Lattice enthalpy of* M^+Cl^-, $-\Delta H^{\ominus}_{lat}$ / $kJ\,mol^{-1}$	*Enthalpy of formation of* M^+Cl^-, $\Delta H^{\ominus}_{f}$ / $kJ\,mol^{-1}$
Li	+161	+525	+121	−370	−846	−409
Na	+109	+500	+121	−370	−771	−411
K	+90	+424	+121	−370	−701	−436
Rb	+86	+408	+121	−370	−675	−430
Cs	+79	+382	+121	−370	−645	−433

For the chlorides, there is a gradual change in crystal structure from face-centred cubic to primitive cubic as the metallic ion gets larger (see page 38).

Table 8.9

Metallic ion	*Ionic radius/ nm*	*Ionic radius of Cl^- ion/nm*	r_{M^+}/r_{Cl^-}	*Theoretical co-ordination number*	*Actual co-ordination number*	*Structure of chloride*
Li^+	0.060	0.181	0.33	4	6	Face-centred cubic
Na^+	0.095	0.181	0.52	6	6	Face-centred cubic
K^+	0.133	0.181	0.73	6	6	Face-centred cubic
Rb^+	0.148	0.181	0.82	8	6 or 8	Face-centred or primitive cubic
Cs^+	0.169	0.181	0.93	8	8	Primitive cubic

The halides of lithium possess certain anomalous properties. The chloride forms a dihydrate, $LiCl \cdot 2H_2O$, whereas the chlorides of sodium and potassium are always anhydrous. It is slightly hydrolysed by hot water and is also deliquescent, whereas the chlorides of sodium and potassium are not hydrolysed and do not deliquesce.

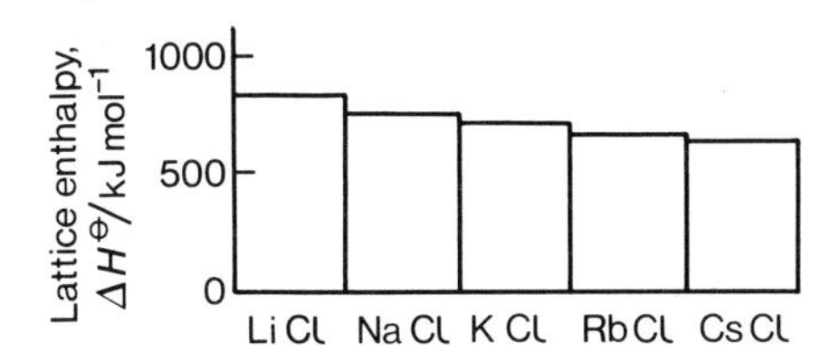

Fig. 8.5 Lattice enthalpies of the alkali metal chlorides

Carbonates

In the anhydrous form, all the carbonates are white solids and, with the exception of lithium carbonate, are soluble in water and stable to heat. The thermal stability is attributable to the high electropositive character of the metals. In this respect, the comparative instability of lithium carbonate closely resembles its diagonally related counterpart, magnesium carbonate (see page 129).

With dilute acids, the carbonates react to liberate carbon dioxide,

$$CO_3^{2-} + 2H^+(aq) \rightarrow CO_2(g) + H_2O(l)$$

and in aqueous solution they hydrolyse to give an alkaline solution:

$$CO_3^{2-} + H_2O(l) \rightleftharpoons HCO_3^-(aq) + OH^-(aq)$$

The most important carbonates are those of sodium and potassium. In addition to the anhydrous form, a number of hydrates exist, the most common being $Na_2CO_3 \cdot 10H_2O$ ('washing soda') which exists in the form of colourless crystals. On exposure to the atmosphere, these crystals effloresce to the monohydrate, a white powder:

$$Na_2CO_3 \cdot 10H_2O(s) \rightarrow Na_2CO_3 \cdot H_2O(s) + 9H_2O(l)$$

Anhydrous sodium and potassium carbonates are frequently used at an elementary level in volumetric analysis for standardizing acids, and in qualitative analysis for precipitating insoluble carbonates. The majority of these reactions proceed in accordance with the equation:

$$2M^+(aq) + CO_3^{2-}(aq) \rightarrow M_2CO_3(s)$$

although in certain cases the basic carbonate is precipitated, notably for magnesium, copper, zinc, and lead.

e.g. $$2Mg^{2+}(aq) + 2OH^-(aq) + CO_3^{2-}(aq) \rightarrow MgCO_3 \cdot Mg(OH)_2(s)$$

Formation of the basic carbonate can be avoided by using aqueous sodium hydrogencarbonate.

e.g. $$Mg^{2+}(aq) + 2HCO_3^-(aq) \rightarrow MgCO_3(s) + H_2O(l) + CO_2(g)$$

Sodium carbonate is also used in qualitative analysis for converting insoluble salts, e.g. barium sulphate, into the corresponding soluble sodium compound before testing for the presence of the anion.

Hydrogencarbonates

With the exception of lithium, the alkali metals are the only elements capable of forming stable, solid hydrogencarbonates. These are obtained in the laboratory by passing carbon dioxide through a concentrated solution of the carbonate and crystallizing out:

$$CO_3^{2-}(aq) + CO_2(g) + H_2O(l) \underset{\text{heat}}{\overset{\text{crystallization}}{\rightleftharpoons}} 2HCO_3^-(s)$$

This process can be reversed by gentle heating.

Hydrogen bonding between the hydrogencarbonate anions plays a significant rôle in binding the ions within the crystal lattice.

Hydrogencarbonates liberate carbon dioxide with dilute acids:

$$HCO_3^- + H^+(aq) \rightarrow H_2O(l) + CO_2(g)$$

and are hydrolysed by water to yield a slightly alkaline solution.

$$HCO_3^- + H_2O(l) \rightleftharpoons H_2CO_3(aq) + OH^-(aq)$$

Nitrates

The alkali metal nitrates, $NaNO_3$ to $CsNO_3$, decompose on heating to yield the nitrite and oxygen.

$$2NO_3^- \rightarrow 2NO_2^- + O_2(g)$$

Lithium nitrate, on the other hand, decomposes to the oxide, showing yet another analogy with the less electropositive metals of Group II (see page 130).

$$4LiNO_3(s) \rightarrow 2Li_2O(s) + 2N_2O_4(g) + O_2(g)$$

The ease of thermal decomposition of the nitrates decreases on descending the group, $LiNO_3$ to $CsNO_3$, in accordance with the increasing electropositive nature of the metal.

Lithium and sodium nitrates both deliquesce although lithium nitrate is the only nitrate of the group to actually crystallize out as a hydrate, $LiNO_3{\cdot}3H_2O$.

Sulphates

The sulphates of potassium, rubidium, and caesium are all anhydrous whereas those of lithium and sodium also exist in the hydrated form, $Li_2SO_4{\cdot}H_2O$ and $Na_2SO_4{\cdot}10H_2O$ (Glauber's salt).

The solubility of sodium sulphate is of particular interest, reaching a maximum at 32.4 °C, at which temperature the solubility decreases, with the anhydrous salt separating out as fine crystals (Figure 8.6).

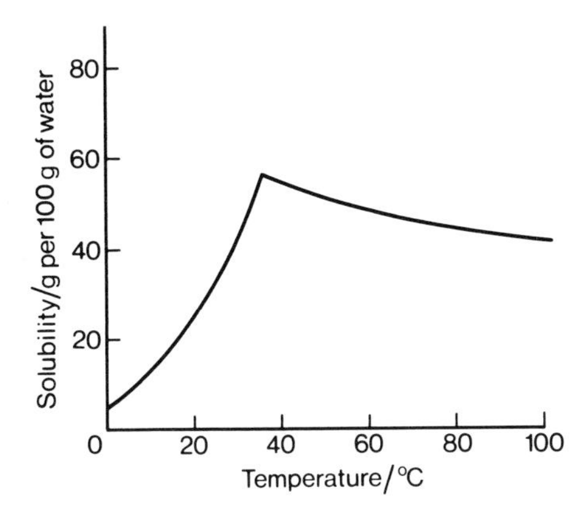

Fig. 8.6 Solubility curve for $Na_2SO_4 \cdot 10H_2O/Na_2SO_4$

Hydrogensulphates

These can be isolated in the solid state, sodium hydrogensulphate crystallizing out as the monohydrate, $NaHSO_4{\cdot}H_2O$.

In aqueous solution they exhibit slightly acidic properties owing to hydrolysis:

$$HSO_4^- + H_2O(l) \rightleftharpoons H_3O^+(aq) + SO_4^{2-}(aq)$$

The solid hydrogensulphate salts of sodium and potassium decompose at 300–350 °C to give the disulphate(VI) (pyrosulphate), $S_2O_7^{2-}$:

$$2HSO_4^- \xrightarrow{300\text{–}350\,^\circ C} S_2O_7^{2-} + H_2O(l)$$

At the elevated temperature of 800–850 °C, further decomposition occurs to yield the sulphate and sulphur(VI) oxide (sulphur trioxide):

$$S_2O_7^{2-} \xrightarrow{800\text{–}850\,^\circ C} SO_4^{2-} + SO_3(g)$$

9 Group II: The Alkaline Earth Metals

Table 9.1

	Beryllium	*Magnesium*	*Calcium*	*Strontium*	*Barium*
Symbol	Be	Mg	Ca	Sr	Ba
Outer electronic structure	$2s^2$	$3s^2$	$4s^2$	$5s^2$	$6s^2$
Principal oxidation state	+2	+2	+2	+2	+2

Electronic structure and size of atoms

The Group II, or alkaline earth, elements possess an outer electronic structure of ns^2, exhibiting an oxidation state of +2.

Owing to the increase in nuclear charge, the atoms are significantly smaller than the preceding alkali metal atoms and consequently have higher densities. The relatively poor screening effect of the *d* electrons (see page 78), enabling the outer electrons to be pulled in and held more firmly, causes strontium, barium, and radium to have considerably higher densities than beryllium, magnesium, and calcium, none of which contains any *d* electrons (Table 9.2).

Table 9.2

	Be	*Mg*	*Ca*	*Sr*	*Ba*
Atomic (covalent) radius/nm	0.089	0.136	0.174	0.191	0.198
Ionic radius, M^{2+}/nm	0.031	0.065	0.099	0.113	0.135
Density/g cm^{-3}	1.85	1.74	1.54	2.62	3.51

The last element in the group, radium, is more renowned for its radioactive properties than for its chemical behaviour. The most stable isotope of radium is ^{226}Ra.

Ionization energies and electropositive character

The smaller size and greater electronegativity of beryllium enables it to form compounds which possess considerably more covalent character than the other elements in the group. The electronegativity of magnesium is also sufficiently high to allow most of its compounds to show at least some degree of covalent nature. Like the alkali metals, the chemistry of calcium, strontium, and barium is essentially that of their ions, in this case the divalent ion, M^{2+}.

Table 9.3

	Be	*Mg*	*Ca*	*Sr*	*Ba*
Electronegativity	1.5	1.2	1.0	1.0	0.9

As a result of the smaller size of the atoms and increased nuclear charge, the first ionization energies of the Group II elements are higher than those of the Group I elements. The second ionization energies are obviously greater than the first, but in the formation of compounds this extra energy is offset by other factors. In the case of compounds in the solid state, the major factor is the increase in lattice enthalpy (for $MgCl_2$, $\Delta H^{\ominus}_{lat} = +2493\,kJ\,mol^{-1}$ compared with a value of $+902\,kJ\,mol^{-1}$ for NaCl), and for compounds in aqueous solution, it is the increased hydration enthalpy of the ions (for Mg^{2+}, $\Delta H^{\ominus}_{hyd} = -1920\,kJ\,mol^{-1}$ compared with a value of $-406\,kJ\,mol^{-1}$ for Na^+).

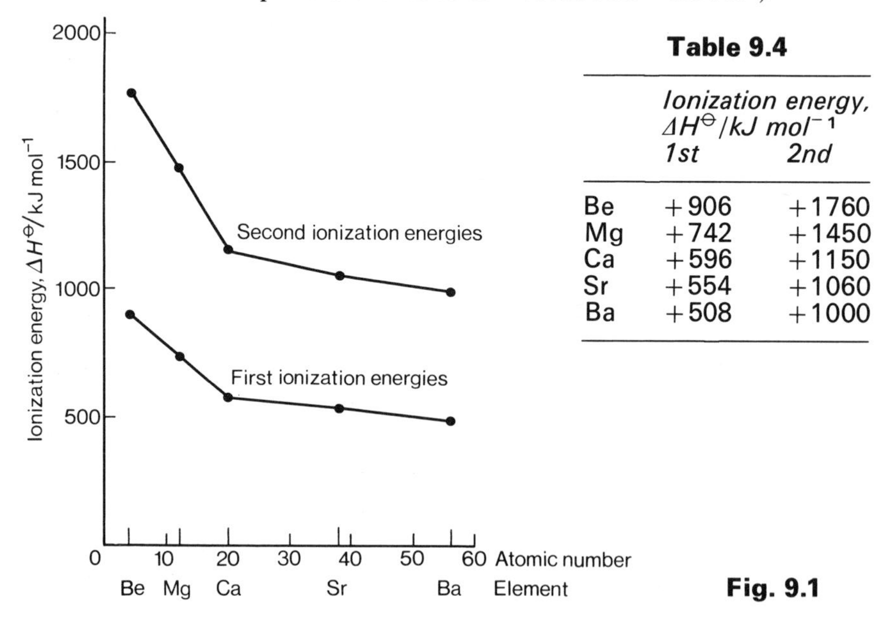

Table 9.4

	Ionization energy, $\Delta H^{\ominus}$/kJ mol^{-1} 1st	2nd
Be	+906	+1760
Mg	+742	+1450
Ca	+596	+1150
Sr	+554	+1060
Ba	+508	+1000

Fig. 9.1

The smaller size and the increased charge of the ions of the Group II elements cause them to have higher hydration enthalpies than their Group I counterparts (Table 9.5 and Figure 9.2). This is indicated by the fact they form a much greater number of hydrated crystalline solids, e.g. $MgCl_2 \cdot 6H_2O$ $CaCl_2 \cdot 6H_2O$, $BaCl_2 \cdot 2H_2O$, etc.

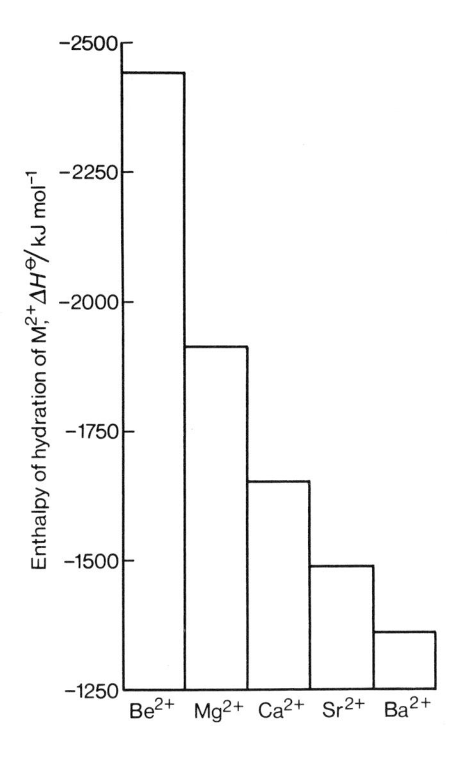

Table 9.5

	Enthalpy of hydration, $\Delta H^{\ominus}$ /kJ mol^{-1}
Be^{2+}	−2450
Mg^{2+}	−1920
Ca^{2+}	−1650
Sr^{2+}	−1480
Ba^{2+}	−1360

Fig. 9.2

General

Magnesium and calcium are by far the most commonly encountered elements in this group, although barium and to a lesser extent strontium compounds are in everyday use.

The M^{2+} ions are all diamagnetic, and compounds containing simple anions are typically white or colourless.

Like the alkali metals, the more electropositive members of this group, namely calcium, strontium, and barium, give distinguishing flame tests:

calcium: brick-red
strontium: crimson
barium: apple green

The availability of two electrons for metallic bond formation enables the atoms to bind more strongly, causing the elements to be harder and have appreciably higher melting and boiling points, than the corresponding alkali metals (Figure 9.3 and Table 9.6).

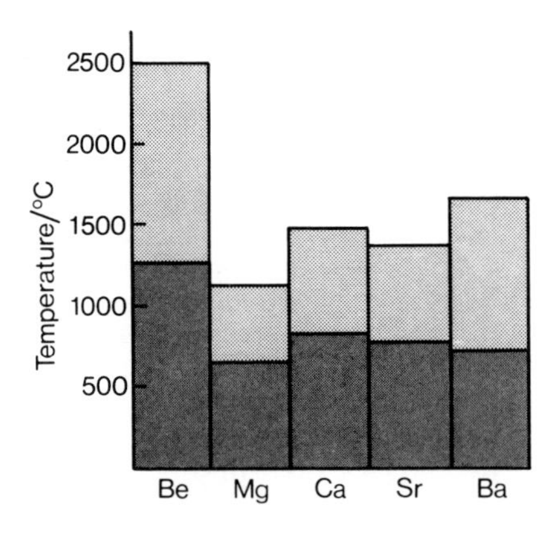

Table 9.6

	M.p./°C	*B.p./°C*
Be	1280	2477
Mg	650	1110
Ca	850	1487
Sr	768	1380
Ba	714	1640

Fig. 9.3 Melting and boiling points of the Group II elements

The irregular patterns for the melting and boiling points are attributable to the different metallic crystal structures of the elements (Table 9.7).

Table 9.7

Element	*Crystal structure*
Be	Hexagonal close-packed
Mg	Hexagonal close-packed
Ca	Hexagonal close-packed and face-centred cubic
Sr	Face-centred cubic
Ba	Body-centred cubic

Whereas the alkali metals react vigorously or violently with cold water, the alkaline earths behave much more mildly:

$$Mg(s) + H_2O(l) \rightarrow MgO(s) + H_2(g)$$
$$M(s) + 2H_2O(l) \rightarrow M(OH)_2(aq) + H_2(g)$$

Beryllium does not react to any appreciable extent even with steam, and any small amount of product formed would be the oxide, BeO, rather than the hydroxide. Magnesium reacts slowly with water but readily with steam, yielding the oxide. Calcium, strontium, and barium react with cold water with progressively greater ease, yielding the hydroxide.

All of the elements react with dilute acids to liberate hydrogen, the ease of reactivity increasing with increasing metallic character from Be to Ba.

$$M(s) + 2H^+(aq) \rightarrow M^{2+}(aq) + H_2(g)$$

Beryllium reacts only slowly.

The amphoteric nature of beryllium is indicated by its ability to react with sodium and potassium hydroxides to form salts containing the beryllate ion (beryllates) and liberate hydrogen. (Compare with aluminium, its diagonally related neighbour.)

Like the alkali metal ions, the alkaline earth metal ions *form comparatively few complexes*, beryllium being the most active in this respect forming, for example, the $[BeF_4]^{2-}$ ion.

Magnesium reacts with both nitrogen and ammonia to form the nitride (cf. lithium):

$$3Mg(s) + N_2(g) \xrightarrow{\text{heat}} Mg_3N_2(s)$$

$$3Mg(s) + 2NH_3(g) \xrightarrow{\text{heat}} Mg_3N_2(s) + 3H_2(g)$$

Magnesium nitride is a pale yellow solid which liberates ammonia when treated with water:

$$Mg_3N_2(s) + 3H_2O(l) \rightarrow 3MgO(s) + 2NH_3(g)$$

The metals (or their oxides) react directly with carbon in an electric furnace to yield ionic carbides, $M^{2+}(C{\equiv}C)^{2-}$.

e.g.

$$Mg(s) + 2C(s) \xrightarrow{500\,^{\circ}C} MgC_2(s)$$

Beryllium forms Be_2C with carbon, although BeC_2 can be obtained by reacting the element with ethyne (acetylene). Of these carbides, calcium dicarbide is probably the best known owing to its use in the production of ethyne with water or dilute acid:

$$CaC_2(s) + 2H_2O(l) \rightarrow Ca(OH)_2(aq) + CH{\equiv}CH(g)$$

Distinctive properties of beryllium

Owing to the small size and relatively high electronegativity and ionization energy values of beryllium by comparison with the other Group II elements, its compounds tend to possess appreciably more covalent character than those of the other members of the group. The unavailability of *d* orbitals limits its coordination number, with the result that beryllium compounds never contain more than four molecules of water of crystallization.

Hydrides

The reactivity of the Group II metals with hydrogen to form hydrides increases on descending the group. Beryllium does not react with hydrogen directly but forms a covalent hydride on reducing beryllium chloride with an ethereal solution of lithium tetrahydridoaluminate(III) (lithium aluminium hydride):

$$2BeCl_2 + LiAlH_4 \xrightarrow{\text{ethoxyethane}} 2BeH_2 + LiCl + AlCl_3$$

The remaining metals in the group all form hydrides by direct combination under comparatively mild conditions.

$$M(s) + H_2(g) \xrightarrow{\text{warm}} MH_2(s)$$

Beryllium and magnesium hydrides are predominantly covalent with bridged polymer-type structures, representing a transition between covalent and ionic hydrides (see also page 106). One possible structure that has been suggested is illustrated in Figure 9.4.

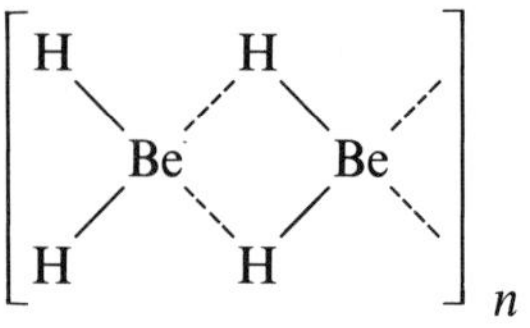

Fig. 9.4

The hydrides of calcium, strontium, and barium are essentially ionic in character, the extent of which increases with increasing metallic character of the alkaline earth metal.

Ca, Sr, and Ba hydrides are capable of reacting with their halides at temperatures of about 900 °C to yield MXH, where X represents a halogen atom:

$$MH_2(s) + MX_2(s) \xrightarrow{900\,^\circ C} 2MXH(s)$$

All of the Group II metal hydrides are *reducing agents* and capable of liberating hydrogen from water under ordinary conditions:

$$MH_2(s) + 2H_2O(l) \rightarrow M(OH)_2(aq) + 2H_2(g)$$

Oxides

All the elements burn in oxygen to form the white solid monoxide, $M^{2+}O^{2-}$,

$$2M(s) + O_2(g) \rightarrow 2MO(s)$$

although these compounds are more usually obtained by the action of heat on the carbonate or nitrate(V).

$$MCO_3(s) \rightarrow MO(s) + CO_2(g)$$
$$2M(NO_3)_2(s) \rightarrow 2MO(s) + 2N_2O_4(g) + O_2(g)$$

Beryllium oxide, BeO, is amphoteric but the other oxides are basic, the basic strength increasing from MgO to BaO (i.e. with increasing metallic character of the alkaline earth metal).

Magnesium oxide is relatively inert, dissolving only *very slightly* in water to form the hydroxide, $Mg(OH)_2$. The other oxides slake exothermically on treatment with water,

e.g. $$CaO(s)+H_2O(l) \rightleftharpoons Ca(OH)_2(s); \quad \Delta H = -65\,kJ\,mol^{-1}$$
'Quicklime' 'Slaked lime'

the amount of heat evolved increasing from CaO to BaO.

The elements Ca to Ba all form peroxides, MO_2, by treating concentrated aqueous solutions of their salts with hydrogen peroxide:

$$M^{2+}(aq)+(O—O)^{2-}(aq) \rightarrow MO_2(aq)$$

The thermal stabilities of the oxides and peroxides increase on descending the group, a property which is reflected in the ability of strontium and barium monoxides to form their respective peroxides by heating in excess oxygen at 400–500 °C.

$$2BaO(s)+O_2(g) \underset{650–700\,°C}{\overset{400–500\,°C}{\rightleftharpoons}} 2BaO_2(s)$$

Hydroxides

Beryllium hydroxide, $Be(OH)_2$, is amphoteric in nature and gives rise to a series of salts containing the *beryllate anion*, $[Be(OH)_4]^{2-}$. (Compare with its diagonally related neighbour, $Al(OH)_3$.)

$$Be(OH)_2(aq)+2OH^-(aq) \rightarrow [Be(OH)_4]^{2-}(aq)$$

Magnesium hydroxide ('milk of magnesia') is virtually insoluble in water and therefore able to furnish only a low hydroxide ion concentration. The hydroxides of the remaining Group II metals are reasonably strong bases although they tend to be only sparingly soluble in water.

The dilute aqueous solutions formed by calcium hydroxide ('lime-water') and barium hydroxide ('baryta water') are often used to detect carbon dioxide.

$$Ca(OH)_2(aq)+CO_2(g) \rightarrow CaCO_3(s)+H_2O(l) \overset{excess\ CO_2}{\rightleftharpoons} Ca(HCO_3)_2(aq)$$
insoluble (white precipitate) — soluble (clear solution)

The basic strength, thermal stability, and solubility of the hydroxides all increase with increasing electropositive character of the elements, i.e. Mg to Ba.

Halides

These can all be formed at a suitable temperature by direct combination of the alkaline earth element and the appropriate halogen,

$$M+X_2 \rightarrow MX_2$$

or by treating the carbonate with the appropriate halogen acid:

$$MCO_3(s)+2HX(aq) \rightarrow MX_2(aq)+H_2O(l)+CO_2(g)$$

Beryllium halides, BeX_2, are covalent in character and fume in air owing to hydrolysis. All the other halides are hygroscopic and form hydrates, e.g. $MgCl_2 \cdot 6H_2O$, $CaCl_2 \cdot 6H_2O$, $BaCl_2 \cdot 2H_2O$, etc., which on further exposure to the atmosphere deliquesce. Anhydrous calcium chloride is commonly employed as a drying agent.

With the exception of the fluorides, which are virtually insoluble in water, the halides are all reasonably soluble. The degree of solubility decreases on descending the group, BeX_2 to BaX_2. The alkaline earth metal halides are generally less soluble in water than the corresponding alkali metal halides This can be attributed to their larger lattice enthalpies which arise as a result of the greater charge on the metal cation, M^{2+}, and its consequent smaller size (Table 9.8 and Figure 9.5).

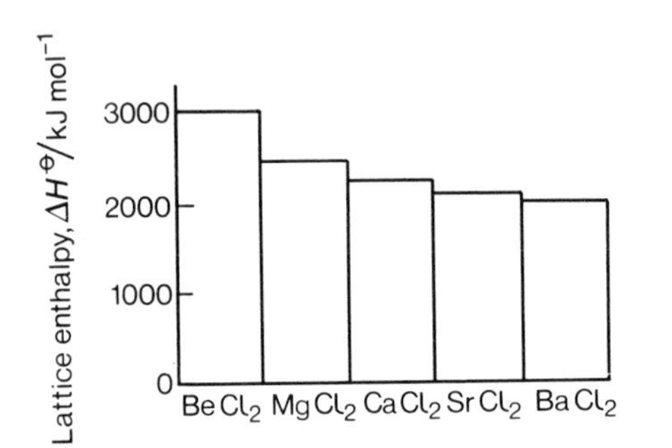

Table 9.8

	$\Delta H^{\ominus}$/kJ mol^{-1}
$BeCl_2$	+3006
$MgCl_2$	+2493
$CaCl_2$	+2237
$SrCl_2$	+2112
$BaCl_2$	+2018

Fig. 9.5 Lattice enthalpies of the alkaline earth metal chlorides

The melting points of the chlorides increase with the increasing electropositive character of the Group II elements, corresponding, in this case, to the higher degree of ionic nature of the bonds (Table 9.9 and Figure 9.6).

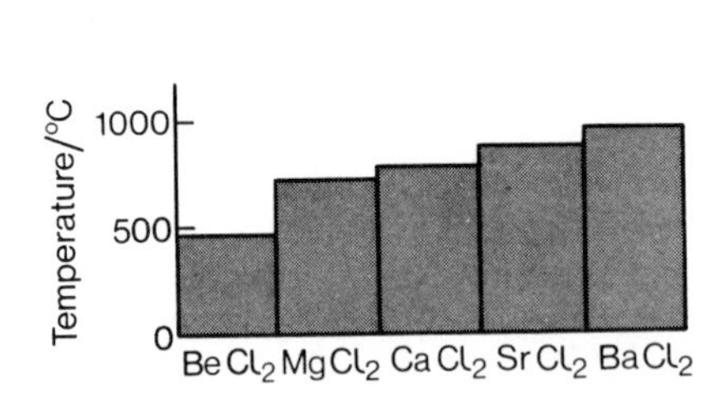

Table 9.9

	M.p./°C
$BeCl_2$	440
$MgCl_2$	708
$CaCl_2$	772
$SrCl_2$	873
$BaCl_2$	962

Fig. 9.6 Melting points of the alkaline earth metal chlorides

For the halides of any particular element, the ionic strength increases with decreasing ionic radius of the anion, X^-, and its corresponding higher electronegativity. This property is reflected in the melting points of the halides of calcium (Table 9.10).

Table 9.10

	CaF_2	$CaCl_2$	$CaBr_2$	CaI_2
Radius of anion, X^-/nm	0.136	0.181	0.195	0.216
M.p./°C	1360	772	730	575

Carbonates

Unlike the alkali metal carbonates (with the exception of lithium carbonate), the Group II metal carbonates decompose on heating and are virtually insoluble in water. The progressive increase in the thermal stabilities of the carbonates $BeCO_3$ to $BaCO_3$ is an indication of the increase in the electropositive character of the metal as the group is descended (Table 9.11). Beryllium carbonate is unstable and can be kept only in an atmosphere of carbon dioxide.

Table 9.11

	$MgCO_3$	$CaCO_3$	$SrCO_3$	$BaCO_3$
Temperature of decomposition/ °C (at 1 atm)	540	900	1290	1360

The solubility of the carbonates in water follows a similar pattern, i.e. it increases from $BeCO_3$ to $BaCO_3$, although even the most soluble of these, namely barium carbonate, does not dissolve appreciably.

Calcium carbonate exists in two crystalline forms (i.e. it is dimorphous). The more stable of these is *calcite*, which has a hexagonal structure and is encountered as limestone, chalk, Iceland spar (pure form), marble, etc. *Aragonite* is metastable and possesses a rhombic structure. This is encountered in shells and coral. Magnesium carbonate conforms to a hexagonal calcite structure, whereas strontium and barium carbonates conform to an aragonite structure.

Hydrogencarbonates

These do not exist in the solid state, but are sufficiently stable to be kept in aqueous solution. Magnesium and calcium hydrogencarbonates are a cause of temporary hardness in hard water. On heating the solution, the hydrogencarbonate decomposes to the carbonate:

$$M(HCO_3)_2(aq) \rightarrow MCO_3(s) + H_2O(l) + CO_2(g)$$

Nitrates

Unlike the more electropositive alkali metal nitrates (except lithium), which decompose on heating only as far as the nitrite, all the alkaline earth metal nitrates decompose to the oxide:

$$2M(NO_3)_2(s) \rightarrow 2MO(s) + 2N_2O_4(g) + O_2(g)$$

Magnesium, calcium, and strontium nitrates crystallize from aqueous solution as hydrates, $Mg(NO_3)_2 \cdot 6H_2O$, $Ca(NO_3)_2 \cdot 4H_2O$, and $Sr(NO_3)_2 \cdot 4H_2O$ whereas barium nitrate separates in the anhydrous form.

Sulphates

The sulphates are of particular interest with regard to hydration. The hydrated beryllium sulphate exists in two forms: $BeSO_4 \cdot 4H_2O$, which is stable below 89 °C, and $BeSO_4 \cdot 2H_2O$, which is stable above this temperature.

Magnesium sulphate has a number of hydrates, corresponding to 1, $1\frac{1}{4}$, $1\frac{1}{2}$, 4, 5, 6, and 7 molecules of water of crystallization. The heptahydrate, $MgSO_4 \cdot 7H_2O$, exists in two allotropic forms, rhombic (Epsom salts) and monoclinic.

Gypsum, $CaSO_4 \cdot 2H_2O$, undergoes ready partial dehydration at 125 °C to the hemihydrate, plaster of Paris, $(CaSO_4)_2 \cdot H_2O$ or $CaSO_4 \cdot \frac{1}{2}H_2O$, which in turn is dehydrated to the anhydrous salt (anhydrite) at a temperature of 140–200 °C:

$$CaSO_4 \cdot 2H_2O(s) \xrightarrow{125\,^\circ C} CaSO_4 \cdot \tfrac{1}{2}H_2O(s) \xrightarrow{140\text{–}200\,^\circ C} CaSO_4(s)$$

Strontium and barium sulphates exist only as the anhydrous form.

The solubility of these compounds decreases quite markedly on descending the group, beryllium and magnesium sulphates being readily soluble in water whereas those of strontium and barium are only very sparingly soluble. Magnesium and calcium sulphate are a cause of 'permanent' hardness in water.

10 Group III

Table 10.1

	Boron	*Aluminium*	*Gallium*	*Indium*	*Thallium*
Symbol	B	Al	Ga	In	Tl
Outer electronic structure	$2s^22p^1$	$3s^23p^1$	$4s^24p^1$	$5s^25p^1$	$6s^26p^1$
Principal oxidation states	**3**	**3**	1, **3**	1, **3**	**1**, 3

Nature of the elements and general properties

The first element in the group, **boron**, is essentially non-metallic and forms compounds which are predominantly covalent. Boron is comparatively rare as an element and difficult to obtain in a high state of purity. It is, therefore, generally encountered as a dark brown, slightly impure, amorphous solid. However, when pure, it is an extremely hard, black, crystalline solid which is chemically unreactive and has poor electrical properties.

Metal borides, such as CaB_6, AlB_2, TiB_2, and FeB, can be prepared by various techniques, one of which involves heating the constituent elements together in the powdered form at a high temperature in a vacuum. The structures of these compounds resemble those of interstitial carbides and nitrides and hence the formulae cannot be accounted for simply in terms of oxidation states.

Aluminium, on the other hand, is the third most abundant element in the Earth's crust and although it displays certain non-metallic properties, such as forming salts with alkalis, it is better known as a metal. In appearance, aluminium is silver-grey and is both malleable and ductile. On account of its low density, it is widely used in alloys where both strength and lightness are essential, e.g. *duralumin* (95 per cent Al; 4 per cent Cu; 0.5 per cent Mg; 0.5 per cent Mn), *magnalium* (90 per cent Al; 10 per cent Mg). Like most metals, aluminium is a good conductor of heat and electricity.

Gallium is a soft, bluish-white metal but, like aluminium, it dissolves in hot alkali. Structurally, each atom has only one other gallium atom near to it, although each is associated with six others at a greater distance. It is this peculiarity of structure which is largely responsible for the remarkably low melting point of the element (Figure 10.1 and Table 10.2) and its soft nature – it is even softer than lead.

Indium and **thallium** are both fairly soft, whitish-grey metals, which give no reaction with alkalis.

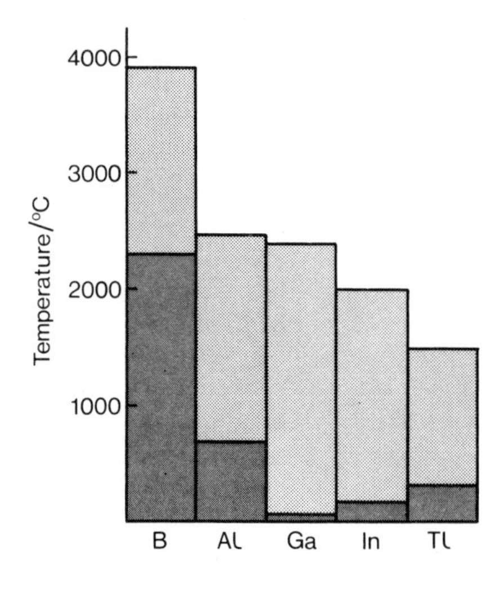

Table 10.2

	M.p./°C	*B.p./°C*
B	2300	3930
Al	660	2470
Ga	29.8	2400
In	157	2000
Tl	304	1460

Fig. 10.1 Melting and boiling points of the Group III elements

Size and density of the atoms

The elements Ga, In, and Tl, which follow immediately after the *d*-block elements, and therefore possess inner *d* electrons, have considerably higher densities than B and Al which possess only *s* and *p* electrons (Table 10.3). This phenomenon is attributed to the poor shielding effect of the inner *d* electrons, which allows the outer electrons to be pulled in closer to the nucleus.

Table 10.3

	B	*Al*	*Ga*	*In*	*Tl*
Atomic (covalent) radius/nm	0.080	0.125	0.125	0.150	0.155
Ionic radius, A^{3+}/nm	(0.020)	0.050	0.062	0.081	0.095
Density/g cm^{-3}	2.34	2.70	5.91	7.30	11.85

Electronic structure and ionization energy

The Group III elements all possess an outer electronic structure of ns^2np^1. The high values of the sum of the first three ionization energies for each of these elements (Table 10.4) is compatible with the fact that, in most of their compounds, the Group III elements exhibit at least some covalent character.

Table 10.4

	Ionization energies, $\Delta H^\ominus$/kJ mol^{-1}		
	1st	*2nd*	*3rd*
B	+805	+2420	+3660
Al	+583	+1820	+2740
Ga	+583	+1980	+2960
In	+562	+1820	+2700
Tl	+596	+1970	+2870

Inspection of the values for the ionization energies indicates that the outer unpaired *p* electron is the one most easily removed.

After aluminium, there is a greater tendency for the *s* electrons to remain paired (*inert pair effect*) and in the case of thallium, the +1 oxidation state is more stable than the +3 state usually encountered for the other elements.

Electronegativity

The pattern of electronegativity values of this group is somewhat irregular (Table 10.5). Instead of following the usual trend of decreasing as the group is descended, the values show a gradual increase from Al to Tl. Because of the poor shielding effect of the inner *d* electrons, the outer electrons of the elements Ga, In, and Tl are more firmly held than might otherwise be expected, and consequently these elements have slightly higher electronegativities.

Table 10.5

	B	*Al*	*Ga*	*In*	*Tl*
Electronegativity	2.0	1.5	1.6	1.7	1.8

The extraction of aluminium

Aluminium is extracted by the electrolysis of alumina, which is obtained from bauxite. Although the formula of bauxite is usually written as $Al_2O_3 \cdot 2H_2O$ it is, in fact, a mixture of the monohydrate and the trihydrate.

In order to remove iron(III) oxide and silicon(IV) oxide (silica) impurities, the ore is first treated with aqueous sodium hydroxide. The aluminium oxide forms a solution of aluminate ions, and the iron(III) oxide, which does not dissolve, is separated by filtration.

$$Al_2O_3(s) + 2OH^-(aq) + 3H_2O(l) \rightarrow 2[Al(OH)_4]^-(aq)$$

After concentration of this solution, the hydroxide is precipitated by one of two methods:

(1) bubbling carbon dioxide through the solution:

$$[Al(OH)_4]^-(aq) + CO_2(g) \rightarrow Al(OH)_3(s) + HCO_3^-(aq)$$

(2) seeding with a little solid aluminium oxide:

$$[Al(OH)_4]^-(aq) \rightleftharpoons Al(OH)_3(s) + OH^-(aq)$$

The silicon(IV) oxide remains in solution.

Pure alumina is then obtained by heating the hydroxide in kilns:

$$2Al(OH)_3(s) \rightarrow Al_2O_3(s) + 3H_2O(l)$$

The alumina is made into a conducting solution by fusing it with sodium hexafluoroaluminate(III) (cryolite), Na_3AlF_6. A little fluorspar, CaF_2, is added to the electrolyte in order to lower the melting point. A small quantity of aluminium fluoride is also added in order to reduce the solubility of the molten aluminium, when formed, in the molten electrolyte. The electrolysis is usually carried out at 800–900 °C.

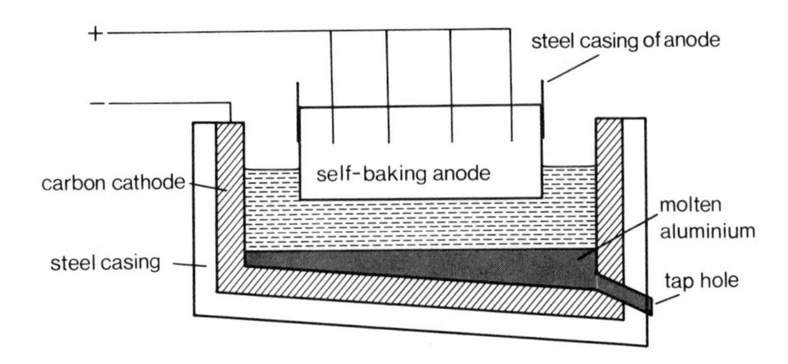

Fig. 10.2 Electrolytic cell for the extraction of aluminium

The cell is made of steel and lined with graphite which acts as the cathode (Figure 10.2). The anodes are also graphite, but since ordinary graphite is rapidly oxidized to carbon dioxide by the oxygen evolved at the anode, a special self-baking Söderberg type of anode is used nowadays. This consists of a mixture of pitch with ground anthracite and/or coke contained in a steel casing from which it can be lowered. When immersed in the molten electrolyte, the mixture is baked into a graphite-type material.

During electrolysis, aluminium is discharged at the cathode and collects in the molten form at the bottom of the cell, from where it can be run off.

General chemistry of aluminium

On exposure to the atmosphere, aluminium slowly acquires a surface coating of the oxide which then protects the element from further oxidation.

The coating of a film of oxide on the surface of aluminium renders the metal much more resistant to corrosion. The electrolytic oxidation of aluminium is known as *anodizing* and is achieved by making the metal the positive electrode (anode) during the electrolysis of chromic(VI) acid. Oxygen is evolved at the surface of the metal and forms a very thin film of the oxide on its surface. The oxide layer has the ability to absorb coloured dyes, and anodized aluminium is widely used for decorative work and making small ornamental objects.

At elevated temperatures, aluminium has a tremendous affinity for oxygen, and at about 800 °C burns vigorously in air to form the oxide, together with some nitride:

$$2Al(s) + \tfrac{3}{2}O_2(g) \rightarrow Al_2O_3(s); \quad \Delta H_f^\ominus = -1669\ kJ\ mol^{-1}$$
$$Al(s) + \tfrac{1}{2}N_2(g) \rightarrow AlN(s); \quad \Delta H_f^\ominus = -\ 241\ kJ\ mol^{-1}$$

The highly exothermic value for the enthalpy of formation of Al_2O_3 provides the basis of the *thermite process* for welding iron and steel,

$$Fe_2O_3(s) + 2Al(s) \rightarrow 2Fe(s) + Al_2O_3(s); \quad \Delta H^\ominus = -840\ kJ\ mol^{-1}$$

and also for reducing the oxides of manganese and chromium.

On heating, aluminium also reacts exothermically with other non-metals such as chlorine, sulphur (at 1200 °C), nitrogen (at 800 °C), and carbon (above 1000 °C) to form the corresponding chloride, $AlCl_3$, sulphide, Al_2S_3, nitride, AlN, and carbide, Al_4C_3.

With moderately concentrated hydrochloric acid, aluminium dissolves, liberating hydrogen,

$$2Al(s) + 6HCl(aq) \rightarrow 2AlCl_3(aq) + 3H_2(g)$$

and with hot, moderately concentrated sulphuric acid, it evolves sulphur dioxide:

$$2Al(s) + 6H_2SO_4(aq) \rightarrow Al_2(SO_4)_3(aq) + 6H_2O(l) + 3SO_2(g)$$

It gives virtually no reaction, however, with either dilute or concentrated nitric acid, forming a so-called *passive* layer.

The non-metallic behaviour of aluminium is shown by the fact that it readily dissolves in hot, dilute sodium or potassium hydroxide, evolving hydrogen and forming the aluminate ion, $[Al(OH)_4]^-$:

$$2Al(s) + 2OH^-(aq) + 6H_2O(l) \rightarrow 2[Al(OH)_4]^-(aq) + 3H_2(g)$$

The hydrated aluminium ion

Because of its small size and high charge, the trivalent aluminium ion has a high hydration energy (see page 59),

$$Al^{3+}(g) + 6H_2O(l) \rightarrow [Al(H_2O)_6]^{3+}(aq); \quad \Delta H_{hyd}^\ominus = -4690\ kJ\ mol^{-1}$$

and exists in aqueous solution as a hexaaqua complex, $[Al(H_2O)_6]^{3+}$, in which the aluminium ion is octahedrally surrounded by six water molecules (Figure 10.3).

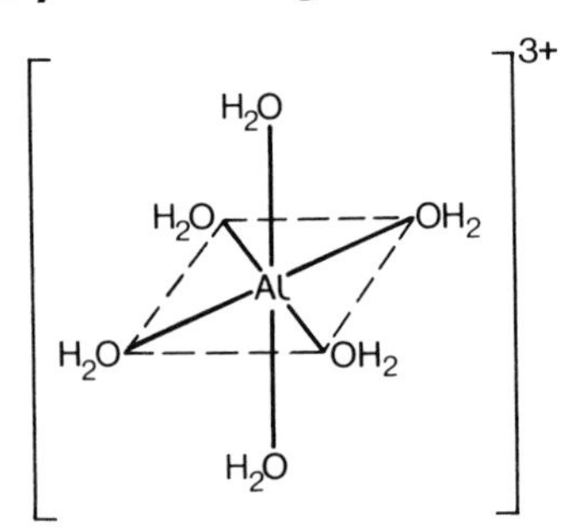

Fig. 10.3

The complex cation functions as a triprotic acid, releasing protons to form hydrated protons with water molecules:

$$[Al(H_2O)_6]^{3+}(aq) \rightleftharpoons [Al(H_2O)_5OH]^{2+}(aq) + H^+(aq)$$
$$[Al(H_2O)_5OH]^{2+}(aq) \rightleftharpoons [Al(H_2O)_4(OH)_2]^+(aq) + H^+(aq)$$
$$[Al(H_2O)_4(OH)_2]^+(aq) \rightleftharpoons [Al(H_2O)_3(OH)_3](s) + H^+(aq)$$

The acidic strength of the solution is such that it can be titrated against sodium hydroxide and, consequently, salts of weak acids, such as the carbonate and sulphide, cannot be prepared in solution as the anions behave as strong bases, removing two protons from the $[Al(H_2O)_5OH]^{2+}$ cation and precipitating the hydroxide:

$$[Al(H_2O)_5OH]^{2+}(aq) + CO_3^{2-}(aq) \rightleftharpoons [Al(H_2O)_3(OH)_3](s) + H_2O(l) + CO_2(g)$$

In view of the comparatively high value for the sum of the first three ionization energies, aluminium has a surprisingly high *electrode potential*, which is a characteristic generally associated with the more reactive metals. This phenomenon is accounted for in terms of the high hydration energy of the aluminium ion, enabling it to be formed readily in solution.

Hydrides

Boron and aluminium both form covalent hydrides. Among the Group III elements, boron is peculiar in that it forms a range of compounds with hydrogen. These are mostly volatile substances known as *boranes*. There are about 14 well-known boranes, although the simple BH_3 monomer does not exist. These include (Table 10.6):

Table 10.6

B_nH_{n+4}	B_nH_{n+6}		
Diborane(6), B_2H_6	Tetraborane(10), B_4H_{10}	gases	
Pentaborane(9), B_5H_9	Pentaborane(11), B_5H_{11}	(unstable)	liquids
Hexaborane(10), B_6H_{10}	Hexaborane(12), B_6H_{12}		liquids

The formulae of these compounds fall into two distinct series, corresponding to the general formulae, B_nH_{n+4} and B_nH_{n+6} (less stable). The number included in brackets following the name denotes the number of hydrogen atoms in the molecule.

One of the most interesting features of these compounds is the uncertainty of their molecular structures. They are all electron-deficient, in as much as there are insufficient valency electrons to enable each of the bonded atoms to be bound by electron pairs, and they are diamagnetic. The most-studied of these compounds is the simplest member, diborane, B_2H_6. In this case, the eight atoms are bound together by only twelve valency electrons.

Electron diffraction studies suggest that two of the hydrogen atoms are distinctive and lie in a plane at right angles to the other four. Further supporting evidence is provided by the fact that when diborane undergoes methylation only four hydrogen atoms are substituted.

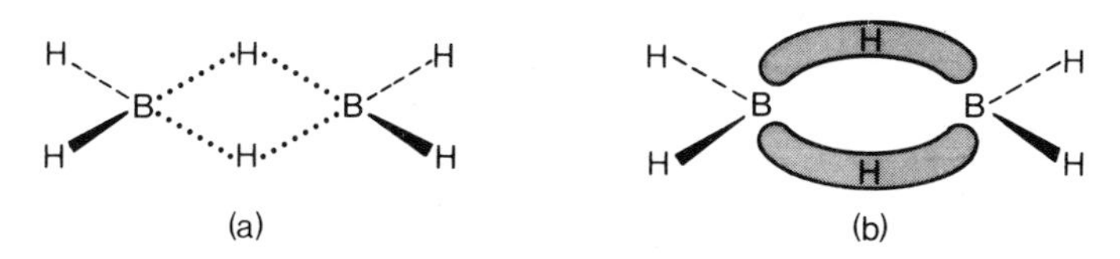

Fig. 10.4

The molecular orbital approach, in which each boron atom is considered to be sp^3 hybridized, provides one of the more satisfactory theories for the structure of this molecule. Two hybrid orbitals are envisaged as forming ordinary covalent bonds with two hydrogen atoms, and the other two forming 'banana' bonds (Figure 10.4(b)) between the boron atoms. The latter are considered to incorporate the bridging hydrogen atoms, each of which contributes one of the two bonding electrons. Similar, but more complex, difficulties are encountered when considering the higher boranes and, again, the concept of bridged structures provides the best explanation.

Aluminium hydride is a white solid possessing a polymer-like structure, $(AlH_3)_n$, although its precise nature is unknown.

None of the hydrides of the Group III elements can be prepared by direct synthesis from the elements.

Oxides

Boron oxide (boric oxide or sesquioxide), B_2O_3, is a colourless, glassy solid formed by heating amorphous boron in air or, more usually, by dehydrating boric (orthoboric) acid, H_3BO_3 or $B(OH)_3$, of which it is the acid anhydride, at low red heat:

$$\underset{\text{boric acid}}{2B(OH)_3(s)} \xrightarrow[-2H_2O]{100\,^\circ C} 2HBO_2(s) \xrightarrow[-H_2O]{\text{low red heat}} B_2O_3(s)$$

X-ray analysis shows the crystalline form of this oxide to be composed of linked spiral chains of BO_4 tetrahedra units which are of two types. Although it is more acidic than aluminium oxide, boron oxide is amphoteric. In alkalis, it predictably forms borate compounds, and on heating with boron to about 1350 °C, it yields a vapour consisting of mainly B_2O_2 molecules which solidify on cooling to form a polymer-type structure.

Aluminium oxide (alumina), Al_2O_3, is obtained by methods typical of those for preparing metallic oxides (i.e. heating the element in air or by dehydration of the hydrated oxides), although it is in fact amphoteric in nature. It is polymorphic, the α- (corundum) and γ- forms being the most usual. When traces of certain metals are present, aluminium oxide provides the basis of a number of precious stones, e.g. ruby (Cr^{3+}), sapphire (Fe^{2+}, Fe^{3+}, and Ti^{4+}), amethyst, and emerald. In the aluminium oxide structure, the oxygen atoms are close-packed with the aluminium atoms occupying two-thirds of the octahedral holes. Consequently, each aluminium is octahedrally surrounded by oxygen atoms and each oxygen by four aluminium atoms.

The M_2O_3 oxides of the remaining elements become progressively more basic as the group is descended. Ga_2O_3, like Al_2O_3, displays amphoteric properties whereas In_2O_3 and Tl_2O_3 are essentially basic.

Hydroxo compounds

Because of the essentially non-metallic character of boron, it is not capable of forming true hydroxides.

Boric acid, H_3BO_3

Boric acid is a soft, flaky, white solid which may be prepared by adding dilute hydrochloric or sulphuric acid to a hot saturated solution of disodium tetraborate–10–water (borax) until the solution is strongly acidic. On cooling, boric acid separates out:

$$B_4O_7^{2-}(aq) + 2H^+(aq) + 5H_2O(l) \xrightarrow{\text{crystallization}} 4B(OH)_3(s)$$

Boric acid is a well-known mild antiseptic, being best known as an eye wash, and is commonly referred to as 'boracic acid'. It functions as a very weak monobasic acid and is better considered as such in terms of the Lewis concept (i.e. an acceptor of an electron pair) rather than the Brønsted–Lowry concept (i.e. a donor of protons):

$$B(OH)_3(aq) + H_2O(l) \rightleftharpoons [B(OH)_4]^-(aq) + H^+(aq); \quad pK_a = 9.0$$

At higher concentrations, the acidic strength is enhanced by the formation of polymeric species:

$$3B(OH)_3(aq) \rightleftharpoons [B_3O_3(OH)_4]^-(aq) + H^+(aq) + 2H_2O(l); \quad pK_a = 6.84$$

The structure of boric acid is complex. It has a planar sheet-like structure (hence the flaky texture) in which the triangular BO_3 units are linked together by means of hydrogen bonds (Figure 10.5).

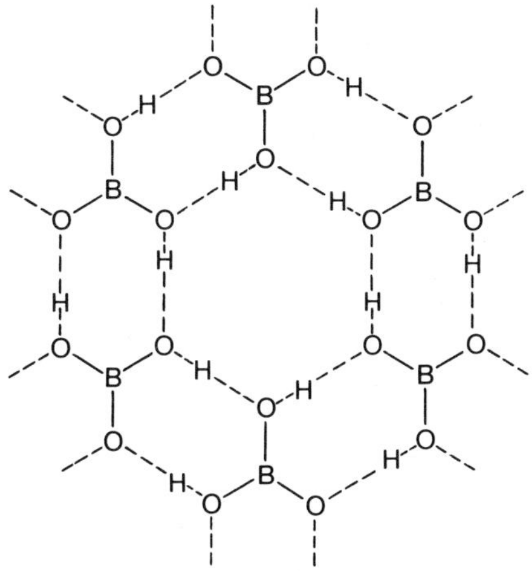

Fig. 10.5

Borates

These are the salts of boric acid. *Simple borate* compounds (*orthoborates*), such as $ScBO_3$, formed by trivalent metals, all contain discrete triangular BO_3^{3-} units. *Metaborates* consist of BO_3^- units linked together into polymeric chains or ring structures (Figures 10.6 and 10.7).

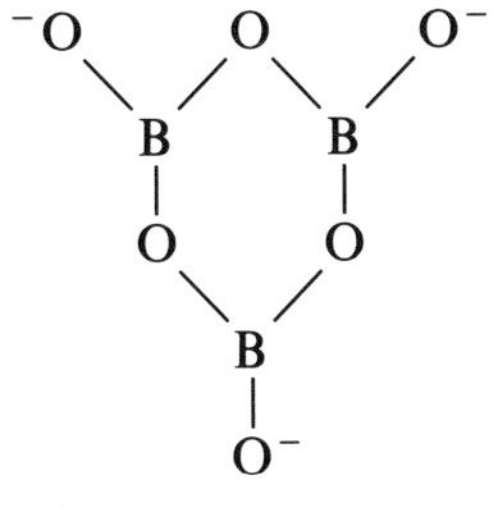

Fig. 10.6 Cyclic metaborate ion, as in $K_3B_3O_6$ and $Na_3B_3O_6$

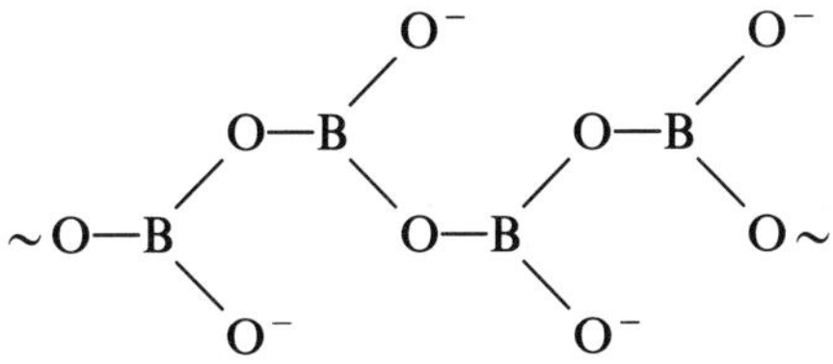

Fig. 10.7 Polymeric metaborate, as in $(Ca(BO_2)_2)_n$

Disodium tetraborate–10–water (borax)

Disodium tetraborate–10–water is probably the best known compound of boron, and contains the hydrated anion $[B_4O_5(OH)_4]^{2-}$ or $[B_4O_7{\cdot}2H_2O]^{2-}$ (Figure 10.8).

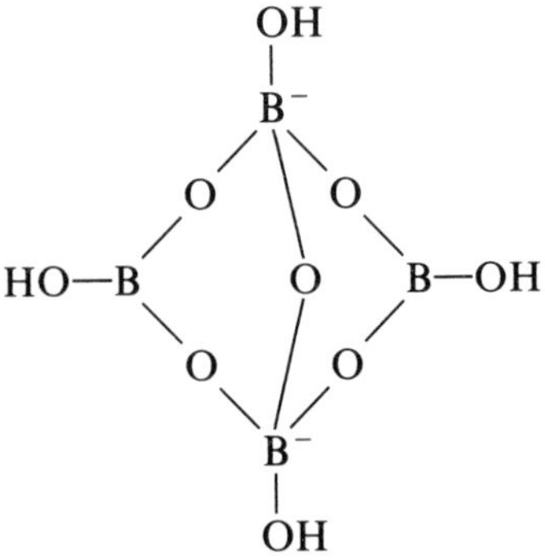

Fig. 10.8

On heating, the clear hydrated crystals lose their water of crystallization and form a glass-like mass of the anhydrous salt. Fused disodium tetraborate–10–water can dissolve metallic oxides to form metaborates, which in certain cases have distinctive colours (borax bead test).

e.g.
$$Na_2B_4O_7(s) + CoO(s) \rightarrow 2NaBO_2(s) + \underset{\text{blue}}{Co(BO_2)_2(s)}$$

As boric acid is such a weak acid, disodium tetraborate–10–water is readily hydrolysed in aqueous media to yield a solution which is distinctly basic. It is suitable for use at an elementary level as a primary standard for titration against acids, methyl orange and methyl red being suitable indicators.

$$B_4O_7^{2-} + 2H^+(aq) + 5H_2O(l) \rightarrow 4B(OH)_3(aq)$$

Hydroxides of the other elements

Aluminium hydroxide, $[Al(H_2O)_3(OH)_3]$ or simply $Al(OH)_3$, is obtained by adding aqueous ammonia to a solution of an aluminium salt, from which the hydroxide separates as a white gelatinous precipitate.

$$[Al(H_2O)_6]^{3+}(aq) + 3OH^-(aq) \rightarrow Al(OH)_3(s) + 6H_2O(l)$$

The gelatinous nature of the precipitate is attributed to hydrogen bonding with water molecules.

Like Al_2O_3, the hydroxide displays amphoteric properties, acting both as a base, forming salts with an acid,

$$Al(OH)_3(s) + 3H^+(aq) + 3H_2O(l) \rightleftharpoons [Al(H_2O)_6]^{3+}(aq)$$

and less typically) as an acid ($pK_a = 12.2$), yielding aluminate ions in the presence of excess OH^- ions:

$$Al(OH)_3(s) + OH^-(aq) \rightleftharpoons [Al(OH)_4]^-(aq)$$

Physical evidence confirms that the $[Al(OH)_4]^-$ anion is in fact hydrated with two water molecules, i.e. $[Al(H_2O)_2(OH)_4]^-$, and has an octahedral structure (Figure 10.9) similar to that of the hexaaqua complex.

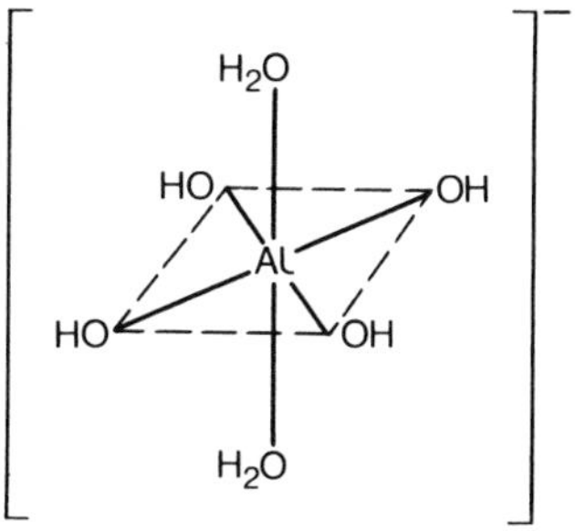

Fig. 10.9

Aluminate ions can be isolated from solution in the crystalline form as the sodium or potassium salts. $Al(OH)_3$ can be re-obtained by bubbling carbon dioxide through the basic solution of aluminate ions.

Gallium hydroxide, $Ga(OH)_3$, resembles $Al(OH)_3$ and is also obtained as a gelatinous, amphoteric solid, although it is predictably more basic than its aluminium counterpart.

By the time thallium is reached in this group, the inert pair effect is such that the only stable hydroxide of this element is TlOH. This is a yellow solid and a base strong enough to absorb carbon dioxide from the air.

Halides

Trihalides (fluorides, chlorides, bromides, and iodides) are formed by all the elements in Group III, although in the case of thallium the monohalide is the more common. However, because of their greater relative importance at this level, our interest is confined to the halides of boron and aluminium.

Boron has a greater affinity for fluorine than for any other element, a fact which is reflected in the values shown in Table 10.7.

Table 10.7

Bond	*B—F*	*B—Cl*	*B—Br*
Bond enthalpy, $\Delta H^\ominus$/kJ mol^{-1}	+620	+456	+377

Under ordinary conditions, boron trifluoride (b.p. −101 °C) and boron trichloride (b.p. 12 °C) are both gases, the tribromide (b.p. 91 °C) is a liquid, and the iodide (m.p. 43 °C) is a solid. In the vapour phase, all are covalent and monomeric.

In addition to the trihalides, boron forms a number of other halides, including the dihalide, B_2X_4, and others such as B_4Cl_4.

With the exception of BF_3, the boron trihalides are all readily hydrolysed by water to boric acid,

$$BX_3 + 3H_2O(l) \rightarrow B(OH)_3(aq) + 3HX(aq)$$

the mechanism involving the progressive substitution of the halogen atoms by hydroxyl groups:

$$BX_3 + H_2O \rightleftharpoons \underset{HX}{\overset{BX_2OH}{+}} \xrightleftharpoons{+H_2O} \underset{HX}{\overset{BX(OH)_2}{+}} \xrightleftharpoons{+H_2O} \underset{HX}{\overset{B(OH)_3}{+}}$$

In these reactions the BX_3 molecule, being electron-deficient, functions as a Lewis acid (strengths decreasing in the order: $BCl_3 > BBr_3 > BI_3$), accepting an electron pair from the oxygen atom of the water molecule. Largely as a result of the strong B–F bond, BF_3 does not undergo substitution in this way but is, nonetheless, a powerful Lewis acid and forms complexes with water,

$$BF_3(g) + H_2O(l) \rightleftharpoons H^+[BF_3OH]^-(aq)$$

and with other species capable of donating lone pairs of electrons, e.g. ammonia (see pages 163 and 165), amines, phosphine, alcohols, etc.

With the exception of the fluoride, which is essentially ionic, the aluminium trihalides are covalent when anhydrous. The most important of these compounds is the chloride which resembles BF_3 in its acidic properties.

Anhydrous aluminium chloride is usually prepared by the action of dry chlorine or hydrogen chloride on aluminium turnings. Vapour density measurements at temperatures up to 400 °C indicate a dimeric structure, Al_2Cl_6, which in the crystalline form is part of a layer lattice.

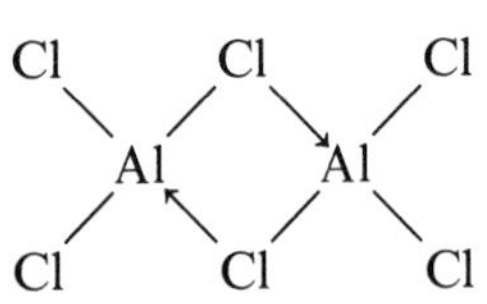

```
Cl     Cl     Cl
  \   /  \   /
   Al      Al
  /   \  /   \
Cl     Cl     Cl
```

Fig. 10.10

Above this temperature, dissociation into the monomeric form starts to take place, the process being complete at about 800 °C.

Anhydrous aluminium chloride is extremely hygroscopic, and, like other aluminium salts, is readily hydrolysed by water, yielding the acidic $[Al(H_2O)_6]^{3+}$ cation:

$$Al_2Cl_6(s) + 12H_2O(l) \rightarrow 2[Al(H_2O)_6]^{3+}(aq) + 6Cl^-(aq)$$

The bromide and iodide of aluminium behave in a similar way although they are slightly more covalent in character.

Aluminium halides resemble BF_3 in their ability to function as Lewis acids and form complexes, the anion, $[AlCl_4]^-$, being involved in organic syntheses, notably in the Friedel-Craft reactions.

Alums

These are *double salts* (see page 217) possessing the general formula $M^IM^{III}(SO_4)_2 \cdot 12H_2O$, where M^I is a large univalent cation (usually Na^+, K^+, or NH_4^+) and M^{III} is a small trivalent cation (usually Al^{3+}, Cr^{3+}, or Fe^{3+}). The three alums most commonly encountered are:

$KAl(SO_4)_2 \cdot 12H_2O$	Aluminium potassium sulphate–12–water (potash alum)
$NH_4Fe(SO_4)_2 \cdot 12H_2O$	Ammonium iron(III) sulphate–12–water (iron alum)
$KCr(SO_4)_2 \cdot 12H_2O$	Chromium(III) potassium sulphate–12–water (chrome alum).

A general method of preparation is to add equimolar quantities of the two metal sulphates to the minimum amount of hot water needed to dissolve them, and allow to crystallize.

Alums all separate from solution as octahedral crystals, although differences in their internal structures exist arising from the differences in the size of the component ions.

11 Group IV

Table 11.1

	Carbon	*Silicon*	*Germanium*	*Tin*	*Lead*
Symbol	C	Si	Ge	Sn	Pb
Outer electronic structure	$2s^2 2p^2$	$3s^2 3p^2$	$4s^2 4p^2$	$5s^2 5p^2$	$6s^2 6p^2$
Principal oxidation states	(2), **4**	(2), **4**	2, **4**	2, **4**	**2**, 4

Nature of the elements

The changeover from non-metallic to metallic character is clearly illustrated by the elements of this group. Carbon and silicon are both non-metals, germanium exhibits some metallic character, and tin and lead are both distinctly metallic. The properties of carbon are unique, showing little resemblance to any of the other elements in the group.

Carbon exists in two allotropic forms, diamond and graphite, both of which have unusually high melting points, especially for a non-metallic element. Their structures are discussed on pages 48 and 49. Silicon (m.p. 1410 °C) and germanium (m.p. 937 °C) both have a diamond-type lattice. Tin (m.p. 232 °C) exists in three polymorphic forms:

$$\text{Grey } (\alpha) \text{ tin} \xrightleftharpoons{13\,^\circ\text{C}} \text{White } (\beta) \text{ tin} \xrightleftharpoons{161\,^\circ\text{C}} \text{Brittle tin}$$

(diamond-type structure)   (body-centred cubic structure)   (rhombic structure)

Lead (m.p. 327 °C) has a typically metallic cubic close-packed lattice.

Electronic structure and bond type

The elements of Group IV possess an outer electronic configuration of ns^2np^2 and all possess an oxidation state of +4. Compounds containing the elements exhibiting this state are generally covalent, the covalency being achieved by promotion of an s electron to the p level:

	ns	np		
Group IV element, ground state:	↿⇂	↿	↿	

Group IV element, excited state: 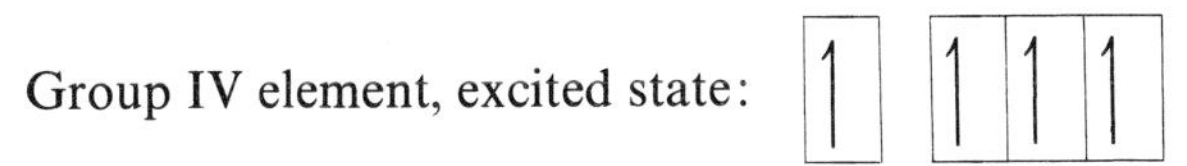

On descending the group, the tendency of the *s* electrons to remain paired (*inert pair effect*) increases, and is accompanied by a decrease in the stability of the +4 state. However, the elements have little tendency to form the M^{4+} ion because of the high third and fourth ionization energies (Table 11.2).

Table 11.2

	Ionization energies, $\Delta H^\ominus$/kJ mol^{-1} 1st	2nd	3rd	4th
C	+1090	+2350	+4610	+6220
Si	+792	+1580	+3230	+4360
Ge	+768	+1540	+3300	+4390
Sn	+713	+1410	+2940	+3930
Pb	+722	+1450	+3080	+4080

Carbon exhibits an oxidation state of four in most inorganic compounds. This is also generally the case for silicon, although in SiO, which is itself highly unstable, silicon displays an oxidation state of +2. Ge(+4) is more stable than Ge(+2), the latter possessing strong reducing properties since it easily loses two electrons to form the +4 state. For tin and lead, both the +2 and the +4 states are important, the higher one being the more stable for tin and the lower the more stable for lead. Like Ge(+2), Sn(+2) has strong reducing properties. Lead(IV) compounds are essentially covalent and good oxidizing agents, gaining two electrons in being converted into the more stable +2 state.

The contrast in the relative stabilities of the +4 and +2 states for tin and lead is illustrated by the marked difference in the standard electrode potentials of the M^{4+}/M^{2+} couples:

$$Sn^{4+}(aq) + 2e \rightleftharpoons Sn^{2+}(aq); \quad E^\ominus = +0.15\,V$$
$$Pb^{4+}(aq) + 2e \rightleftharpoons Pb^{2+}(aq); \quad E^\ominus = +1.69\,V$$

The increase in stability of the lower oxidation state on descending the group is accompanied by an increase in the tendency to form ionic bonds.

Electronegativity

The general pattern of electronegativity values is slightly irregular (Table 11.3). This behaviour was also observed in Group III as being typical of a group of elements which follow the *d*-block elements and contain inner *d* electrons.

Table 11.3

	C	*Si*	*Ge*	*Sn*	*Pb*
Electronegativity	2.5	2.0	1.8	1.8	1.8

Distinctive properties of carbon

Unlike the other elements in the group, the carbon atom has no *d* orbitals available to accept lone pairs of electrons, and is limited to a coordination number of four.

Carbon atoms possess, to a unique extent, the ability to form multiple bonds between themselves, $>C=C<$ and $-C\equiv C-$ (see pages 24 and 25), and also with the atoms of certain other elements such as oxygen, $>C=O$, sulphur, $>C=S$, and nitrogen, $-C\equiv N$.

Catenation

Carbon is also one of the few elements for which *catenation* (the ability to form long chains of identical atoms) is an essential feature of its chemistry; indeed it can catenate better than any other element. These chains may exist as short or long open systems or, alternatively, as closed ring systems. Each different arrangement corresponds to a different compound with its own distinctive properties. In order to be able to catenate, an element must have a valency of at least two and be able to form fairly strong covalent bonds with itself. It is this property, coupled with its ability to form multiple bonds, which enables carbon to form the phenomenal range of compounds which comprise organic chemistry.

Catenation also occurs in the chemistry of silicon, although, by comparison with carbon, the chains are limited in length and the range of compounds is much more restricted. Unlike carbon, silicon does not form multiple bonds with itself. The great difference between the numbers of compounds that these two elements form through catenation is attributable to the much greater strength of the carbon-carbon bond as compared to that of the silicon-silicon bond (Table 11.4). Germanium, tin, and lead do not catenate, and this can be interpreted similarly in terms of their lower bond strengths.

Table 11.4

Bond	*Bond enthalpy,* $\Delta H^{\ominus}/kJ\ mol^{-1}$
C—C	+348
C=C	+612
C≡C	+837
Si—Si	+176
Ge—Ge	+168
Sn—Sn	+155

Carbides

The electronegativity values of the Group IV elements are not sufficiently high to enable them to form anions, although ionic dicarbides, containing the

$(C{\equiv}C)^{2-}$ ion, are formed with the elements of Groups I and II and yield ethyne (acetylene), $CH{\equiv}CH$, on hydrolysis.

The carbides of beryllium, Be_2C, and aluminium, Al_4C_3, are both partly covalent in character and yield methane on hydrolysis. Silicon(IV) carbide, SiC carbordundum), and boron carbide, B_4C, are almost entirely covalent and have giant structures. They are both chemically stable and extremely hard, possessing strong abrasive properties. With the transition metals, carbon forms hard, high melting point, interstitial (see page 241) carbides in which the carbon atoms occupy the octahedral holes (see page 42) in the crystal lattices, and do not significantly affect the electrical conductivity of the metals.

The carbon cycle

Despite the vast quantity of carbon dioxide produced by the combustion of fuels, the percentage of the gas in the atmosphere remains fairly constant (approximately 0.03 per cent). This is maintained by the natural operation of the *carbon cycle*, which is outlined in a simplified form in Figure 11.1.

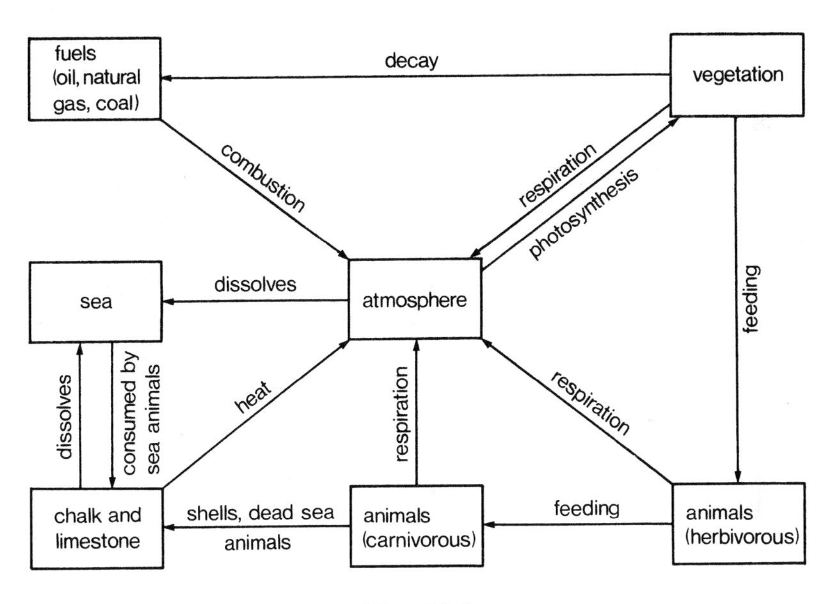

Fig. 11.1

Hydrides

All the elements of this group form covalent hydrides. The hydrides of carbon provide the basis of organic chemistry, which is far too extensive to attempt even to summarize here. Both silicon and germanium form a number of hydrides, referred to as silanes and germanes respectively, which correspond to the general formulae Si_nH_{2n+2} and Ge_nH_{2n+2} (cf. alkanes, C_nH_{2n+2}), the highest members having n equal to 7 or 8. Tin and lead form only the tetrahydrides SnH_4 (stannane) and PbH_4 (plumbane).

The thermal stability and volatility of the tetrahydrides decreases from CH_4 to PbH_4, the latter being highly unstable. This decrease in thermal stability is reflected by the progressive decrease in the A—H bond enthalpy from CH_4 to PbH_4.

Table 11.5

	C	*Si*	*Ge*	*Sn*	*Pb*
A—H bond enthalpy, $\Delta H^\ominus/\text{kJ mol}^{-1}$	+412	+318	+289	+251	+205

CH_4 is very stable and exhibits weak reducing properties (e.g. it is capable of reducing certain heated metal oxides, such as CuO, to the metal), whereas all the other tetrahydrides are strong reducing agents, increasing in power on descending the group. This contrast in properties is probably attributable to the difference in the polarities of the bonds in the different hydrides. In CH_4, the hydrogen atom is the more electropositive element, the bonds being polarized $C^{\delta-}—H^{\delta+}$, whereas in the hydrides of the elements Si to Pb it is the more electronegative, $M^{\delta+}—H^{\delta-}$, the degree of polarization increasing as the group is descended.

SiH_4, together with smaller quantities of other silanes, can be prepared by adding dilute hydrochloric acid to magnesium silicide, Mg_2Si, which is obtained by heating magnesium and silicon together in the absence of air,

$$2Mg(s) + Si(s) \rightarrow Mg_2Si(s)$$
$$Mg_2Si(s) + 4HCl(aq) \rightarrow 2MgCl_2(aq) + SiH_4(g) \quad \text{(plus other silanes)}$$

Mg_2Si can also be obtained in small amounts in the laboratory by heating magnesium powder with silicon(IV) oxide (silica).

The hydrides of the elements Si, Ge, and Sn can all be obtained by the reduction of the appropriate halides using lithium tetrahydridoaluminate(III) in ethoxyethane.

$$MX_4 + LiAlH_4 \xrightarrow{\text{dry ethoxyethane}} MH_4 + LiAlX_4$$

In contrast to CH_4, which is stable to hydrolysis, silanes are readily hydrolysed by dilute aqueous alkali (a trace of alkali is sufficient).

e.g. $$SiH_4(g) + 4H_2O(l) \rightarrow SiO_2{\cdot}xH_2O(s) + 4H_2(g)$$

The hydrides of germanium and tin undergo hydrolysis only slowly.

Halides

The halides of carbon are numerous because of the ability of carbon to catenate. Chlorofluorohydrocarbons (Freons) are used as refrigerants and as propellants in aerosols, and poly(tetrafluoroethene) (PTFE or Teflon) is widely used in the manufacture of non-stick cooking utensils because of its chemical inertness and very low coefficient of friction. Silicon forms a few halides which correspond to the general formula, Si_nX_{2n+2}. The chlorides form the longest chains, the highest being Si_6Cl_{14}.

With the exception of PbI_4, all the tetrahalides of the elements of this group are known, although the existence of $PbBr_4$ is also questionable. That PbI_4 should not exist is understandable in view of the powerful reducing properties of I^- and the strong oxidizing properties of Pb^{4+}. SnF_4 and PbF_4 are predictably ionic, but the other tetrahalides are essentially covalent and volatile in character and possess a tetrahedral structure. The thermal stability of the tetrahalides generally decreases from Si to Pb.

$$PbCl_4(l) \xrightarrow{\text{heat}} PbCl_2(s) + Cl_2(g)$$

The physical state (at room temperature) of the tetrahalides is summarized as follows:

gases – CF_4, SiF_4, GeF_4
liquids – CCl_4, $SiCl_4$, $GeCl_4$, $SnCl_4$, $PbCl_4$, $SiBr_4$, $GeBr_4$, $SnBr_4$
solids – CBr_4, CI_4, SiI_4, GeI_4, SnI_4, SnF_4, PbF_4

The tetrahalides of Si, Ge, Sn, and Pb can be obtained by the action of the appropriate halogen on the element,

$$M + 2X_2 \rightarrow MX_4$$

although the fluorides and the chlorides are often obtained by the reaction of hydrofluoric acid (obtained from CaF_2 and concentrated H_2SO_4) and concentrated hydrochloric acid respectively on the dioxide.

$$CaF_2(s) + H_2SO_4(l) \rightarrow CaSO_4(s) + 2HF(l)$$
$$MO_2 + 4HF(l) \rightarrow MF_4 + 2H_2O(l)$$
$$MO_2 + 4HCl\,(\text{conc. aq}) \rightarrow MCl_4 + 2H_2O(l)$$

By utilizing available *d* orbitals, tetrafluorides of the elements Si to Pb dissolve in excess hydrofluoric acid to form the complex $[MF_6]^{2-}$ ion. In excess concentrated hydrochloric acid, the tetrachlorides of tin and lead yield the complex anions, $[SnCl_6]^{2-}$ and $[PbCl_6]^{2-}$.

With the exception of the carbon tetrahalides (carbon being incapable of extending its coordination number because of the lack of availability of suitable *d* orbitals), all the tetrahalides undergo hydrolysis with water to yield the dioxides or hydrated dioxides. The tetrahalides of silicon are hydrolysed rapidly, forming a hydrated oxide sometimes referred to as silicic acid, which appears as a colloidal solution or gel.

e.g.
$$SiX_4 + 4H_2O \rightarrow \underset{\substack{\text{hydrated oxide} \\ \text{(silicic acid)}}}{SiO_2 \cdot xH_2O} + 4HX$$

The hydrolysis probably takes place via the formation of intermediates of the type $Si(OH)X_3$,........., $Si(OH)_4$. $GeCl_4$ and $GeBr_4$ undergo this reaction less readily, and the hydrolysis of $SnCl_4$ and $PbCl_4$ is slow and can be prevented altogether by the presence of dilute acid.

Ge, Sn, and Pb also form dihalides. As is usual for the lower positive oxidation states of those elements exhibiting more than one state, the dihalides are more ionic in character than the tetrahalides and can function as reducing agents, the reducing powers decreasing from GeX_2 to PbX_2.

The ease with which Sn(+2), particularly in the form of an aqueous solution of $SnCl_2$, is converted into the more stable Sn(+4) state can be utilized for a number of reduction processes. Some of these include the reactions:

$$I_2(aq) + Sn^{2+}(aq) \rightleftharpoons 2I^-(aq) + Sn^{4+}(aq); \quad E = +0.39\,V$$
$$2Fe^{3+}(aq) + Sn^{2+}(aq) \rightleftharpoons 2Fe^{2+}(aq) + Sn^{4+}(aq); \quad E = +0.62\,V$$
$$2Hg^{2+}(aq) + Sn^{2+}(aq) \rightleftharpoons Hg_2^{2+}(aq) + Sn^{4+}(aq); \quad E = +0.76\,V$$
$$Hg_2^{2+}(aq) + Sn^{2+}(aq) \rightleftharpoons 2Hg(l) + Sn^{4+}(aq); \quad E = +0.64\,V$$

In organic chemistry, tin(II) chloride, formed during the course of the reaction from tin and concentrated hydrochloric acid, is used in reducing nitrobenzene to phenylamine (aniline).

Silicon also forms a dichloride, $SiCl_2$, but this only exists in significant quantities above 1100 °C.

Oxides of carbon

Of the five most stable oxides of carbon, CO, CO_2, C_3O_2, C_5O_2, and $C_{12}O_9$, the first two are by far the most common and important.

Carbon monoxide

This is a colourless and odourless gas (b.p. −190 °C, m.p. −205 °C). It is highly poisonous owing to its ability to combine with the haemoglobin in the blood more easily and more firmly than oxygen, thus preventing the haemoglobin from acting in its normal capacity as an oxygen carrier.

Despite its preparation in the laboratory from ethanedioic (oxalic) acid or methanoic (formic) acid, carbon monoxide is too insoluble in water to be considered as an acidic oxide solely on the basis of its solubility. However, it does form a salt, sodium methanoate (formate) with sodium hydroxide (see page 114).

$$\underset{\text{methanoic acid}}{HCOOH(l)} \xrightarrow{\text{conc. } H_2SO_4} CO(g) + H_2O(l)$$

$$\underset{\text{ethanedioic acid}}{(COOH)_2{\cdot}2H_2O(s)} \xrightarrow{\text{conc. } H_2SO_4} CO(g) + CO_2(g) + 3H_2O(l)$$

Commercially, it is produced in the form of water and producer gas.

$$2C(s) + \underbrace{O_2(g) + 4N_2(g)}_{\text{air}} \longrightarrow \underbrace{2CO(g) + 4N_2(g)}_{\text{producer gas}} \quad \text{EXOTHERMIC}$$

$$\underset{\text{steam}}{C(s) + H_2O(g)} \xrightarrow{>1000\,°C} \underbrace{CO(g) + H_2(g)}_{\text{water gas}} \quad \text{ENDOTHERMIC}$$

Carbon monoxide is an important fuel:

$$2CO(g) + O_2(g) \rightarrow 2CO_2(g); \quad \Delta H^\ominus = -566\,kJ\,mol^{-1}$$

It is also a reducing agent, being formed and utilized in the blast furnace to reduce the iron oxide ore.

$$FeO(s) + CO(g) \rightarrow Fe(l) + CO_2(g)$$
$$(Fe_2O_3(s) + 3CO(g) \rightarrow 2Fe(s) + 3CO_2(g))$$

Refer to page 252.

Nowadays, its reaction with hydrogen is used in manufacturing methanol:

$$2H_2(g) + CO(g) \underset{}{\overset{Cr_2O_3/ZnO\text{ catalyst}}{\underset{350\text{–}400\,^\circ C,\text{ high pressure}}{\rightleftharpoons}}} CH_3OH(l)$$

In bright sunlight, it reduces both chlorine and bromine to the highly poisonous carbonyl halides, COX_2,

$$CO(g) + Cl_2(g) \xrightarrow{\text{sunlight}} \underset{\text{carbonyl chloride (phosgene)}}{COCl_2(g)}$$

The lone pair of electrons on the carbon atom (see page 27) enables carbon monoxide to function as an important ligand, combining with a number of transition metals to form complex *carbonyl compounds* (see page 224).

Carbon monoxide is oxidized quantitatively to carbon dioxide by iodine(V) oxide (iodine pentoxide) (see page 206) and the iodine liberated can be estimated volumetrically by titrating with sodium thiosulphate.

Carbon dioxide

This is a colourless, odourless gas (m.p. $-78\ ^\circ C$) and possesses a slightly sharp taste. It is significantly denser than air and does not support combustion. Under ordinary conditions, carbon dioxide is the most stable oxide of carbon.

Commercially, the gas is obtained as a by-product of other processes, e.g. from lime kilns and furnace gases and from the fermentation of monosaccharides (e.g. glucose, fructose).

$$C_6H_{12}O_6(aq) \xrightarrow{\text{enzyme}} 2C_2H_5OH(aq) + 2CO_2(g)$$

It is usually stored under pressure as either dry ice ('Drikold'), or as a liquid. At ordinary pressure, the solid sublimes.

It is prepared in the laboratory by the action of dilute acids on carbonates.

It is an acidic oxide, although it is only sparingly soluble in water, forming 'carbonic acid, H_2CO_3', which is probably more correctly represented as $CO_2{\cdot}H_2O$.

$$H_2O(l) + CO_2(g) \rightleftharpoons H^+(aq) + HCO_3^-(aq);$$

The acidic properties of the solution are very weak as less than 1 per cent of the dissolved gas is in fact converted into the acid with the greater part only loosely hydrated. At room temperature and a pressure of one atmosphere the pH of a saturated solution is about 3.7. On warming, however, the gas is expelled from the solution and its acidic strength diminishes.

With the strong alkalis, carbon dioxide forms two series of salts, carbonates and hydrogencarbonates:

$$2OH^-(aq)+CO_2(g) \rightarrow \underset{\text{carbonate}}{CO_3^{2-}(aq)}+H_2O(l)$$

$$OH^-(aq)+\underset{\text{excess}}{CO_2(g)} \rightarrow \underset{\text{hydrogencarbonate}}{HCO_3^-(aq)}$$

These salts are hydrolysed in water, making the solution weakly basic.

$$CO_3^{2-}+H_2O(l) \rightleftharpoons OH^-(aq)+HCO_3^-(aq)$$
$$HCO_3^-+H_2O(l) \rightleftharpoons OH^-(aq)+H_2CO_3(aq)$$

Carbon dioxide is usually detected by its action with a solution of the sparingly soluble hydroxides of calcium ('lime-water') or barium ('baryta-water'), which involves the initial formation of the white insoluble carbonate followed, on passing excess of the gas, by the formation of a clear solution of the soluble hydrogencarbonate.

$$Ca(OH)_2(aq)+CO_2(g) \rightarrow \underset{\text{white, insoluble}}{CaCO_3(s)}+H_2O(l)$$

$$CaCO_3(s)+H_2O(l)+CO_2(g) \rightleftharpoons \underset{\text{colourless, soluble}}{Ca(HCO_3)_2(aq)}$$

Carbon dioxide is also involved in two essential biological processes, photosynthesis and respiration. In photosynthesis, it combines with water in the presence of sunlight to form glucose, and oxygen is evolved:

$$6H_2O(l)+6CO_2(g) \underset{\text{respiration}}{\overset{\text{photosynthesis}}{\rightleftharpoons}} \underset{\text{glucose}}{C_6H_{12}O_6(s)}+6O_2(g) \quad \text{(simplified equation)}$$

Respiration is the reverse process, the oxygen combining with sugars in the body to form carbon dioxide, water, and energy.

Oxides of silicon

Silicon forms two oxides, the monoxide, SiO, and the dioxide, SiO_2. Unlike the oxides of carbon, these do not exist in the gaseous, monomeric forms but as three-dimensional crystal structures.

Silicon(II) oxide

This oxide is believed to be formed by the high temperature reduction of silicon(IV) oxide by silicon,

$$SiO_2(s)+Si(s) \rightarrow 2SiO(s)$$

but its existence at ordinary temperatures is questionable.

Silicon(IV) oxide

Silicon(IV) oxide, SiO_2, or *silica* as it is more generally known, exists in three crystalline forms, quartz, tridymite, and crystobalite, which themselves have subsidiary polymorphic forms.

$$\begin{array}{ccccc} \alpha\text{-quartz} & & \alpha\text{-tridymite} & & \alpha\text{-crystobalite} \\ \upharpoonleft\!\downharpoonright 573\,^{\circ}\mathrm{C} & & \upharpoonleft\!\downharpoonright 140\,^{\circ}\mathrm{C} & & \upharpoonleft\!\downharpoonright 240\,^{\circ}\mathrm{C} \\ \beta\text{-quartz} & \underset{}{\overset{870\,^{\circ}\mathrm{C}}{\rightleftharpoons}} & \beta\text{-tridymite} & \overset{1470\,^{\circ}\mathrm{C}}{\rightleftharpoons} & \beta\text{-crystobalite} \end{array}$$

Above 1710 °C, crystobalite exists as liquid silica which, on cooling, supercools to form silica-quartz glass. A further type of naturally-occurring silicon(IV) oxide is *kieselguhr* (a fossil mineral), which is a hydrated amorphous solid possessing great absorptive properties. Each of these polymorphs exists as a three-dimensional lattice in which each silicon atom is bonded tetrahedrally to four oxygen atoms, each oxygen atom being common to two tetrahedra.

Fusion of silicon(IV) oxide with excess sodium carbonate yields sodium silicate which, in aqueous solution, is known as *water glass*.

$$SiO_2(s) + Na_2CO_3(s) \rightarrow Na_2SiO_3(s) + CO_2(g)$$

The above equation greatly over-simplifies the change involved.

Silicon(IV) oxide is an *acidic oxide* which, although insoluble in water, dissolves in aqueous alkali, forming the silicate anion:

$$SiO_2(s) + 2OH^-(aq) \rightleftharpoons SiO_3^{2-}(aq) + H_2O(l)$$

With hydrogen fluoride, silicon(IV) oxide is converted into silicon tetrafluoride,

$$SiO_2(s) + 4HF(g) \rightarrow SiF_4(g) + 2H_2O(l)$$

which on reacting with water, is hydrolysed to the hydrated oxide, $SiO_2 \cdot xH_2O$, which appears as a gelatinous suspension. *Pure, anhydrous* silicon(IV) oxide can be obtained from this gel by filtering, drying, and then heating it. This anhydrous form, known as *silica gel*, absorbs water when exposed to a moist atmosphere, and is a useful drying agent.

Silicates

Although sodium silicate is soluble, the great majority of silicates are not, largely because of the strength of the Si—O bond (bond enthalpy, +452 kJ mol^{-1}), which, despite being covalent, is highly polar.

Because of the similarity in size of their cationic radii, replacement of silicon ($r_{Si^{4+}}$ = 0.041 nm) by aluminium ($r_{Al^{3+}}$ = 0.050 nm) to form aluminosilicates is quite common. The difference in oxidation state of these cations requires the presence of other cations (e.g. Na^+) to balance the charges.

The simple silicate ion, SiO_4^{4-}, is tetrahedral (Figure 11.2)

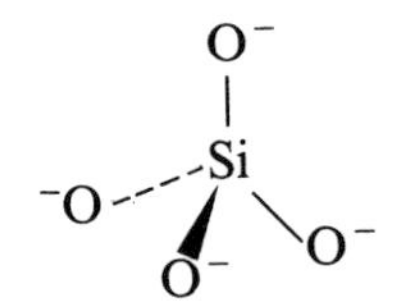

Fig. 11.2

but, as a result of the different ways in which these fundamental units can be bound together, a group of compounds is formed which almost rivals the chemistry of carbon and hydrogen in complexity. This is very briefly summarized as follows:

1. *Discrete anions*
These contain the individual tetrahedral anion, or a limited number of them, in which one or two oxygens are shared by each tetrahedron, and include:

(a) *orthosilicates*, SiO_4^{4-}, e.g. willemite, Zn_2SiO_4, fosterite, Mg_2SiO_4

(b) *pyrosilicates*, $Si_2O_7^{6-}$, e.g. hemimorphite, $Zn_4(OH)_2Si_2O_7 \cdot H_2O$

(△ represents the tetrahedral SiO_4 units)

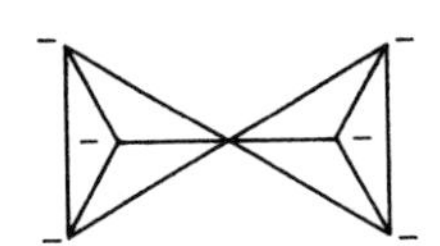

Fig. 11.3 $Si_2O_7^{6-}$ anion

(c) *cyclic silicates*, $Si_3O_9^{6-}$, e.g. wollastonite, $Ca_3(Si_3O_9)$

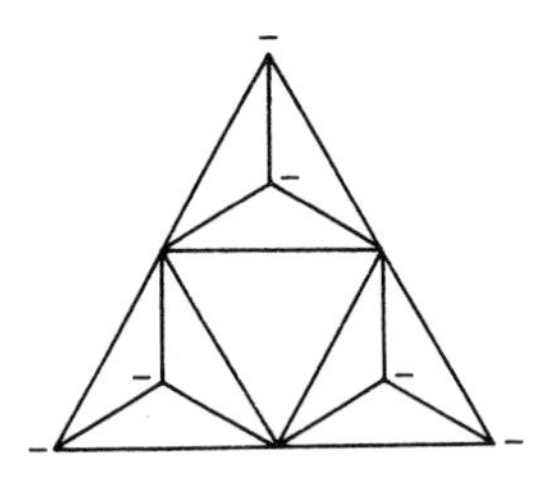

Fig. 11.4 $Si_3O_9^{6-}$ anion

and $Si_6O_{18}^{12-}$, e.g. beryl, $Be_3Al_2(Si_6O_{18})$.

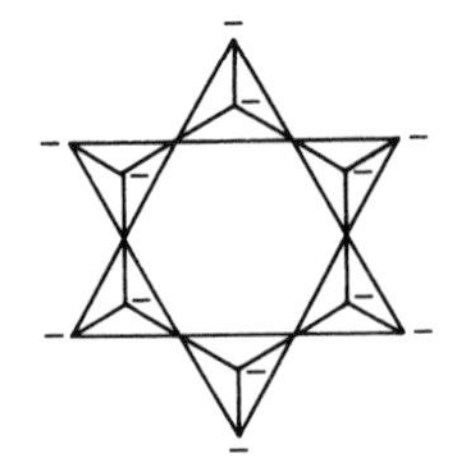

Fig. 11.5 $Si_6O_{18}^{12-}$ anion

2. *Extended anions*

These contain the tetrahedral units linked as chains or sheets of indeterminate size and include:

(a) *Chain silicates*, of which there are two types:

(i) *metasilicates (pyroxenes)*, $(SiO_3)_n^{2n-}$, which have a single-chain structure, e.g. estatite, $MgSiO_3$, jadeite, $NaAl(SiO_3)_2$.

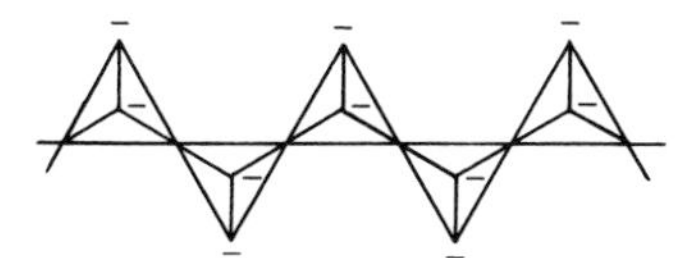

Fig. 11.6 Metasilicate anion structure, $(SiO_3)_n^{2n-}$

(ii) *metatetrasilicates (amphiboles)*, $(Si_4O_{11})_n^{6n-}$, which have a double-chain structure. These are sometimes referred to as fibrous silicates, e.g. tremolite, $(OH)_2Ca_2Mg_5(Si_4O_{11})_2$. Cleavage readily occurs in directions parallel to the strong $(Si_4O_{11})_n$ chains. This property is typified by the various asbestos minerals.

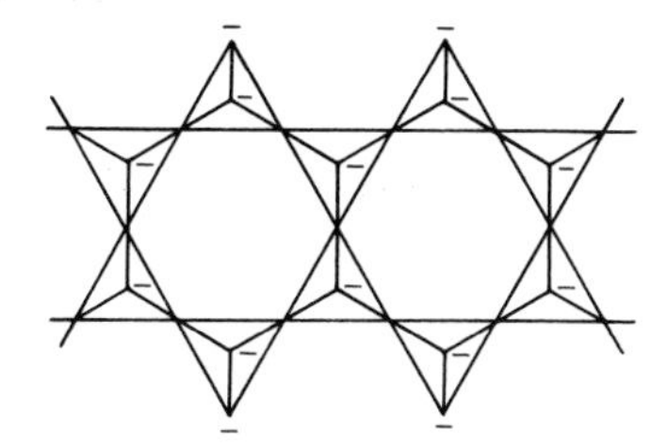

Fig. 11.7 Metatetrasilicate anion structure, $(Si_4O_{11})_n^{6n-}$

(b) *Sheet silicates*, $(Si_2O_5)_n^{2n-}$. In these, three oxygens of each tetrahedron are shaired, e.g. kaolin, $Al_2(OH)_4(Si_2O_5)$, talc, $Mg_3(OH)_2(Si_2O_5)_2$ and micas.

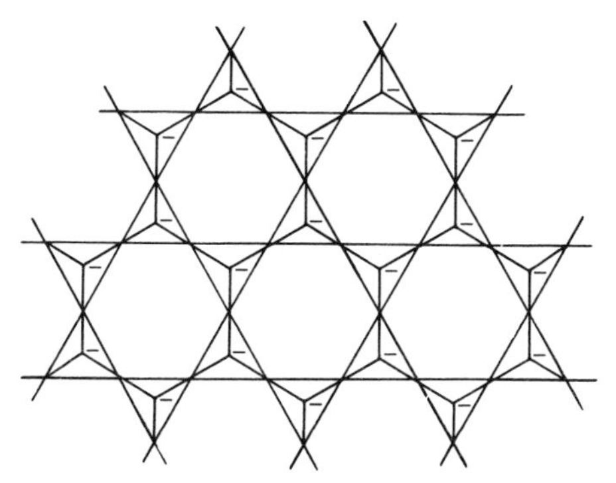

Fig. 11.8 $(Si_2O_5)_n^{2n-}$ anion

3. *Three-dimensional silicates*
In these, all four oxygens of each tetrahedron are shared and the tetrahedral units are linked together in three dimensions. If no cationic replacement of silicon occurs, they have the formula SiO_2 (i.e. quartz, tridymite, or cristobalite). Replacement of silicon, however, results in feldspars, zeolites, and ultramarines.

Silicones

Silicone polymers are manufactured by hydrolysing alkyl-substituted chlorosilanes, the nature of which determines the type of polymer produced, followed by dehydration. The polymers may possess either fairly simple or very complex structures.

Trialkylmonochlorosilanes, R_3SiCl, where the R groups are generally methyl but may also be higher alkyl or even phenyl groups, undergo hydrolysis-dehydration to yield a simple, mobile, and volatile compound, hexaalkylsiloxane.

$$2R_3SiCl + H_2O \rightarrow \underset{\text{hexaalkylsiloxane}}{R_3Si{-}O{-}SiR_3} + 2HCl$$

Dialkyldichlorosilanes, R_2SiCl_2, polymerize to yield long straight-chain molecules,

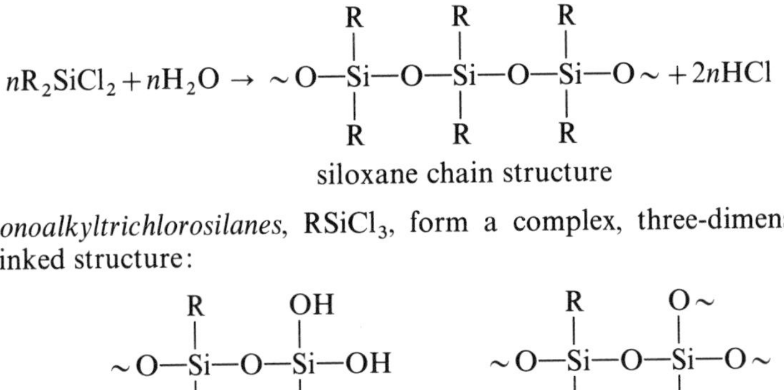

$$nR_2SiCl_2 + nH_2O \rightarrow \sim O{-}\underset{R}{\overset{R}{Si}}{-}O{-}\underset{R}{\overset{R}{Si}}{-}O{-}\underset{R}{\overset{R}{Si}}{-}O\sim + 2nHCl$$

siloxane chain structure

and *monoalkyltrichlorosilanes*, $RSiCl_3$, form a complex, three-dimensional, cross-linked structure:

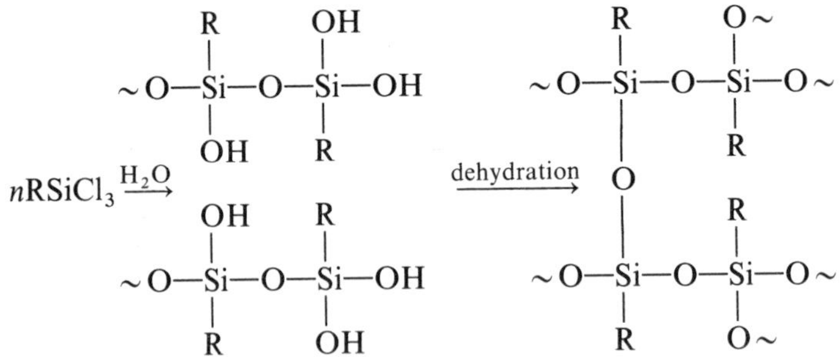

$$nRSiCl_3 \xrightarrow{H_2O} \begin{matrix} \sim O{-}\underset{OH}{\overset{R}{Si}}{-}O{-}\underset{R}{\overset{OH}{Si}}{-}OH \\ \\ \sim O{-}\underset{R}{\overset{OH}{Si}}{-}O{-}\underset{OH}{\overset{R}{Si}}{-}OH \end{matrix} \xrightarrow{\text{dehydration}} \begin{matrix} \sim O{-}\overset{R}{Si}{-}O{-}\underset{R}{\overset{O\sim}{Si}}{-}O\sim \\ | \\ O \\ | \\ \sim O{-}\underset{R}{Si}{-}O{-}\underset{O\sim}{\overset{R}{Si}}{-}O\sim \end{matrix}$$

siloxane cross-linked three-dimensional structure

The degree of cross-linking and the extent of polymerization of monoalkyltrichlorosilanes can be controlled by adding calculated quantities of R_3SiCl and R_2SiCl_2. Similarly, the length of the polymeric chain resulting from the hydrolysis-dehydration of dialkyl compounds can be limited by the addition of measured amounts of R_3SiCl to the reaction mixture, in order to block the ends of the polymeric structures.

Since in practice a rather complex mixture of products is yielded by the above hydrolysis processes, the reactions have been outlined very simply in order to illustrate the formation of the major products.

For straight-chain polymers, boiling points and viscosity increase with increasing relative molecular mass. One of the great assets of these compounds is their resistance to heat, most of them being stable to temperatures of at least 200 °C. This is most significant in phenyl- and, to a lesser degree, methyl-substituted polymers. On heating in air to temperatures approaching 400 °C, the polymers become brittle and crack owing to cleavage of the polymeric molecules into smaller chains, some of which adopt cyclic structures.

Silicon polymers are used for manufacturing silicon rubbers, such as methylsilicone rubber, which is far more resistant to chemicals than natural and other synthetic rubbers and is capable of retaining its elasticity over a much wider range of temperature (−90 to +250 °C). Silicone resins are used in paints and varnishes, and for water-repellent surface coatings. Linear polymers are employed as lubricants and have the advantage over hydrocarbon oils in that their viscosity changes only slowly with temperature. In addition, they find applications in waterproofing textiles, as electrical insulators, in polishes, in glassware, and in the manufacture of numerous other materials. One unfortunate drawback is that they tend to be fairly expensive to produce.

The oxides of Ge, Sn, and Pb

The monoxides, MO

The monoxides, GeO, SnO, and PbO, are all amphoteric, although the acidic properties of SnO and PbO are comparatively weak. With alkalis, these oxides dissolve to form germanate(II) ions, stannate(II) ions, and plumbate(II) ions, which probably contain the species $[Ge(OH)_6]^{4-}$, $[Sn(OH)_6]^{4-}$, and $[Pb(OH)_6]^{4-}$ respectively. These solutions are quite powerful reducing agents, the M(+2) state being readily converted into the M(+4) state.

e.g.

$$[Sn(OH)_6]^{4-}(aq) \longrightarrow [Sn(OH)_6]^{2-}(aq) + 2e$$

stannate(II) ion (hexahydroxostannate(II) ion) → stannate(IV) ion (hexahydroxostannate(IV) ion)

Germanium(II) oxide, GeO, is a black solid and is the most acidic of this group of oxides.

Tin(II) oxide, SnO, is a blue–black powder when anhydrous, but is more commonly obtained as the white hydrated oxide, $SnO{\cdot}xH_2O$. The anhydrous form possesses a giant molecular-type planar structure in which each metal atom is attached to four oxygens, the planes being bound by metal-metal bonds. When freshly prepared, it glows on standing in air and *rapidly oxidizes* to tin(IV) oxide.

Lead(II) oxide, PbO, is much more common than its tin counterpart. The yellow powder obtained on heating lead in air is called *massicot*. On heating above about 500 °C, however, this is converted into a red-brown fused crystal form, known as *litharge*. This oxide has a structure similar to tin(II) oxide, and is widely used in the glass and pottery industries.

The higher oxides

As is common among the higher oxidation states of elements displaying more than one state, the MO_2 oxides are more covalent and slightly more acidic in character than their monoxide counterparts. The degree of acidic character of the MO_2 oxides predictably decreases on descending the group. GeO_2 is essentially acidic in its behaviour, whereas SnO_2 and PbO_2 are both amphoteric, although SnO_2 is predominantly acidic and PbO_2 is predominantly basic. On reacting with fused or concentrated alkali, these oxides yield *germanate(IV)*, $[Ge(OH)_6]^{2-}$, *stannate(IV)*, $[Sn(OH)_6]^{2-}$, and *plumbate(IV)*, $[Pb(OH)_6]^{2-}$ ions respectively; the ease with which the reactions occur diminishing as the basic character of the oxides increases.

Tin(IV) oxide, SnO_2, is a soft grey–white powder, which occurs naturally as cassiterite, and can be prepared in the laboratory by dissolving tin in concentrated nitric(V) acid and igniting the product.

Lead(IV) oxide, PbO_2, is a dark brown solid which can be obtained by treating dilead(II) lead(IV) oxide (red lead), Pb_3O_4, with dilute nitric acid (see below).

It functions as a strong oxidizing agent, liberating chlorine from concentrated hydrochloric acid:

$$PbO_2(s) + 4HCl(l) \rightarrow PbCl_2(s) + 2H_2O(l) + Cl_2(g)$$

and oxidizing sulphur to sulphur dioxide,

$$PbO_2(s) + S(s) \rightarrow Pb(s) + SO_2(g)$$

which then reacts with the lead(IV) oxide to give lead(II) sulphate:

$$PbO_2(s) + SO_2(g) \rightarrow PbSO_4(s)$$

On heating to about 300 °C, the lead(IV) oxide decomposes to the monoxide:

$$2PbO_2(s) \rightarrow 2PbO(s) + O_2(g)$$

If a jet of hydrogen sulphide is directed on to the solid oxide, instantaneous ignition occurs with the Pb(+4) state being reduced to the Pb(+2) state:

$$PbO_2(s) + 2H_2S(g) \rightarrow PbS(s) + S(s) + 2H_2O(g)$$

The comparative ease with which all of these oxidation processes take place is an indication of the greater stability of the Pb(+2) state compared with the Pb(+4).

Both PbO_2 and SnO_2 are readily reduced to the metal by heating with carbon or with carbon monoxide.

$$PbO_2(s) + C(s) \rightarrow Pb(s) + CO_2(g)$$
$$SnO_2(s) + 2CO(g) \rightarrow Sn(s) + 2CO_2(g)$$

Dilead(II) lead(IV) oxide (red lead), Pb_3O_4, is a bright red powder which is widely used as an ingredient of anti-rust paints. It is formulated as $(Pb^{II})_2Pb^{IV}O_4$, (although this is often simplified to $PbO_2{\cdot}2PbO$) and is prepared by heating the monoxide in air at temperatures above 550 °C.

With hydrochloric and sulphuric acids, dilead(II) lead(IV) oxide gives white precipitates of lead(II) chloride and lead(II) sulphate respectively.

$$Pb_3O_4(s) + 8HCl(aq) \rightarrow 3PbCl_2(s) + 4H_2O(l) + Cl_2(g)$$
$$2Pb_3O_4(s) + 6H_2SO_4(aq) \rightarrow 6PbSO_4(s) + 6H_2O(l) + O_2(g)$$

With nitric acid, however, the dark brown lead(IV) oxide is precipitated:

$$Pb_3O_4(s) + 4HNO_3(aq) \rightarrow PbO_2(s) + 2Pb(NO_3)_2(aq) + 2H_2O(l)$$

12 Group V

Table 12.1

	Nitrogen	*Phosphorus*	*Arsenic*	*Antimony*	*Bismuth*
Symbol	N	P	As	Sb	Bi
Outer electronic structure	$2s^2 2p^3$	$3s^2 3p^3$	$4s^2 4p^3$	$5s^2 5p^3$	$6s^2 6p^3$
Principal oxidation states	1, 2, **3**, 4, **5**	**3**, **5**	**3**, **5**	**3**, **5**	**3**, 5

General chemistry and nature of the elements

The transition from non-metallic to metallic character on descending the group occurs rather more gradually than in Groups III and IV. Nitrogen and phosphorus are non-metals, arsenic and antimony have some metallic character, and bismuth is a metal.

Nitrogen is gaseous, diatomic, and, as a result of its high bond enthalpy ($+944\,kJ\,mol^{-1}$), is a very stable molecule. It is usually prepared in the laboratory by warming an aqueous solution of ammonium chloride and sodium nitrite:

$$NH_4^+(aq) + NO_2^-(aq) \rightarrow 2H_2O(l) + N_2(g)$$

On an industrial scale, it is obtained by the fractional distillation of liquid air.

Below 700 °C, **phosphorus** vapour consists of tetrahedral P_4 molecules which condense to a white crystalline solid comprising the same molecular form. Above 700 °C, the gaseous molecules dissociate into P_2 units and, at still higher temperatures, into single atoms. There are at least four other forms of phosphorus which are more stable than the metastable white allotrope, the most common being the relatively chemically inert red phosphorus.

Arsenic is trimorphic with yellow, black, and grey forms. **Antimony** is dimorphic, having yellow and grey forms. In both cases the grey allotrope is the more stable and the most commonly encountered form. **Bismuth**, the last member of the group, is also dimorphic but, of the two allotropes, the white form is by far the more common. Despite the possession of at least some metallic character, all of these latter three solid elements are poor conductors of heat and electricity.

Electronic structure and bond type

The outer electronic structure of ns^2np^3 causes the Group V elements to exhibit principal oxidation states of 3 and 5. With the exception of nitrogen, which has no available d orbitals, the maximum covalency of 5 is attained by utilizing all five outer electrons in bond formation. This involves unpairing the s electrons and promoting to the d level.

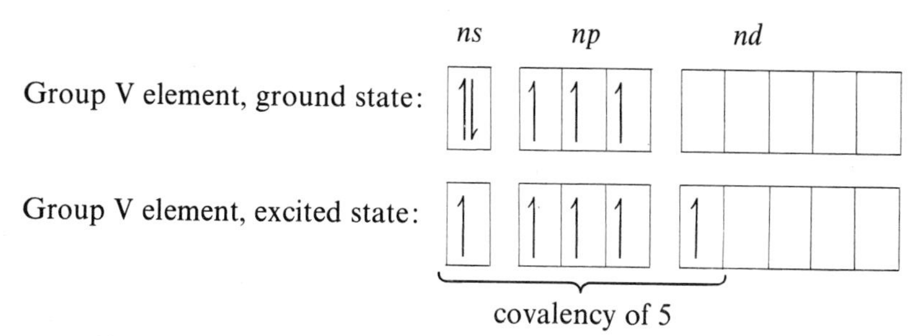

The equivalence of all five of these electrons can be explained in terms of sp^3d hybridization. The absence of available d orbitals prevents the nitrogen atom from accepting electron pairs.

As the group is descended, the tendency for the s electrons to remain as an inert pair increases and by the time bismuth is reached, this is virtually always the case. Consequently, an oxidation state of 3 is the one generally associated with bismuth. This is exemplified by the strong oxidizing power of the Bi(+5) state in, say, sodium bismuthate(V), $NaBiO_3$, which is capable of converting Mn(+2) to Mn(+7) in acidic solution. In this process the Bi(+5) is reduced to the more stable Bi(+3) state.

In the majority of their compounds, the bonds formed by these elements are generally predominantly covalent, a tendency which increases with increasing oxidation state.

The formation of positively charged ions is made difficult by the high ionization energies (Table 12.2), the only element to have any real tendency to form trivalent cations, A^{3+}, being bismuth; and even here the ion is not particularly stable.

Table 12.2

	Ionization energies, $\Delta H^\ominus$/kJ mol^{-1}				
	1st	*2nd*	*3rd*	*4th*	*5th*
N	+1400	+2860	+4590	+7480	+9440
P	+1060	+1900	+2920	+4960	+6280
As	+972	+1950	+2730	+4850	+6020
Sb	+839	+1590	+2440	+4270	+5360
Bi	+780	+1610	+2460	+4350	+5400

The acquisition of three electrons is equally difficult and only the most electronegative element of the group, nitrogen, has any appreciable tendency to form simple A^{3-} anions. There are, in fact, only two elements in the Periodic Table (fluorine and oxygen) which possess a higher electronegativity than nitrogen, and this electronegativity is a significant factor in nitrogen's ability to form hydrogen bonds (see page 106).

Table 12.3

	N	*P*	*As*	*Sb*	*Bi*
Electronegativity	3.0	2.1	2.0	1.9	1.9

With the electropositive elements of Groups I and II, Group V elements form metallic nitrides, phosphides, arsenides, antimonides, and bismuthides, which tend to have giant molecular structures although they do possess some degree of ionic character.

The nitrogen cycle

Nitrogen compounds, particularly in the form of proteins, play vital rôles in the growth of vegetable matter and animals. The *nitrogen cycle* (Figure 12.1) describes a continuous interchange of the element between the atmosphere, which acts as an enormous inexhaustible reservoir, and living organisms.

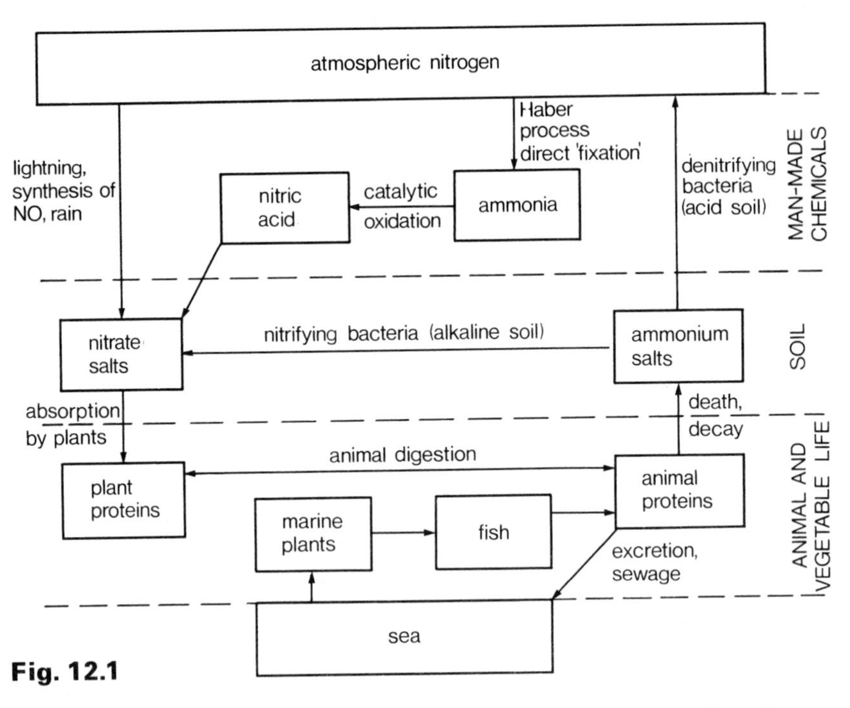

Fig. 12.1

Plants remove nitrogen compounds from the soil during growth and convert it into proteins, while animals acquire their proteins by eating plants. The loss of nitrogen compounds by the soil is compensated by the decay of plants, the decay of animals, and the addition of nitrogenous fertilizers (in the form of nitrate salts and ammonium salts). In addition, certain leguminous plants exist symbiotically with bacteria which are capable of direct conversion of atmospheric nitrogen to ammonia and ammonium compounds. This direct fixation of nitrogen is the subject of intense research in both chemistry and genetics, since it provides a possible solution to problems caused by shortages of food, especially proteins.

Hydrides

All the members of the group form gaseous hydrides, AH_3, which have pyramidal structures, exemplified by the ammonia molecule (see page 29).

The ability of the Group V element in these hydrides to donate its lone pair of electrons and function as a Lewis base decreases quite markedly as the group is descended. Ammonia is a typical weak base, accepting protons from mineral acids and from hydrogen chloride gas:

$$NH_3(g) + HCl(g) \rightarrow NH_4Cl(s)$$

In aqueous solution, the hydrated form undergoes partial dissociation to yield ammonium ions (see below). Phosphine, on the other hand, is virtually insoluble in water.

The basic properties displayed by ammonia are not so prominent in the behaviour of phosphine, PH_3. Although it forms phosphonium salts, e.g. PH_4Cl, with mineral acids, these are readily decomposed by water:

$$PH_4^+ + aq \rightarrow PH_3(g) + H^+(aq)$$

Whereas ammonia requires red heat to reduce certain metal oxides (see below), phosphine is a fairly strong reducing agent, being capable of precipitating copper and silver from aqueous solutions of their salts:

$$4Cu^{2+}(aq) + PH_3(g) + 4H_2O(l) \rightarrow 4Cu(s) + H_3PO_4(aq) + 8H^+(aq)$$

The thermal stability of the hydrides is reflected in the decreasing bond enthalpies (Table 12.4). NH_3 is reasonably stable to heat, PH_3 and AsH_3 both decompose on heating, and SbH_3 and BiH_3 are both unstable at room temperature.

Table 12.4

Bond	*N—H*	*P—H*	*As—H*	*Sb—H*
Bond enthalpy, $\Delta H^\ominus$/kJ mol^{-1}	+388	+322	+247	+256

As the strength of the A—H bond diminishes, the easier it is for other atoms or groups to replace the hydrogen atoms in the molecule.

In addition to ammonia, nitrogen forms two other well defined hydrides, hydrazine, NH_2NH_2, and hydrogen azide (hydrazoic acid), HN_3. Although the latter is stable in aqueous solution and behaves as a weak acid, it is explosive when pure. Phosphorus also forms phosphorus dihydride, P_2H_4, which is a colourless, volatile liquid. Its vapour is unstable and spontaneously flammable in air. The remaining members of the group form only those hydrides mentioned above, namely, arsine, AsH_3, stibine, SbH_3, and bismuth(III) hydride, BiH_3.

Ammonia

Of all the common gases, ammonia is by far the most soluble in water (one volume of water at 0 °C dissolving 1300 volumes of ammonia), forming a solution which functions as a weak base ($pK_b = 4.75$). Physical evidence suggests that this solution contains hydrated ammonia, $NH_3{\cdot}H_2O$, in which the nitrogen is probably hydrogen-bonded to the water molecule, rather than existing as discrete ammonium hydroxide molecules, NH_4OH, which cannot be successfully accounted for in terms of the number of unpaired nitrogen electrons.

$$NH_3(g)+H_2O(l) \rightleftharpoons NH_3{\cdot}H_2O(aq) \rightleftharpoons NH_4^+(aq)+OH^-(aq)$$

The gas is readily expelled from solution on boiling.

Pure liquid ammonia is one of the most studied of the non-aqueous solvents. In this state, the molecules are linked by means of hydrogen bonds and establish the equilibrium:

$$2NH_3 \rightleftharpoons \underset{\text{Lewis acid}}{NH_4^+} + \underset{\text{Lewis base}}{NH_2^-}$$

$$(\text{cf. } H_2O(l) \rightleftharpoons H^+(aq)+OH^-(aq))$$

The position of the equilibrium lies very much in favour of the undissociated molecule.

Ammonia is produced industrially by the direct synthesis of its constituent elements (Haber process); this is the basis of the fertilizer industry, essential for adequate food production (see page 176). Hydrogen, obtained from methane (see page 99) or naphtha (a mixture of hydrocarbons), and nitrogen, obtained from air, are passed over a heated, finely divided iron catalyst promoted by traces of aluminium oxide and potassium oxide. The reaction is reversible and exothermic:

$$N_2(g)+3H_2(g) \rightleftharpoons 2NH_3(g); \quad \Delta H^\ominus = -92\,\text{kJ mol}^{-1}$$

As the reaction is exothermic, elevating the temperature will inhibit the formation of ammonia and promote the backward reaction, as this will have the effect of absorbing heat and lowering the temperature (Le Châtelier's principle). Thus, in order to obtain a high yield, it is theoretically desirable to adopt the lowest possible working temperature. In practice, however, such conditions retard the rate of formation of the required product and a compromise has to be sought. An optimum temperature of about 350–400 °C is usually adopted, although this varies from plant to plant. Furthermore, since the reaction proceeds with a decrease in volume, it is favoured by a high pressure. Pressures of 250–350 atmospheres are usually employed in the U.K., although higher pressures are often employed elsewhere. High pressures require expensive steel vessels to contain them. Research is in progress to find alternative low energy methods of converting nitrogen into ammonia. Certain phosphine complexes of molybdenum and tungsten have given promising results; they absorb nitrogen, and ammonia can then be released by suitable treatment.

In the laboratory, ammonia is usually prepared by heating ammonium chloride with a slurry of calcium hydroxide:

$$2NH_4Cl(s) + Ca(OH)_2(aq) \rightarrow 2NH_3(g) + CaCl_2(aq) + 2H_2O(l)$$

Because it reacts with concentrated sulphuric acid (forming ammonium sulphate) and with calcium chloride (forming a complex, $CaCl_2 \cdot 8NH_3$), it is dried by passing through calcium oxide (quicklime).

Ammonia has many chemical reactions, including:

(1) the formation of amides when passed over heated potassium, sodium, or calcium:

e.g. $$2K(s) + 2NH_3(g) \rightarrow \underset{\text{potassamide}}{2KNH_2(s)} + H_2(g)$$

(2) the reduction of many heated metal oxides at red heat:

e.g. $$3CuO(s) + 2NH_3(g) \rightarrow 3Cu(s) + N_2(g) + 3H_2O(g)$$

(3) combustion in oxygen (but not air),

$$4NH_3(g) + 3O_2(g) \rightarrow 6H_2O(g) + 2N_2(g)$$

in which it burns with a yellow–green flame.
The presence of a catalyst, in the form of a red hot spiral of platinum, however, directs the course of the oxidation differently.

$$4NH_3(g) + 5O_2(g) \xrightleftharpoons[]{\text{red hot Pt spiral, } 800\text{–}900\,^\circ\text{C}} 4NO(g) + 6H_2O(g)$$

This is a key reaction in the manufacture of nitric acid.

(4) Its reaction with chlorine depends upon the amount of chlorine present:

Excess ammonia: $8NH_3(g) + 3Cl_2(g) \rightarrow N_2(g) + 6NH_4Cl(s)$

Excess chlorine: $2NH_3(g) + 6Cl_2(g) \rightarrow 2NCl_3(l) + 6HCl(g)$

NCl_3 is dangerously explosive.

Owing to its ability to donate a lone-pair of electrons, ammonia is a commonly encountered ligand, forming complex ammines with transition metals (refer to Chapters 16 and 17).

Hydrazine and hydroxylamine are two important derivatives of ammonia and both are weaker bases than ammonia. Hydrazine has been used as a rocket fuel, and is manufactured by oxidizing ammonia with sodium chlorate(I) (hypochlorite) (Raschig process):

$$NH_3(g) + NaOCl(aq) \rightleftharpoons NH_4Cl(aq) + NaOH(aq)$$

$$\big\updownarrow +NH_3(g)$$

$$NH_2NH_2(aq) + NaCl(aq) + H_2O(l)$$

Ammonium compounds are thermally unstable, the nature of the decomposition products being dependent on the type of anion. If the anion is resistant to reduction, ammonia is formed, e.g.

$$NH_4Cl(s) \rightleftharpoons NH_3(g) + HCl(g)$$

but if it possesses strong oxidizing powers, an oxidation product is formed, e.g.

$$NH_4NO_3(s) \rightleftharpoons N_2O(g) + 2H_2O(g)$$

Ammonium compounds are often anhydrous and soluble in water, undergoing salt hydrolysis to yield an acidic solution:

$$NH_4Cl(s) + aq \rightarrow NH_4^+(aq) + Cl^-(aq)$$

$$NH_4^+(aq) \rightleftharpoons NH_3(g) + H^+(aq)$$

Halides

Trihalides

All the Group V elements form trivalent halides which are mainly covalent in character. All possible trihalides of the elements of this group exist with the exception of NBr_3 and NI_3, although complexes of these units with ammonia, $NBr_3 \cdot 6NH_3$ and $NI_3 \cdot NH_3$, are well known.

Of these, *phosphorus trichloride*, PCl_3, which is a colourless, fuming liquid, is probably the most useful in the laboratory and is sometimes used as a chlorinating agent, especially in organic chemistry.

All of the phosphorus trihalides undergo hydrolysis with water to yield phosphonic (phosphorous) acid and the halogen acid.

$$PX_3 + 3H_2O(l) \rightarrow H_3PO_3(aq) + 3HX(aq)$$

Ease of hydrolysis decreases with increasing ionic nature, i.e. $PI_3(s) > PBr_3(l) > PCl_3(l) > PF_3(g)$.

The trihalides of As, Sb, and Bi possess progressively more ionic character than their nitrogen and phosphorus counterparts and tend to be either liquids or, more generally, solids.

Pentahalides

Because of the absence of available *d* orbitals, nitrogen does not form pentahalides, but the other elements in the group give rise to many examples of these trigonal bipyramidal structures.

The most common pentahalide is phosphorus pentachloride, which, like the trihalide, is a useful chlorinating agent in organic chemistry and is reduced by certain metals on warming.

e.g. $$PCl_5(s) + Cd(s) \rightarrow PCl_3(l) + CdCl_2(s)$$

On warming, phosphorus pentachloride undergoes thermal dissociation into the trichloride,

$$PCl_5(s) \rightleftharpoons PCl_3(g) + Cl_2(g)$$

Its reaction with water is violent and takes place in two stages:

$$PCl_5(s) + H_2O(l) \longrightarrow POCl_3(l) + 2HCl(aq)$$

$$POCl_3(l) + 3H_2O(l) \xrightarrow{\text{boil}} \underset{\text{phosphoric(V) acid}}{H_3PO_4(aq)} + 3HCl(aq)$$

In the crystalline form, PCl_5 has an ionic structure, $[PCl_4]^+[PCl_6]^-$, in which the phosphorus is respectively tetrahedrally and octahedrally surrounded. The $[PCl_6]^-$ and the $[PF_6]^-$ anions, are the only known examples in which phosphorus displays a coordination number of six.

Oxides

Nitrogen forms five stable oxides, N_2O, NO, N_2O_3, $N_2O_4(NO_2)$, and N_2O_5, in which the nitrogen atom adopts the oxidation states of $+1$, $+2$, $+3$, $+4$, and $+5$ respectively. Those oxides corresponding to the lower states are neutral whereas those corresponding to the higher states ($+3$, $+4$, $+5$) are acidic. With the exception of N_2O_5, which is a solid of low melting point, they are all gases at room temperature. Three of these oxides, N_2O, NO, and $N_2O_4(NO_2)$, are particularly well-known and are discussed in more detail below.

As one might expect, the main oxides of the other Group V elements are the tri- and pentavalent compounds.

Relative molecular mass and electron diffraction measurements indicate that the two main oxides of phosphorus have the formulae P_4O_6 and P_4O_{10}. The physical properties of the oxides of phosphorus differ quite considerably from those of the oxides of nitrogen, largely as a result of the tendency of phosphorus to form polymeric structures with oxygen.

Phosphorus(III) oxide (phosphorus trioxide), P_4O_6, is a white, waxy solid. Its structure in the vapour state corresponds to four phosphorus atoms at the apices of a tetrahedron (similar to the elemental structure) with the three oxygen atoms forming bridges between the two phosphorus atoms (Figure 12.2(a)).

Phosphorus(V) oxide (phosphorus pentoxide), P_4O_{10}, is polymorphic, existing in three crystalline forms and at least one amorphous form. Of the crystalline polymorphs, one is hexagonal and the other two are orthorhombic.

The hexagonal form comprises separate P_4O_{10} molecules which are structurally similar to the P_4O_6 units, with the exception that each phosphorus atom is bonded to an additional oxygen atom (Figure 12.2(b)). In the gaseous state, the pentoxide always contains some discrete P_4O_{10} units.

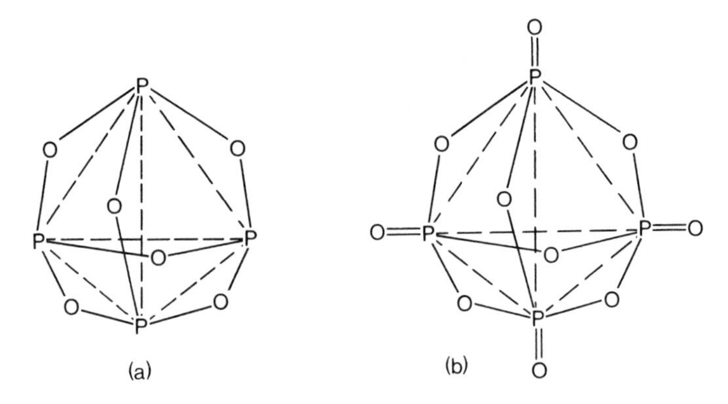

Fig. 12.2 Structures of (a) the P_4O_6 and (b) the P_4O_{10} units

Phosphorus(III) oxide reacts with cold water to give phosphonic (phosphorous) acid, H_3PO_3, which, on treating with hot water, yields several products, among them being phosphine and phosphoric(V) acid, H_3PO_4.

$$P_4O_6(s)+6H_2O(l) \xrightarrow{\text{cold}} 4H_3PO_3(aq) \xrightarrow{\text{hot}} PH_3(g)+3H_3PO_4(aq)$$

On heating P_4O_6 in vacuo at 210 °C, it is converted into a polymeric form, $(PO_2)_n$.

Phosphorus(V) oxide reacts vigorously with water to form phosphoric(V) acid (of which it is the acid anhydride) together with some HPO_3 and $H_4P_2O_7$. P_4O_{10} has an especially high affinity for water and finds use as a drying and dehydrating agent.

Arsenic(III) oxide (arsenious oxide) exists up to temperatures of 800 °C as As_4O_6 molecules, at which point it dissociates into the monomeric form. It has reducing properties and in aqueous solution acts as a weak acid, arsenious acid, H_3AsO_3. The pentavalent oxide, As_4O_{10}, unlike its phosphorus counterpart, has oxidizing powers.

Antimony(III) oxide (antimonous oxide), Sb_2O_3, and bismuth(III) oxide, Bi_2O_3, have giant molecular structures which are transitional in character between those of typical covalent and ionic lattices.

Dinitrogen oxide, N_2O

Dinitrogen oxide (nitrous oxide) is usually prepared by the thermal decomposition of ammonium nitrate:

$$NH_4NO_3(s) \rightarrow N_2O(g)+2H_2O(g)$$

If the heating is too rapid, side reactions occur which proceed with great rapidity and may cause an explosion. An alternative procedure is to heat a mixture of ammonium chloride and sodium nitrate.

Dinitrogen oxide is a neutral oxide and is sometimes used as a mild anaesthetic ('laughing gas'). It possesses a linear structure, infra-red spectroscopy showing the molecule to be unsymmetrical with the structure probably being a resonance hybrid of two canonical forms.

$$N^-{=}N^+{=}O \leftrightarrow N{\equiv}N^+{-}O^-$$

The gas decomposes at about 500–600 °C into nitrogen and oxygen:

$$2N_2O(g) \rightarrow 2N_2(g)+O_2(g); \quad \Delta H^\ominus = -163\,\text{kJ mol}^{-1}$$

Once this decomposition has been initiated, it supports the combustion of a number of elements such as C, S (rapidly extinguished if only feebly burning), P, Cu, Fe, and Pb.

e.g.

$$C(s)+2N_2O(g) \rightarrow CO_2(g)+2N_2(g)$$
$$Cu(s)+N_2O(g) \rightarrow CuO(s)+N_2(g)$$

Like oxygen, the gas is capable of relighting a glowing splint as a result of a sufficiently high localized temperature. The two gases can be easily distinguished by the fact that dinitrogen oxide is reduced to nitrogen on passing over heated copper turnings, is slightly more soluble in water, and gives no reaction with nitrogen oxide (see below).

Nitrogen oxide, NO

Nitrogen oxide (nitric oxide) is a neutral oxide, obtained on an industrial scale by direct synthesis of the component elements, and, in the laboratory, usually by the action of 50 per cent nitric acid on copper turnings.

$$3Cu(s)+8HNO_3(aq) \rightarrow 3Cu(NO_3)_2(aq)+2NO(g)+4H_2O(l)$$
(approximate equation)

The two atoms comprising the monomer possess between them eleven valency electrons, with the result that one of them is necessarily unpaired giving the molecule paramagnetic (see page 240) properties. Such structures are rarely encountered and are referred to as **odd electron molecules**. The bonding in nitrogen oxide is represented as containing a *three electron bond*:

:N$\overset{\cdots}{=}$O:

In the solid state, the oxide dimerizes to N_2O_2 in which the magnetic properties of the unpaired electrons effectively nullify each other causing it to display diamagnetism.

The NO molecule loses an electron from the three electron bond with relative ease to form the *nitrosyl (nitrosonium) cation*, NO^+. This strengthens the nitrogen-oxygen linkage, and reduces the bond length from 0.115 nm in the NO molecule to 0.016 nm in the NO^+ ion. A number of salts containing the nitrosyl ion are known, including: $NO^+HSO_4^-$, $NO^+BF_4^-$, and $NO^+ClO_4^-$.

Odd electron compounds are usually highly reactive, and by comparison with others NO is remarkably stable. Nonetheless, it reacts instantaneously with air (and oxygen) forming brown fumes of dinitrogen tetraoxide.

$$2NO(g) + O_2(g) \rightarrow N_2O_4(g)$$

It combines directly with the halogens to form the nitrosyl halides, NOX.

e.g. $$2NO(g) + Cl_2(g) \rightarrow 2NOCl(g)$$

It is the most thermally stable oxide of nitrogen, beginning to decompose only above 1000 °C. Consequently, it only supports the combustion of brightly burning elements such as phosphorus and magnesium.

One of the more familiar reactions of NO is with iron(II) sulphate in the 'brown ring test' for nitrate ions. The 'brown ring' compound, $[Fe(H_2O)_5NO]^{2+}$, is extremely unstable and rapidly decomposes with a small increase in temperature. Expulsion of NO from the complex provides a means of acquiring small quantities of the gas in a pure state. Nitrogen oxide can also replace a cyanide ion in the hexacyanoferrate(III) ion, $[Fe(CN)_6]^{3-}$, to form the pentacyanonitrosylferrate(II) ion (the nitroprusside ion), $[Fe(CN)_5NO]^{2-}$ which is used to detect sulphur (see page 256).

Dinitrogen tetraoxide, N_2O_4

Dinitrogen tetraoxide is manufactured by allowing nitrogen oxide to react with air, and in the laboratory it is obtained by the action of heat on a number of metal nitrates.

e.g. $$2Pb(NO_3)_2(s) \rightarrow 2PbO(s) + 2N_2O_4(g) + O_2(g)$$

N_2O_4 solidifies to a colourless solid which melts at −9.3 °C to a pale yellow liquid. The yellow colour is attributable to the presence of some brown dioxide, NO_2, molecules. Usually present is some N_2O_3 (resulting from the combination of NO and NO_2) which is blue. This causes the liquid usually obtained in the laboratory to appear green. At the boiling point of 21.2 °C, vapour density measurements indicate that the N_2O_4 molecules are about 15 per cent dissociated into NO_2 molecules. On raising the temperature further, dissociation continues and is accompanied by a gradual increase in intensity of colour until dissociation is complete at about 140 °C. Above 140 °C, the dioxide molecules dissociate into colourless nitrogen oxide and oxygen. As a result the brown colour gradually becomes paler, until at 620 °C the process is complete and the mixture is colourless.

$$N_2O_4(s) \underset{}{\overset{-9.3\,^\circ C}{\rightleftharpoons}} N_2O_4(l) \overset{21.2\,^\circ C}{\rightleftharpoons} N_2O_4(g) \overset{140\,^\circ C}{\rightleftharpoons} 2NO_2(g) \overset{620\,^\circ C}{\rightleftharpoons} 2NO(g) + O_2(g)$$

The N_2O_4 dimer has a planar structure and is diamagnetic. The weakness of the N—N linkage in the dimer is ascribed to its relatively long bond length (0.175 nm).

Like NO, the NO_2 monomer is an *odd electron molecule* (17 valency electrons, Figure 12.3) and has paramagnetic properties. The N_2O_4 molecule probably results from pairing of the unpaired electrons of the NO_2 molecules:

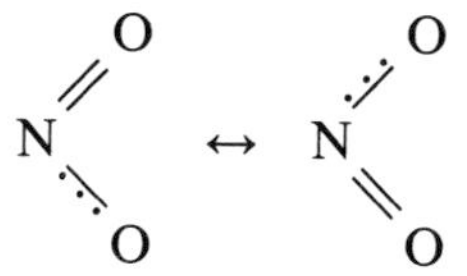

Fig. 12.3

In the liquid state, N_2O_4 tends to ionize:

$$N_2O_4 \rightleftharpoons \underset{\text{Lewis acid}}{NO^+} + \underset{\text{Lewis base}}{NO_3^-}$$

and finds considerable use as a non-aqueous solvent.

Dinitrogen tetraoxide is the mixed anhydride of nitric and nitrous acids.

$$N_2O_4(g) + H_2O(l) \rightleftharpoons HNO_3(aq) + HNO_2(aq)$$

It is a strong oxidizing agent, oxidizing a solution of iodide ions to iodine,

$$4I^-(aq) + N_2O_4(g) + 2H_2O(l) \rightarrow 2I_2(s) + 2NO(g) + 4OH^-(aq); \quad E = +0.49\ V$$

and hydrogen sulphide to sulphur:

$$2H_2S(g) + N_2O_4(g) \rightarrow 2NO(g) + 2H_2O(l) + 2S(s)$$

However, with stronger oxidizing agents, such as acidified manganate(VII) (permanganate) solution, it behaves as a reducing agent, being converted into the nitrate ion and removing the purple colour.

$$2MnO_4^-(aq) + 5N_2O_4(g) + 2H_2O(l) \rightarrow 2Mn^{2+}(aq) + 10NO_3^-(aq) + 4H^+(aq); \quad E = +0.71\ V$$

It supports the combustion of phosphorus, carbon, and magnesium to form their oxides and is also capable of relighting a glowing splint.

Oxoacids of nitrogen

Nitric acid, HNO_3

When pure, nitric acid is a colourless, fuming liquid, although it frequently appears yellow because of dissolved nitrogen dioxide produced by photolytic decomposition. It is manufactured on an industrial scale by the catalytic oxidation of ammonia by excess air to nitrogen oxide. This combines with the excess oxygen to give dinitrogen tetraoxide, which is then absorbed in water in the presence of oxygen.

The platinum catalyst is heated electrically to start the reaction, but, once started, the process is maintained at a temperature of 800–900 °C by the heat produced by the reaction.

$$4NH_3(g) + 5O_2(g) \xrightleftharpoons[]{\text{heated Pt catalyst}} 4NO(g) + 6H_2O(g); \Delta H^\ominus = -906\,\text{kJ mol}^{-1}$$

$$2NO(g) + O_2(g) \rightarrow N_2O_4(g)$$

$$2N_2O_4(g) + 2H_2O(l) + O_2(g) \rightleftharpoons 4HNO_3(aq)$$

Ordinary commercial concentrated nitric acid is obtained by distillation from water as a constant boiling mixture containing 68 per cent acid. This can be further concentrated by distillation with concentrated sulphuric acid.

In the laboratory, nitric acid is prepared by distilling potassium nitrate with concentrated sulphuric acid.

$$KNO_3(s) + H_2SO_4(l) \rightarrow KHSO_4(s) + HNO_3(g)$$

Physical evidence indicates that the nitric acid molecule is planar and has the structure (Figure 12.4).

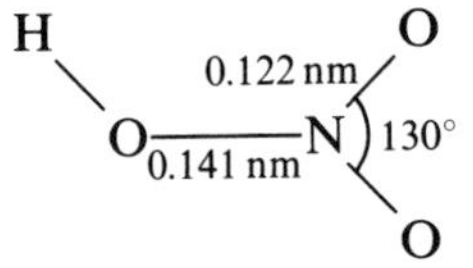

H, O, O, N, 0.122 nm, 0.141 nm, 130°

Fig. 12.4

In addition to functioning as a strong monobasic acid, of which the salts, the nitrates, are nearly all soluble in water, nitric acid is a powerful oxidizing agent. It has the ability to oxidize certain non-metals (e.g. P, S, I) to their oxo-acids, tin to white hydrated tin(IV) oxide, tin(II) chloride to tin(IV) chloride, and turpentine to carbon.

The nitrogen atom of the nitrate anion, NO_3^-, may be considered to be sp^2 hybridized, giving the anion a symmetrical planar arrangement of atoms. The structure can be represented by the canonical forms of Figure 12.5.

$O{=}N^+(O^-)(O^-) \leftrightarrow O^-{-}N^+({-}O^-)({=}O) \leftrightarrow O^-{-}N^+({=}O)({-}O^-)$

Fig. 12.5

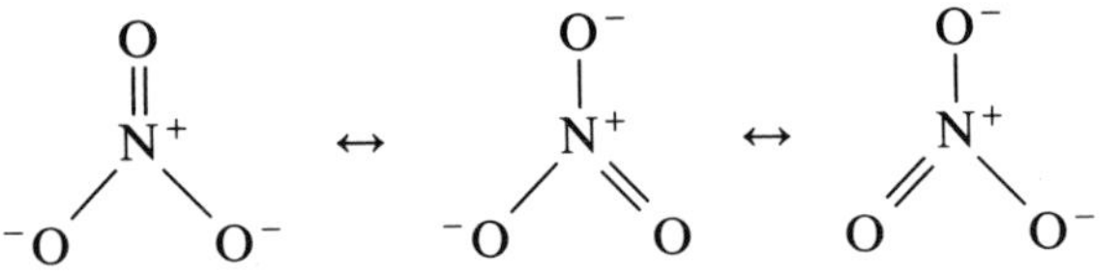

Although nitric acid behaves as a typical strong acid, its reactions are made more complex as a result of its oxidizing power.

With weaker acids, nitric acid acts as a proton donor,

e.g. $$HNO_3(aq) \rightleftharpoons H^+(aq) + NO_3^-(aq)$$

and with stronger acids, as a proton acceptor (i.e. as a base). With chloric(VII) acid, the following equilibria are established:

$$HNO_3(aq) + HClO_4(aq) \rightleftharpoons [(HO)_2NO]^+(aq) + ClO_4^-(aq)$$

$$\rightleftharpoons NO_2^+(aq) + H_2O(l)$$

A mixture of concentrated sulphuric and nitric acids, commonly referred to as a *nitrating mixture*, is of immense importance in organic chemistry as a source of the *nitryl* (nitronium) cation, NO_2^+.

$$HNO_3(aq) + 2H_2SO_4(aq) \rightleftharpoons H^+(aq) + NO_2^+(aq) + 2HSO_4^-(aq) + H_2O(l)$$

Action with metals

Pt, Au, Rh, and a few other metals have no action at all with nitric acid. Al, Fe, Co, and Cr adopt the so-called 'passive' state, probably owing to the formation of a protective layer of the oxide. Most of the remaining metals react readily, the more weakly electropositive ones (e.g. Sn, As, Sb, W, and Mo) forming the metal oxide and the more electropositive ones forming nitrate ions. During the course of these reactions, various nitrogen oxides, sometimes together with elemental nitrogen, are evolved.

A summary of the main reduction products of nitric acid with certain metals is given in Table 12.5.

Table 12.5

	Metal	*Concentration of acid*
$4H^+(aq) + 2NO_3^-(aq) + 2e \rightarrow 2H_2O(l) + N_2O_4(g)$	Pb, Cu, Ag, Hg	Conc.
$4H^+(aq) + NO_3^-(aq) + 3e \rightarrow 2H_2O(l) + NO(g)$	Pb, Cu, Ag, Hg	Mod. conc.
$10H^+(aq) + 2NO_3^-(aq) + 8e \rightarrow 5H_2O(l) + N_2O(g)$	Mg, Zn, Fe	Mod. conc.
$10H^+(aq) + NO_3^-(aq) + 8e \rightarrow 3H_2O(l) + NH_4^+(aq)$	Sn	Dilute

These equations represent a very much over-simplified form of extremely complex processes, and even for the metals and approximate concentrations of acid stated, variations in the composition of the products will occur depending upon the precise concentration of the acid and the temperature of reaction. Furthermore, the reaction products themselves, particularly N_2O_4, may catalyse side-reactions.

Nitrous (nitric(III)) acid, HNO_2

Nitrous acid is obtained by treating a solution containing nitrite ions with a dilute acid. The resulting solution (which is pale blue because of the presence of N_2O_3) is unstable and decomposes to yield nitric acid:

$$3HNO_2(aq) \rightleftharpoons HNO_3(aq) + 2NO(g) + H_2O(l) \quad \text{(simplified equation)}$$

Pure nitrous acid has not been isolated.

Aqueous solutions of nitrous acid ($pK_a = 3.34$) can be employed as both an oxidizing and a reducing agent.

It reduces strong oxidizing agents, such as potassium manganate(VII) (permanganate) in acidic but not alkaline solution, potassium dichromate(VI), hydrogen peroxide, and bromine water.

$$2MnO_4^-(aq) + 5NO_2^-(aq) + 6H^+(aq) \rightarrow 2Mn^{2+}(aq) + 5NO_3^- + 3H_2O(l); \quad E = +0.58\,V$$

$$Br_2(aq) + H_2O(l) + NO_2^-(aq) \rightarrow 2Br^-(aq) + NO_3^-(aq) + 2H^+(aq); \quad E = +0.13\,V$$

As an oxidizing agent, it reacts with acidified potassium iodide solution,

$$2HI(aq) + 2HNO_2(aq) \rightarrow 2H_2O(l) + 2NO(g) + I_2(aq); \quad E = +0.45\,V$$

with Sn^{2+} ions to give Sn^{4+} ions,

$$2Sn^{2+}(aq) + 2HNO_2(aq) + 4H^+(aq) \rightarrow 2Sn^{4+}(aq) + N_2O(g) + 3H_2O(l); \quad E = +0.84\,V$$

with Fe^{2+} ions to give Fe^{3+} ions, with hydrogen sulphide precipitating sulphur, and with sulphur dioxide to give sulphuric acid.

Nitrous acid forms nitrites, mainly with the more electropositive metals of Groups I and II. Nitrite ions may be detected in aqueous solution by the liberation of iodine from acidified potassium iodide.

One of the most important industrial applications of nitrous acid is in the preparation of diazonium compounds used in the manufacture of azo dyes.

Nitric(I) (hyponitrous) acid, $H_2N_2O_2$

This is a solid, weak, dibasic acid, prepared by the action of nitrous acid on hydroxylamine:

$$HNO_2(aq) + NH_2OH(s) \rightarrow HON{=}NOH(s) + H_2O(l)$$

Nitrate(I) salts (hyponitrites) are reducing agents, being oxidized by acidified potassium manganate(VII) solution to nitric acid and by an alkaline solution of manganate(VII) to the nitrite.

Oxoacids of phosphorus

There are several well-characterized oxoacids of phosphorus, but only two of these, phosphonic (phosphorous) acid, H_3PO_3, and phosphoric(V) acid, H_3PO_4, have been studied in the crystalline state. In both of them, the phosphorus atom is tetrahedrally surrounded; but in phosphonic acid there is a P—H linkage which enables the acid to exhibit reducing properties, whereas in phosphoric(V) acid there are only P—O linkages (Figure 12.6).

Fig. 12.6 (a) phosphonic acid (b) phosphoric(V) acid

Phosphonic acid, H_3PO_3, is a white, deliquescent, crystalline compound. It is dibasic, forming two series of acid salts, dihydrogenphosphonates, $H_2PO_3^-$, and hydrogenphosphonates, HPO_3^{2-}. It is a moderately strong acid in aqueous solution (Table 12.6) but is better known as a reducing agent:

$$H_3PO_4(aq) + 2H^+(aq) + 2e \rightleftharpoons H_3PO_3(aq) + H_2O(l); \quad E^\ominus = -0.20\,V$$

On warming, it decomposes to phosphine and phosphoric(V) acid.

Table 12.6

	pK_{a_1}	pK_{a_2}	pK_{a_3}
H_3PO_3	2.00	6.58	—
H_3PO_4	2.15	7.21	12.36

Phosphoric(V) acid is a colourless, deliquescent, crystalline compound of similar acid strength to phosphonic acid. In this case, however, all three hydrogen atoms are replaceable. It gives rise to two series of acid salts containing the dihydrogenphosphate(V) ion, $H_2PO_4^-$, and the hydrogenphosphate(V) ion, HPO_4^{2-}, and a normal salt containing the phosphate(V) ion, PO_4^{3-}. In all three cases, the anions are tetrahedral in shape.

In aqueous solution, the $H_2PO_4^-$ ion is weakly acidic,

$$H_2PO_4^-(aq) \rightarrow HPO_4^{2-}(aq) + H^+(aq)$$

whereas the HPO_4^{2-} and PO_4^{3-} ions both undergo hydrolysis to yield a basic solution:

$$HPO_4^{2-}(aq) + H_2O(l) \rightleftharpoons H_2PO_4^-(aq) + OH^-(aq) \quad \text{(weakly basic)}$$

$$PO_4^{3-}(aq) + H_2O(l) \rightarrow HPO_4^{2-}(aq) + OH^-(aq) \quad \text{(strongly basic)}$$

Crystalline H_3PO_4 has a layer structure and, within each layer, the individual PO_4 groups are attached to six others by hydrogen bonds. This structure is also present in concentrated solution and probably accounts for the syrupy nature of the liquid.

Phosphates(V)

Calcium phosphate(V) occurs naturally as *rock phosphate* and is the main source of phosphorus. It is too insoluble to be used as a *phosphatic fertilizer* but can be converted into the well-known 'superphosphate' by treating with 70 per cent sulphuric acid:

$$Ca_3(PO_4)_2(s) + 2H_2SO_4(aq) \rightarrow \underbrace{Ca(H_2PO_4)_2(s) + 2CaSO_4(s)}_{\text{'superphosphate'}}$$

With aqueous phosphoric(V) acid it yields 'triple phosphate':

$$Ca_3(PO_4)_2(s) + 4H_3PO_4(aq) \rightarrow \underset{\text{'triple phosphate'}}{3Ca(H_2PO_4)_2(s)}$$

and, with nitric acid, it yields 'nitrophos', which is a mixed phosphatic and nitrogenous fertilizer:

$$Ca_3(PO_4)_2(s) + 4HNO_3(aq) \rightarrow \underbrace{Ca(H_2PO_4)_2(s) + 2Ca(NO_3)_2(s)}_{\text{'nitrophos'}}$$

Sodium hexametaphosphate(V) ('Calgon'), $(NaPO_3)_n$, is a polymeric compound obtained by heating sodium dihydrogenphosphate(V). It is widely used as a water softener, removing Ca^{2+} and Mg^{2+} ions by sequestration (see page 23).

Organic phosphates provide some of the most essential chemicals involved in life processes. These include the genetic compounds, DNA (deoxyribonucleic acid) and RNA (ribonucleic acid). In biological systems, the compound adenosine triphosphate(V), ATP, is formed in photosynthesis and is the end product of respiration. It can be used as an energy source in the chemical reactions of a body cell, the energy arising as it is hydrolysed to the diphosphate(V), ADP.

13 Group VI

Table 13.1

	Oxygen	*Sulphur*	*Selenium*	*Tellurium*	*Polonium*
Symbol	O	S	Se	Te	Po
Outer electronic structure	$2s^2 2p^4$	$3s^2 3p^4$	$4s^2 4p^4$	$5s^2 5p^4$	$6s^2 6p^4$
Principal oxidation states	2	2, 4, **6**	2, **4**, **6**	2, **4**, **6**	2, **4**, 6

Nature and general chemistry of the elements

The general, transitional trend of the previous two groups is again observed here. However, the non-metallic character of each of the first four elements is now quite pronounced, being predictably most evident in oxygen and sulphur, and they are collectively known as the *chalcogens* ('ore-forming elements'). The fifth member of the group, polonium, behaves primarily as a metal but, as it is radioactive and has a short half-life (^{210}Po; $t_{\frac{1}{2}} = 138.4$ days), its compounds are not well characterized.

All five elements have allotropic forms; the more familiar forms of oxygen and sulphur are discussed below.

Oxygen

Oxygen is the only gaseous member of Group VI. It has three naturally occurring isotopes, ^{16}O (99.76 per cent), ^{17}O (0.04 per cent), and ^{18}O (0.2 per cent) and is the most common element in the Earth's crust, accounting for 46.6 per cent by mass of the components of the crust, 89 per cent of water, and about 21 per cent of the atmosphere.

Vast quantities of oxygen are manufactured in industry by the fractional distillation of liquid air, but in the laboratory it is normally prepared by adding hydrogen peroxide to manganese(IV) oxide, or by heating potassium chlorate(V); each time manganese(IV) oxide is used as a catalyst.

$$2H_2O_2(l) \xrightarrow{MnO_2 \text{ catalyst}} 2H_2O(l) + O_2(g)$$

$$2KClO_3(s) \xrightarrow{MnO_2 \text{ catalyst}} 2KCl(s) + 3O_2(g)$$

Oxides are formed with most elements, the oxygen acting as an oxidizing agent by acquiring electrons. Heat is usually needed for these reactions, although the readiness of oxygen to react with white phosphorus and nitrogen oxide is well-known.

Oxygen gives rise to three important ions: the oxide ion, O^{2-}, the peroxide ion, O_2^{2-}, and the superoxide ion, O_2^-; the last two feature mainly in the chemistry of the electropositive metals of Groups I and II.

Trioxygen (*ozone*) is an allotrope of oxygen and is formed by striking a silent electric discharge through an atmosphere of oxygen.

$$3O_2(g) \rightleftharpoons 2O_3(g); \quad \Delta H^\ominus = +284\,\text{kJ}\,\text{mol}^{-1}$$

It occurs in the Earth's stratosphere as a result of the action of ultra-violet light on oxygen. Although present in the atmosphere in only minute amounts, it has the vital role of shielding animal and vegetable life from certain harmful ultra-violet radiations. One of its major uses is in the purification of water.

Gaseous trioxygen is diamagnetic, and spectroscopic and electron diffraction evidence suggests that the molecule has a bond angle of 116.5° and bond lengths of 0.128 nm.

Chemically, its reactions are principally those of a strong oxidizing agent, the trioxygen itself being usually reduced to molecular oxygen:

$$O_3(g) + 2H^+(aq) + 2e \rightleftharpoons H_2O(l) + O_2(g); \quad E^\ominus = +2.07\,\text{V}$$

Typical of its oxidation reactions is the conversion of sulphide to sulphate ions, acidified iodide ions to iodine, iron(II) ions to iron(III) ions, and tin(II) to tin(IV) ions:

$$S^{2-}(aq) + 4O_3(g) \rightarrow SO_4^{2-}(aq) + 4O_2(g); \quad E = +1.70\,\text{V}$$

$$2I^-(aq) + 2H^+(aq) + O_3(g) \rightarrow I_2(aq) + H_2O(l) + O_2(g); \quad E = +1.53\,\text{V}$$

$$2Fe^{2+}(aq) + 2H^+(aq) + O_3(g) \rightarrow 2Fe^{3+}(aq) + H_2O(l) + O_2(g); \quad E = +1.30\,\text{V}$$

$$3Sn^{2+}(aq) + 6H^+(aq) + O_3(g) \rightarrow 3Sn^{4+}(aq) + 3H_2O(l); \quad E = +1.92\,\text{V}$$

With mercury, trioxygen forms mercury(II) oxide, which has the peculiar property of adhering to the sides of the glass vessel. This is referred to as 'tailing', and is sometimes used for detecting the gas.

In organic chemistry, ozonolysis is used in determining the location of the double bond in alkenes, the analysis depending upon the subsequent identification of the carbonyl compounds resulting from the hydrolysis of the ozonide.

Sulphur

Sulphur exists in various allotropic forms. The most stable allotrope is the yellow, crystalline, *rhombic* (α) form (m.p. 112.8 °C). At the transition temperature of 95.6 °C, rhombic sulphur changes into the *monoclinic* (β) form (m.p. 119.25 °C). The element boils at 444.6 °C.

In both of these allotropic forms, and in the liquid state, the structural unit is the puckered S_8 ring (Figure 13.1). At temperatures just above the melting point, the pale yellow liquid is mobile and composed of S_8 units. On raising the temperature further, the liquid rapidly darkens, and at about 160 °C some of the S_8 rings are broken down forming long spiral chains (μ-sulphur) together with some S_6 units (π-sulphur). Between 160 and 180 °C, the viscosity of the liquid reaches a maximum and it cannot be poured.

The gaseous element contains a mixture of S_8, S_6, S_4, and S_2 units, the precise composition depending upon temperature.

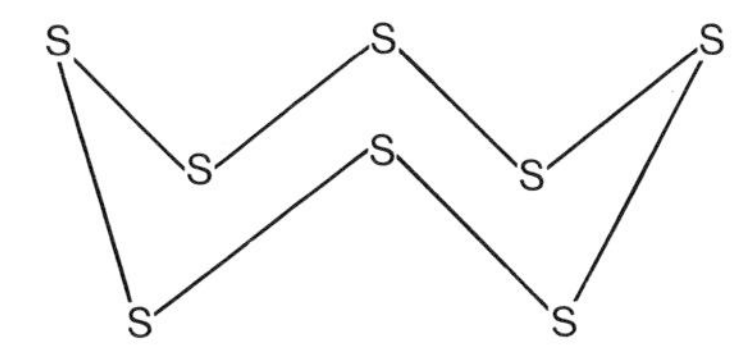

Fig. 13.1

Plastic sulphur, obtained by pouring molten sulphur (ideally at 250–350 °C) into cold water, has a fibrous nature comprising mainly the μ-form. On standing, it gradually becomes rhombic sulphur.

Sulphur is a chemically active element, forming sulphides, usually by direct synthesis of the elements. The more electropositive metals of Groups I and II form mainly ionic sulphides containing the S^{2-} anion, whereas most other elements form essentially covalent sulphides, some of which are giant molecules possessing three-dimensional, layer, or chain structures.

Sulphur is attacked by oxidizing acids,

e.g. $$S(s) + 2H_2SO_4(l) \xrightarrow{\text{heat}} 3SO_2(g) + 2H_2O(l)$$

and by alkalis:

$$3S(s) + 6OH^-(aq) \longrightarrow 2S^{2-}(aq) + SO_3^{2-}(aq) + 3H_2O(l)$$

Electronic structure and bond type

The elements have an outer electronic structure of ns^2np^4. Oxygen is limited to a valency of two as it has no available *d* orbitals. The other elements, having suitable available *d* levels, utilize them in extending their oxidation states to +4 and +6 (e.g. SO_2, SO_3, SF_6). The tendency for the elements to donate electron pairs decreases in the order O to Te. Polonium is analogous to bismuth, the last member of Group V, in the sense that the pair of *s* electrons behaves as an inert pair.

The ease with which the gaseous atoms of these elements fill their outer *p* levels is given by their *electron affinities* (Table 13.2).

Table 13.2

	Electron affinities, $\Delta H^\ominus$*/kJ mol*$^{-1}$	
	1st	*2nd*
O	−148	+850
S	−206	+538
Se	−211	+440
Te	−230	—

Despite the high resistance of O^- to acquiring a second electron, the simple ionic oxides of the more electropositive metals all contain the O^{2-} ion as opposed to the O^- ion. The explanation of this is not to be found in a cursory examination of electron affinity values, but by consideration of a Born-Haber cycle for the formation of the solid oxides. The enthalpy change on forming the solid crystal (i.e. $-\Delta H_{lat}$) more than compensates for the adverse second electron affinity value (Table 13.3).

Table 13.3

Group II metal oxide	*−Lattice enthalpy, $-\Delta H^{\ominus}_{lat}$/kJ mol^{-1}*
MgO	−3889
CaO	−3513
SrO	−3310
BaO	−3152

A lowering of the lattice enthalpies with sulphides, selenides, and tellurides, makes the formation of ionic compounds from these elements progressively more difficult.

The non-metallic character of these elements is reflected by the high values of their *ionization energies*, which prohibit the formation of positive ions (Table 13.4).

Table 13.4

	Ionization energy, $\Delta H^{\ominus}$/kJ mol^{-1}			
	1st	*2nd*	*3rd*	*4th*
O	+1310	+3390	+5320	+7450
S	+1000	+2260	+3390	+4540
Se	+947	+2080	+3090	+4140
Te	+876	+1800	+3010	+3680

The large electronegativity values (Table 13.5) of the group as a whole further enhance the expectation that the elements will be predominantly non-metallic in their behaviour.

Table 13.5

	O	*S*	*Se*	*Te*	*Po*
Electronegativity	3.5	2.5	2.4	2.4	2.0

The ability to catenate is a feature of these elements, particularly sulphur. In addition to the ring and chain structures it adopts in the elemental state, sulphur forms a number of polysulphur compounds, including the polysulphur dichlorides, S_nCl_2(n = 3 to 6), and the thionic acids, $H_2S_nO_6$ (n = 3 to 6). Oxygen has a maximum chain length of three in ozone, and selenium and tellurium give rise to a few short, unstable chains, e.g. Se_2Cl_2. Like the Group IV elements (see page 146), the ability to catenate is closely related to the A—A bond enthalpies (Table 13.6).

Table 13.6

Bond	*O—O*	*S—S*	*Se—Se*	*Te—Te*
Bond enthalpy, $\Delta H^\ominus$/kJ mol^{-1}	+142	+226	+172	+126

Hydrides

The hydrides H_2O, H_2S, H_2Se, and H_2Te are all known and, with the exception of water, are all poisonous and pungent gases.

Because of the high electronegativity of oxygen and its readily available lone pairs of electrons in the H_2O molecule (see page 31), water is highly associated by means of intermolecular hydrogen bonds in the liquid state (see page 106) and has an unusually high boiling point. The melting and boiling points of H_2S, H_2Se, and H_2Te, however, form a graded series and the values increase with increasing relative molecular mass (Table 13.7).

Table 13.7

	H_2O	*H_2S*	*H_2Se*	*H_2Te*
M.p./°C	0	−85.5	−66	−51.2
B.p./°C	100	−60.5	−41.5	−20

Each of these four molecules has an angular arrangement of atoms with the bond angle, α, decreasing in the order H_2O to H_2Te (Table 13.8).

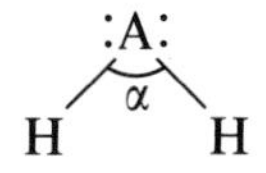

Table 13.8

	H_2O	*H_2S*	*H_2Se*	*H_2Te*
Bond angle, α	104° 30′	92° 20′	91°	89° 30′

As the electronegativity of the Group VI element decreases, the bond pairs of electrons are to be found further away from the central atom, with the result that the repulsive force between them becomes progressively weaker. Interaction between the two sets of lone pairs, and the lone pairs and the bond pairs, consequently causes a progressively greater deviation from the regular tetrahedral angle of 109°28′ (Table 13.8).

One property common to all of these hydrides is their ability to function as weak acids. The relative strengths of these acids can be determined by means of a Born-Haber cycle, similar to that outlined for aqueous solutions of the hydrogen halides (see page 203), and increases from H_2O to H_2Te.

Table 13.9

	H_2O	H_2S	H_2Se	H_2Te
pK_a	14	7	4	3

Oxides, sulphides, selenides, and tellurides may be considered as salt derivatives of these weak acids.

The thermal stability of water is unusually high in comparison with the hydrogen compounds of the other elements of this group. At about 2000 °C it undergoes only slight decomposition into its component elements, whereas hydrogen sulphide decomposes on passing through a red hot tube. This difference in stability is related to the greater strength of the O—H bond.

Table 13.10

Bond	*O—H*	*S—H*
Bond enthalpy, $\Delta H^{\ominus}$/kJ mol^{-1}	+463	+338

Another contrast between water and hydrogen sulphide is their relative reducing powers. This also is attributable to the difference in strength between the O—H and S—H bonds. Whereas hydrogen sulphide is a strong reducing agent,

$$2H^+(aq) + S(s) + 2e \rightleftharpoons H_2S(g); \quad E^{\ominus} = +0.14\,V$$

oxygen is liberated from water only by very powerful oxidizing agents such as fluorine (see page 199).

$$4H^+(aq) + O_2(g) + 4e \rightleftharpoons 2H_2O(l); \quad E^{\ominus} = +1.23\,V$$

Water, H_2O

Solvent properties

Water has remarkable properties as a solvent. In addition to ionic compounds, the solubility of which is related to the relative values of their lattice and hydration enthalpies (see page 60), a vast number of other polar, but essentially covalent, substances have a marked tendency to dissolve in water as a result of their ability to form intermolecular hydrogen bonds with the water molecules (Figure 13.2). Notable examples of these include compounds containing hydroxyl, carbonyl, carboxyl, or amino groups.

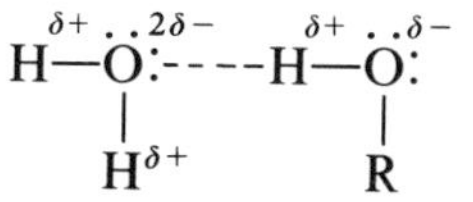

$$\begin{array}{ccc} \overset{\delta+}{\text{H}}-\overset{\cdot\cdot\,2\delta-}{\text{O}}\text{:} & \text{- - - -} & \overset{\delta+}{\text{H}}-\overset{\cdot\cdot\,\delta-}{\text{O}}\text{:} \\ | & & | \\ \text{H}^{\delta+} & & \text{R} \end{array}$$

Fig. 13.2 Intermolecular hydrogen bonding between water and alcohol molecules

Water of crystallization

X-ray analysis provides a clear image of the way in which water of crystallization is bound up in the crystal structure. Although the formula of copper(II) sulphate crystals is usually written as $CuSO_4{\cdot}5H_2O$, X-ray analysis indicates that only four of these water molecules are in fact bound to the copper(II) ion. A more realistic representation of the structure is therefore given by the formula, $[Cu(H_2O)_4]^{2+}SO_4^{2-}{\cdot}H_2O$. This also provides a plausible explanation as to why, when heated to about 100 °C, the salt dehydrates only as far as the monohydrate. The fifth water molecule is lost at about 250 °C, suggesting that it is more strongly bound in the crystal than the other four water molecules.

Reactions

The highly reactive metals, such as K, Na, and Ca, react readily with cold water, liberating hydrogen; other reactive metals, such as Mg, Zn, and Fe, react similarly with steam.

Water is slightly dissociated into ions when pure ($K_w \approx 1 \times 10^{-14}$ at 25 °C),

$$H_2O(l) \rightleftharpoons H^+(aq) + OH^-(aq)$$

and is at the same time both an acid and a base. In the presence of a base, it functions as an acid:

e.g. $$NH_3(g) + H_2O(l) \rightleftharpoons NH_3{\cdot}H_2O(aq) \rightleftharpoons NH_4^+(aq) + OH^-(aq)$$

and in the presence of an acid, it functions as a base:

e.g. $$HNO_3(l) + H_2O(l) \rightleftharpoons {}^*H_3O^+(aq) + NO_3^-(aq)$$

Since it has the ability to both accept and donate protons, water is classified as an *amphiprotic solvent.*

The reversible process by which water reacts with salts (other than those of a strong acid/strong base) to generate the acidic and basic components of the salt is known as *salt hydrolysis.* The process is dependent upon either or both components being weak and interacting with the water. The pH of the resulting solution is determined by the relative strengths of the acidic and basic species.

---

* Representation of the hydrated proton as an oxonium ion, H_3O^+, is known to be an over-simplification, as it undergoes further hydration and is more accurately represented as $[H(H_2O)_x]^+$ or, for simplicity, as $H^+(aq)$. However, in order to illustrate the basic character of water with the acid, H_3O^+ is adopted in this particular context.

For example, an aqueous solution of an alkali metal carbonate (the salt of a strong base and a weak acid) will be basic.

$$CO_3^{2-}(aq)+H_2O(l) \rightleftharpoons OH^-(aq)+HCO_3^-(aq)$$

Conversely, aqueous solutions of salts of weak bases and strong acids, such as ammonium chloride, yield acidic solutions. In this case, the ammonium cation reacts with the OH^- ions yielded by the water and also with the water molecules themselves:

$$NH_4^+(aq)+OH^-(aq) \rightleftharpoons NH_3{\cdot}H_2O(aq)$$
$$NH_4^+(aq)+H_2O(l) \rightleftharpoons NH_3{\cdot}H_2O(aq)+H^+(aq)$$

The concentration of H_3O^+ ions in solution therefore exceeds that of OH^- ions.

The soluble salts of certain metals undergo salt hydrolysis to form the hydrated cation. This is typified by the $[Al(H_2O)_6]^{3+}$ cation (see page 135) and the hydrated cations formed by the transition metals, such as $[Fe(H_2O)_6]^{3+}$ (see page 255).

Hydrogen sulphide, H_2S

Hydrogen sulphide is a colourless, poisonous gas with a distinctive 'bad egg' smell. At s.t.p., one volume of water dissolves 4.7 volumes of the gas. On boiling such a solution, the gas is completely expelled.

In the laboratory, hydrogen sulphide is usually obtained by the action of hydrochloric or sulphuric acid on iron(II) sulphide in Kipp's apparatus:

$$FeS(s)+2H^+(aq) \rightarrow Fe^{2+}(aq)+H_2S(g)$$

In a plentiful supply of oxygen, the gas burns with a blue flame to sulphur dioxide,

$$2H_2S(g)+3O_2(g) \rightarrow 2H_2O(g)+2SO_2(g)$$

and in a limited supply, sulphur is deposited:

$$2H_2S(g)+O_2(g) \rightarrow 2H_2O(l)+2S(s)$$

Being a *powerful reducing agent*, it reduces concentrated sulphuric acid to sulphur dioxide and sulphur,

$$H_2SO_4(l)+H_2S(g) \rightarrow SO_2(g)+2H_2O(l)+S(s)$$

and iron(III) ions to iron(II) ions:

$$2Fe^{3+}(aq)+H_2S(g) \rightarrow 2Fe^{2+}(aq)+2H^+(aq)+S(s); \quad E = +0.63\,V$$

In acidified solution, purple manganate(VII) (permanganate) ions are reduced to the very pale pink (almost colourless) manganese(II) ions,

$$2MnO_4^-(aq)+5H_2S(g)+6H^+(aq) \rightarrow 2Mn^{2+}(aq)+8H_2O(l)+5S(s); \quad E = +1.38\,V$$

and orange dichromate(VI) ions to green chromium(III) ions:

$$Cr_2O_7^{2-}(aq) + 3H_2S(g) + 8H^+(aq) \rightarrow 2Cr^{3+}(aq) + 7H_2O(l) + 3S(s); \quad E = +1.91\,V$$

(Yellow chromate(VI) ions, CrO_4^{2-}, are similarly reduced by hydrogen sulphide to chromium(III) ions via the intermediate formation of dichromate(VI) ions.)

Aqueous solutions of hydrogen sulphide function as a weak dibasic acid, dissociating in two stages:

$$H_2S(aq) \overset{pK_{a_1} = 7.05}{\rightleftharpoons} H^+(aq) + HS^-(aq) \overset{pK_{a_2} = 13.92}{\rightleftharpoons} 2H^+(aq) + S^{2-}(aq)$$

On absorption in alkalis, it forms sulphides, S^{2-}, and hydrogen sulphides, HS^-

$$2OH^-(aq) + H_2S(g) \rightleftharpoons S^{2-}(aq) + 2H_2O(l)$$
$$S^{2-}(aq) + H_2S(g) \rightleftharpoons 2HS^-(aq)$$

The presence of *the gas may be detected* by its action on white lead(II) ethanoate or nitrate paper, which is blackened by the formation of the sulphide, or by the production of a purple colouration in an alkaline solution of sodium pentacyanonitrosylferrate(II) (sodium nitroprusside).

$$\underset{\text{yellow-brown}}{[FeNO(CN)_5]^{2-}(aq)} + HS^-(aq) + OH^-(aq) \rightarrow \underset{\text{purple}}{[Fe(CN)_5NOS]^{4-}(aq)} + H_2O(l)$$

The latter test is very sensitive.

Application of hydrogen sulphide to qualitative analysis

Hydrogen sulphide is used in qualitative inorganic analysis under controlled conditions to bring about precipitation of insoluble metal sulphides, many of which have distinctive colours. In Group 2 (the Group number in analysis has nothing whatsoever to do with the periodic classification) of most elementary analysis schemes, the ions Hg^{2+}, Bi^{3+}, Cu^{2+}, As^{3+}, Sb^{3+}, Cd^{2+}, and Pb^{2+} are isolated in acidic media as their insoluble sulphides. The purpose of the high proton concentration is to inhibit the dissociation of the hydrogen sulphide (an example of Le Châtelier's principle) and consequently reduce the sulphide ion concentration so that only the less soluble sulphides are precipitated.

In Group 4 analysis the sulphides ZnS, MnS, NiS, and CoS, all of which are slightly more soluble than the Group 2 metal sulphides, are precipitated by the passage of hydrogen sulphide through a solution which has been made just alkaline by the addition of aqueous ammonia. The hydroxide ions remove the acidic protons furnished by the hydrogen sulphide to form water and cause the equilibrium position to adjust in favour of the production of more sulphide ions.

$$H_2S(aq) \rightleftharpoons 2H^+(aq) + S^{2-}(aq)$$
$$\Big\Updownarrow 2OH^-(aq)$$
$$2H_2O(l)$$

This increase in sulphide ion concentration ensures that any Group 4 metal sulphides present are precipitated.

Hydrogen peroxide, H_2O_2

In the laboratory, hydrogen peroxide is prepared by the action of an acid on a metal peroxide, usually sulphuric acid and barium peroxide, so that the barium sulphate precipitate can be removed by simple filtration.

$$BaO_2(s) + H_2SO_4(aq) \xrightarrow{\text{ice cold}} BaSO_4(s) + H_2O_2(aq)$$

The resulting aqueous solution may be concentrated by distilling under reduced pressure, or by freezing the water present and removing the ice.

The most important processes for the manufacture of hydrogen peroxide involve the hydrogenation of a 2-alkylanthracene-9,10-dione to the corresponding 2-alkylanthracene-9,10-diol which then undergoes oxidation by oxygen-enriched air yielding hydrogen peroxide and regenerating the dione:

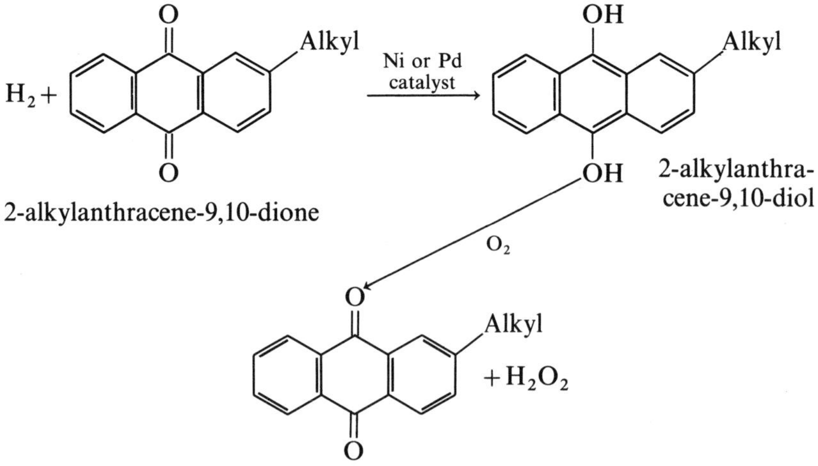

The peroxide is extracted with water and concentrated by vacuum distillation.

Pure hydrogen peroxide is a colourless, slightly viscous liquid with a density of 1.4 g cm^{-3}. It is usually stored in aqueous solution, the concentration of which is usually quoted in terms of its 'volume rating'. This relates to the number of volumes of oxygen, corrected to s.t.p., evolved when one volume of the hydrogen peroxide solution is fully decomposed. For example, one volume of '20 volume' hydrogen peroxide would decompose to give 20 volumes of oxygen at s.t.p.

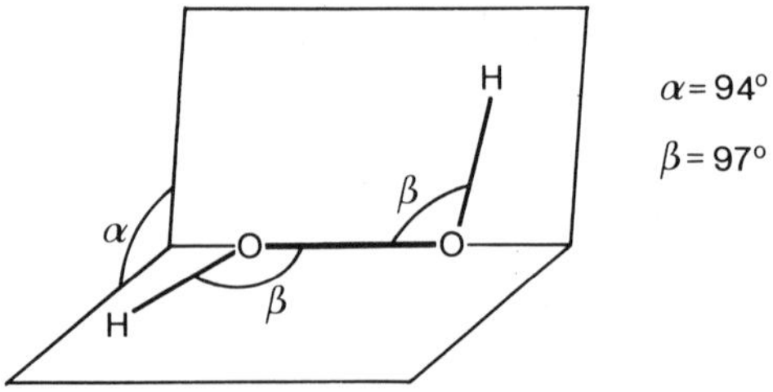

Fig. 13.3

Structurally, the H_2O_2 molecule is referred to as a dihedron as the two —OH groups lie in different planes (Figure 13.3). This represents the most stable configuration, as interaction between the lone pairs on the oxygen atoms is at a minimum.

Hydrogen peroxide is thermodynamically unstable, slowly undergoing decomposition into water and oxygen:

$$2H_2O_2(l) \rightarrow 2H_2O(l) + O_2(g); \quad \Delta H^{\ominus} = -196\,\text{kJ}\,\text{mol}^{-1}$$

The process is accelerated by light and a variety of catalysts, including a number of finely divided metals (e.g. Cu, Fe, Au, Ag, Pt), alkalis, some metal oxides (e.g. MnO_2, PbO_2), and certain organic materials (e.g. blood).

It can behave as both an oxidizing and a reducing agent. In acidic solution, its oxidizing power is enhaced by the reaction:

$$H_2O_2(aq) + 2H^+(aq) + 2e \rightleftharpoons 2H_2O(l); \quad E^{\ominus} = +1.77\,\text{V}$$

As an oxidizing agent, it converts lead(II) sulphide to white lead(II) sulphate (a reaction which is utilized in the restoration of discoloured oil paintings),

$$PbS(s) + 4H_2O_2(aq) \rightarrow PbSO_4(s) + 4H_2O(l)$$

iron(II) salts to iron(III) salts,

$$2Fe^{2+}(aq) + H_2O_2(aq) + 2H^+(aq) \rightarrow 2Fe^{3+}(aq) + 2H_2O(l); \quad E = +1.00\,\text{V}$$

iodide ions to iodine,

$$2I^-(aq) + H_2O_2(aq) + 2H^+(aq) \rightarrow I_2(aq) + 2H_2O(l); \quad E = +1.23\,\text{V}$$

and similarly for many other reactions. Its mild oxidizing action makes it suitable for bleaching hair and for use as a mild antiseptic.

Its reducing properties are in accordance with the partial ionic equation:

$$O_2(g) + 2H^+(aq) + 2e \rightleftharpoons H_2O_2(aq); \quad E^{\ominus} = +0.68\,\text{V}$$

Some of the reactions involving hydrogen peroxide as a reducing agent include:

the reduction of silver(I) oxide to silver,

$$Ag_2O(s) + H_2O_2(aq) \rightarrow 2Ag(s) + O_2(g) + H_2O(l); \quad E = +0.12\,\text{V}$$

trioxygen to oxygen,

$$O_3(g) + H_2O_2(aq) \rightarrow H_2O(l) + 2O_2(g); \quad E = +1.39\,\text{V}$$

chlorine water (chloric(I) acid) to hydrochloric acid,

$$HOCl(aq) + H_2O_2(aq) \rightarrow H_2O(l) + HCl(aq) + O_2(g); \quad E = +0.81\,\text{V}$$

and the acidified manganate(VII) ion to the manganese(II) ion,

$$2MnO_4^-(aq) + 5H_2O_2(aq) + 6H^+(aq) \rightarrow 2Mn^{2+}(aq) + 8H_2O(l) + O_2(g); \quad E = +0.84\,\text{V}$$

Detection of hydrogen peroxide is usually carried out by adding a drop of the liquid to the interface of an ethoxyethane/acidified potassium dichromate(VI) mixture, the formation of a royal blue colour of a chromium complex in the upper ethoxyethane layer indicating its presence.

Oxides and oxoacids

The Group VI elements form several series of oxides (Table 13.11), of which the dioxides and trioxides, particularly those of sulphur, are the most familiar.

Table 13.11

Oxides of sulphur	*Oxides of selenium*	*Oxides of tellurium*	*Oxides of polonium*
S_2O			
	(SeO)	TeO	PoO
S_2O_3			
SO_2	SeO_2	TeO_2	PoO_2
SO_3	SeO_3	TeO_3	
SO_4			

The dioxides SO_2, SeO_2, and TeO_2, can all be obtained by the combustion of the element in air.

Sulphur dioxide is a colourless, pungent gas (b.p. $-10\,^{\circ}C$) and exists as discrete molecules, even in the solid state. The molecule is angular in shape and possesses what are effectively two sulphur-oxygen double bonds (use is made of the *d* orbitals of the sulphur for this purpose) (see Figure 13.4).

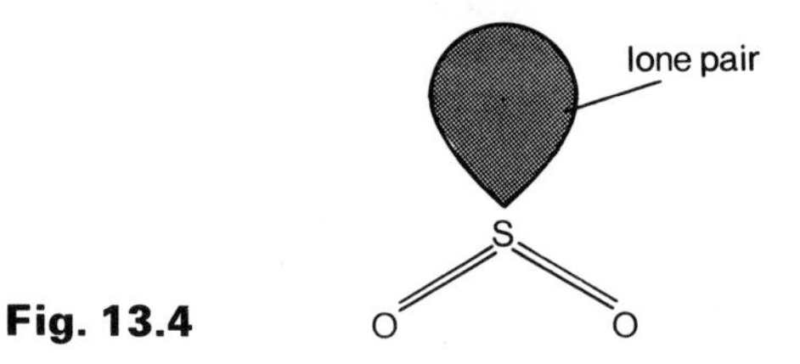

Fig. 13.4

Both selenium dioxide, SeO_2, and tellurium dioxide (two forms), TeO_2, are colourless solids but, whereas SeO_2 exists in the form of infinite, covalent chains:

```
    O       O       O
    ||      ||      ||
    Se      Se      Se
   /  \    /  \    /  \    /
       O       O       O
```

TeO_2 has a more ionic structure. PoO_2 has two predominantly ionic forms, a tetragonal (red) and a face-centred cubic (yellow).

SO_2 is extremely soluble in water (79.8 volumes in one of water at s.t.p.), forming sulphurous acid, H_2SO_3, an acid which cannot be isolated. SeO_2 dissolves to form selenious acid, H_2SeO_3, which is readily obtained in the crystalline state.

Of the trioxides, SO_3 is without doubt the most important. It exists in three distinct solid forms, the α-form (m.p. 17 °C; b.p. 45 °C) which is an ice-like cyclic trimer, the β-form, which has a helical chain structure, and the γ-form, which has a cross-linked chain structure. The β- and γ-forms sublime between 45 and 50 °C. In the vapour phase sulphur(VI) oxide (sulphur trioxide) exists as discrete, symmetrical planar molecules in which the sulphur atom is sp^2 hybridized:

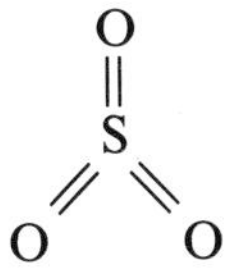

O
‖
S
// \\
O O

Fig. 13.5

SeO_3 (m.p. 118 °C) and TeO_3 are white solids of unknown structure, the latter being thermally very stable.

Sulphur dioxide, SO_2

In the laboratory, sulphur dioxide can be generated by the action of dilute acid on a sulphite,

$$SO_3^{2-}(aq) + 2H^+(aq) \rightarrow H_2O(l) + SO_2(g)$$

or by heating a mixture of concentrated sulphuric acid and copper turnings:

$$Cu(s) + 2H_2SO_4(l) \rightarrow CuSO_4(aq) + 2H_2O(l) + SO_2(g)$$

Industrially it is obtained as a by-product during the roasting of sulphide ore, notably in the manufacture of sulphuric acid (see page 191).

It is used for bleaching wool and paper pulp, and also as a fumigant.

Under compression (3 atmospheres at 20 °C) the gas is easily liquefied and is usually available for small-scale use in small steel cylinders. Liquid sulphur dioxide is a useful non-aqueous solvent for a number of compounds such as iodine, potassium iodide, sulphur, and sulphur dichloride oxide (thionyl chloride), $SOCl_2$. The ionization of the liquid is uncertain but has been postulated as:

$$2SO_2 \rightleftharpoons \underset{\text{Lewis acid}}{SO^{2+}} + \underset{\text{Lewis base}}{SO_3^{2-}}$$

$$(\text{cf. } H_2O(l) \rightleftharpoons H^+(aq) + OH^-(aq))$$

In the presence of charcoal, which acts as a catalyst, it reacts with chlorine to form sulphur dichloride dioxide (sulphuryl chloride), SO_2Cl_2.

$$SO_2(g) + Cl_2(g) \xrightarrow[\text{catalyst}]{\text{charcoal}} SO_2Cl_2(l)$$

Certain dioxides and peroxides are reduced by sulphur dioxide to the sulphate.

e.g.
$$PbO_2(s)+SO_2(g) \rightarrow PbSO_4(s)$$
$$Na_2O_2(s)+SO_2(g) \rightarrow Na_2SO_4(s)$$

Sulphur dioxide also functions as an *oxidizing agent*, supporting the combustion of elements such as magnesium, potassium, tin, and iron. At temperatures above 1100 °C it also supports the combustion of carbon.

e.g.
$$2Mg(s)+SO_2(g) \rightarrow 2MgO(s)+S(s)$$
$$C(s)+SO_2(g) \rightarrow CO_2(g)+S(s)$$

Potassium yields a mixture of the sulphite, K_2SO_3, and the thiosulphate, $K_2S_2O_3$; tin and iron both produce a mixture of the oxide and sulphide.

In the presence of moisture, sulphur dioxide oxidizes hydrogen sulphide to sulphur (refer below to the reactions of the sulphite ion).

Sulphurous acid, H_2SO_3

Most of the reactions of sulphur dioxide take place in aqueous solution, in which it is readily soluble forming the so-called sulphurous acid, although this acid has not been isolated as a pure substance and any attempt at evaporation results in expulsion of the gas from the solution.

The solution behaves as a weak dibasic acid:

$$H_2SO_3(aq) \underset{pK_{a_1} = 1.92}{\rightleftharpoons} H^+(aq)+HSO_3^-(aq) \underset{pK_{a_2} = 7.21}{\rightleftharpoons} 2H^+(aq)+SO_3^{2-}(aq)$$

It forms two series of salts, sulphites, SO_3^{2-}, and hydrogensulphites, HSO_3^-. Sulphites of the alkali metals can be isolated as solids but the hydrogensulphites, on crystallizing out, eliminate water to form the disulphates(IV).

$$2HSO_3^-(aq) \rightleftharpoons S_2O_5^{2-}(aq)+H_2O(l)$$

Aqueous solutions of alkali metal sulphites are alkaline as a result of hydrolysis.

$$Na_2SO_3(s)+aq \rightarrow 2Na^+(aq)+SO_3^{2-}(aq)$$
$$2H_2O(l) \rightleftharpoons 2OH^-(aq)+2H^+(aq)$$
$$\upharpoonleft\downharpoonright$$
$$H_2SO_3(aq)$$

On exposure to atmospheric oxygen, the sulphite ion slowly undergoes oxidation to the sulphate ion:

$$2SO_3^{2-}(aq)+O_2(g) \rightarrow 2SO_4^{2-}(aq)$$

Other reducing reactions of the sulphite ion include:

$$2MnO_4^-(aq)+5SO_3^{2-}(aq)+6H^+(aq) \rightarrow 2Mn^{2+}(aq)+5SO_4^{2-}(aq)+3H_2O(l); \quad E = +1.35\,V$$

$$Cr_2O_7^{2-}(aq)+3SO_3^{2-}(aq)+8H^+(aq) \rightarrow 2Cr^{3+}(aq)+3SO_4^{2-}(aq)+4H_2O(l); \quad E = +1.16\,V$$

$$2Fe^{3+}(aq)+SO_3^{2-}(aq)+H_2O(l) \rightarrow 2Fe^{2+}(aq)+SO_4^{2-}(aq)+2H^+(aq); \quad E = +0.60\,V$$

$$ClO^-(aq)+SO_3^{2-}(aq) \rightarrow Cl^-(aq)+SO_4^{2-}(aq); \quad E = +1.32\,V$$

By contrast, the sulphite ion can also function as an oxidizing agent. For example, it oxidizes hydrogen sulphide to sulphur,

$$2H_2S(g)+SO_3^{2-}(aq)+2H^+(aq) \rightarrow 3H_2O(l)+3S(s); \quad E = +0.31\,V$$

and in *strongly acidic solution* it oxidizes iron(II) ions to iron(III) ions:

$$4Fe^{2+}(aq)+SO_3^{2-}(aq)+6H^+(aq) \rightarrow 4Fe^{3+}(aq)+3H_2O(l)+S(s); \quad E = +0.32\,V$$

Sulphur(VI) oxide (sulphur trioxide), SO_3

Sulphur(VI) oxide is obtained by the catalytic oxidation of sulphur dioxide using a platinized asbestos or vanadium(V) oxide catalyst. It is also formed as a white vapour on heating certain metal sulphates.

e.g. $$Fe_2(SO_4)_3(s) \rightarrow Fe_2O_3(s)+3SO_3(g)$$

It is an exceptionally strong Lewis acid which fumes in moist air and reacts explosively with liquid water to form sulphuric acid, of which it is the anhydride.

$$H_2O(l)+SO_3(s) \rightarrow H_2SO_4(l); \quad \Delta H^\ominus = -87\,kJ\,mol^{-1}$$

Dissolution in water does not occur readily.

With concentrated sulphuric acid it reacts to form oleum (disulphuric(VI) acid or fuming sulphuric acid), $H_2S_2O_7$:

$$H_2SO_4(l)+SO_3(s) \rightarrow H_2S_2O_7(l)$$

With basic oxides, it combines exothermically to form sulphates:

$$BaO(s)+SO_3(s) \rightarrow BaSO_4(s)$$

It also functions as an oxidizing agent and will, for example, liberate bromine from hydrogen bromide.

$$2HBr(g)+SO_3(s) \rightarrow H_2O(l)+Br_2(aq)+SO_2(g)$$

Sulphuric acid, H_2SO_4

Sulphuric acid is commercially the most important acid of sulphur, and about 85 per cent of the United Kingdom production is made by the Contact Process.

Sulphur (the principal raw material) or some other sulphur-containing ore, such as anhydrite or pyrites, is roasted in air to produce sulphur dioxide. This is passed over a catalyst at 450–550 °C. The use of platinum or vanadium(V) oxide as a catalyst has been superseded and the catalyst is now 'promoted' vanadium, which can be thought of as a granule of silica covered with a thin film of liquid alkyl metal pyrosulphate in which some ionic species is present incorporating vanadium. The oxidation state of vanadium is +4 or +5 depending upon the gas composition and the temperature, and it is suggested that the ion carrying the vanadium atoms is a complex ion involving pyrosulphate anions as liquids.

$$2SO_2(g)+O_2(g) \underset{}{\overset{\text{'promoted' vanadium catalyst, } 450\text{–}550\,°C}{\rightleftharpoons}} 2SO_3(g); \quad \Delta H^\ominus = -190\,\text{kJ mol}^{-1}$$

As the reaction is reversible and exothermic, the efficiency of the conversion of the dioxide to sulphur(VI) oxide decreases with rising temperature (cf. Haber process, page 164). Because air is used rather than oxygen, the diminution in volume is considerably less than that implied by the stoichiometric equation, and compression is generally considered uneconomic, although a *slightly* elevated pressure is employed to shift the gases through the system.

After cooling, the resultant sulphur(VI) oxide is absorbed in 98 per cent sulphuric acid to form oleum (disulphuric(VI) acid) which is then diluted with water to give 98 per cent acid.

$$H_2SO_4(l)+SO_3(g) \rightarrow H_2S_2O_7(l) \xrightarrow{+H_2O} 2H_2SO_4(l)$$

Direct absorption of sulphur(VI) oxide in water forms a mist of corrosive vapours and is a generally inefficient reaction.

Because of strong intermolecular hydrogen bonding, sulphuric acid is a viscous liquid with a high boiling point (270 °C).

Concentrated sulphuric acid has a great affinity for water and will actually remove the elements of water from certain organic compounds.

e.g.

$$\underset{\text{sucrose}}{C_{12}H_{22}O_{11}(s)} \xrightarrow{\text{conc. } H_2SO_4} 12C(s)+11H_2O(l)$$

$$\underset{\text{ethanedioic acid}}{(COOH)_2(s)} \xrightarrow[\text{heat}]{\text{conc. } H_2SO_4,} CO(g)+CO_2(g)+H_2O(l)$$

$$\underset{\text{ethanol}}{CH_3CH_2OH(l)} \xrightarrow[\text{heat } 170\,°C]{\text{conc. } H_2SO_4,} CH_2{=}CH_2(g)+H_2O(l)$$

The concentrated acid also behaves as an oxidizing agent, particularly when hot, oxidizing a number of non-metals to their oxides or oxoacids, e.g. carbon to carbon dioxide, sulphur to sulphur dioxide, and phosphorus to phosphoric(V) acid. It also oxidizes certain metals, such as Al, Zn, Cu, Ag, and Hg, to their sulphates, liberating sulphur dioxide in the process.

The *dilute acid*, however, is a typical strong dibasic acid, giving rise to two series of salts, normal sulphates and hydrogensulphates. With the exception of some of the alkaline earth metal sulphates and lead(II) sulphate, most of the normal sulphate compounds are readily soluble in water. Although the Group I and II metal sulphates are stable to heat, most other metal sulphates decompose.

e.g. $$CuSO_4(s) \rightarrow CuO(s) + SO_3(g)$$

Of the other sulphur-containing anions, the *thiosulphate anion*, $S_2O_3^{2-}$, is probably the best known and is usually encountered in the form of sodium thiosulphate, $Na_2S_2O_3 \cdot 5H_2O$. This compound is very soluble in water and is used in volumetric analysis in the quantitative estimation of iodine.

$$2S_2O_3^{2-}(aq) + I_2(aq) \rightarrow 2I^-(aq) + \underset{\text{tetrathionate ion}}{S_4O_6^{2-}(aq)}$$

It is familiar to photographers, who employ it as a fixing agent ('hypo' crystals) to dissolve unaltered silver halides by forming the complex ion, $[Ag(S_2O_3)_2]^{3-}$. Another useful reaction involving the thiosulphate anion is that with dilute acid, a reaction which is widely used in an elementary study of reaction rates:

$$S_2O_3^{2-}(aq) + 2H^+(aq) \rightarrow SO_2(g) + H_2O(l) + S(s)$$

Halides

There are numerous binary halides of the Group VI elements (Table 13.12).

Table 13.12

	F	*Cl*	*Br*	*I*
O	F_2O_2, F_2O	Cl_2O, ClO_2, Cl_2O_6, Cl_2O_7	Br_2O, BrO_2, BrO_3	I_2O_4, I_4O_9, I_2O_5
S	S_2F_2, SF_4, SF_6, S_2F_{10}	S_2Cl_2, SCl_2, SCl_4	S_2Br_2	
Se	Se_2F_2, SeF_4, SeF_6	Se_2Cl_2, $SeCl_4$	Se_2Br_2, $SeBr_4$	
Te	TeF_4, Te_2F_{10} TeF_6	$TeCl_2$, $TeCl_4$	$TeBr_2$, $TeBr_4$	TeI_4
Po		$PoCl_2$, $PoCl_4$	$PoBr_2$, $PbBr_4$	PoI_4

The halides A_2X_2, AX_2, and AX_4 can all be obtained by direct synthesis of the elements. They have molecular covalent structures and are chemically reactive, being hydrolysed by water. S_2Cl_2 has a dihedral structure similar to that of H_2O_2.

Spectral evidence indicates that the ACl_4 molecules have a distorted trigonal bipyramidal structure (Figure 13.6) with one equatorial position occupied by a lone pair.

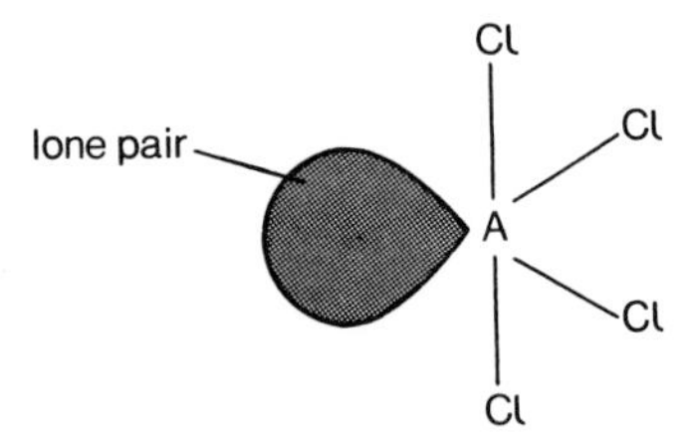

Fig. 13.6

The ionic character of these compounds gradually increases from SX_4 to PoX_4. Molten $TeCl_4$ actually conducts electricity, suggesting a certain degree of self-ionization.

$$2TeCl_4(l) \rightleftharpoons TeCl_3^+(l) + TeCl_5^-(l)$$

SCl_4 is a colourless, unstable liquid; $SeCl_4$ is a colourless solid, subliming at 196 °C; and $TeCl_4$ is a white hygroscopic solid (m.p. 224 °C).

Fluorine tends to bring out the maximum oxidation states (see page 198) and SF_6, SeF_6, and TeF_6 can all be prepared by direct synthesis. They are all colourless gases and fairly inert in comparison with the tetrahalides. Reactivity increases as the electronegativity of the central atoms decreases. TeF_6 is slowly hydrolysed by water to telluric acid, H_6TeO_6, at room temperature.

$$TeF_6(g) + 6H_2O(l) \rightarrow Te(OH)_6(aq) + 6HF(aq)$$

The AF_6 molecules have a regular octahedral structure as illustrated for SF_6 (see page 32).

14 Group VII: The Halogens

Table 14.1

	Fluorine	*Chlorine*	*Bromine*	*Iodine*	*Astatine*
Symbol	F	Cl	Br	I	At
Outer electronic structure	$2s^22p^5$	$3s^23p^5$	$4s^24p^5$	$5s^25p^5$	$6s^26p^5$
Principal oxidation state	−1	−1	−1	−1	

General

The halogens are all too reactive to be found naturally in the elemental state. The melting points and boiling points of the elements increase on descending the group (Table 14.2), a factor which is apparent from their normal physical states; fluorine and chlorine are gases, bromine is a liquid, and iodine is a solid.

Table 14.2

	F	*Cl*	*Br*	*I*
Melting point/°C	−223	−102	−7.3	114
Boiling point/°C	−188	−34.6	59	184

They all form coloured, diatomic molecules, the colours arising from electronic transitions brought about by the absorption of different frequencies from the visible spectrum (Table 14.3). For example, fluorine absorbs violet light (high energy – the electrons in this small atom being firmly held) and appears yellow, whereas bromine absorbs blue–green light (lower energy – the electrons being less firmly held than in fluorine) and appears reddish–brown.

The elements fluorine and iodine each possess only one stable nuclide, ^{19}F

Table 14.3

	F	*Cl*	*Br*	*I*
Colour	Pale yellow	Yellow-green	Reddish-brown	Dark purple, almost black (violet vapour)

and ^{127}I, whereas chlorine and bromine possess two. Naturally occurring chlorine is composed of 75.4 per cent ^{35}Cl and 24.6 per cent ^{37}Cl. Bromine comprises 50.6 per cent ^{79}Br and 49.4 per cent ^{81}Br.

The chemistry of the heaviest element in this group, astatine, At, is of little significance at this level. It is a volatile solid and has about twenty short-lived isotopes. e.g. ^{210}At has a half-life of 8.3 hours and ^{211}At has a half-life of 7.5 hours.

Electronic structure and oxidation states

The halogens have an outer electronic configuration of ns^2np^5. All readily acquire an additional outer electron and most commonly exhibit an oxidation state of -1. Fluorine, being the most electronegative element in the Periodic Table, is limited to this single oxidation state, whereas the others exhibit a range of positive oxidation states which extends to $+7$ for iodine in the interhalogen compound, IF_7, and for chlorine in the oxide, Cl_2O_7. All of the higher states involve the unpairing of *p* electrons and promotion to the *d* level. Despite the existence of positive oxidation states, the halogens have little tendency to form positive ions, as indicated by the high values of the first ionization energies (Table 14.4).

Table 14.4

	F	*Cl*	*Br*	*I*
First ionization energy, $\Delta H^{\ominus}$/kJ mol^{-1}	+1680	+1260	+1140	+1010

Size of atoms and ions

Because of the strong attractive force between the nuclear protons and the other electrons (see page 78), the halogen atoms are the smallest in their respective periods. The acquisition of an additional electron on forming a negative ion, however, brings about a tremendous transformation in size, causing the X^- ion to be almost twice the size of the parent atom (Table 14.5).

Table 14.5

	F	*Cl*	*Br*	*I*
Atomic (covalent) radius/nm	0.072	0.099	0.114	0.133
Ionic radius, X^-/nm	0.136	0.181	0.195	0.216

Trihalide ions

Although I_3^-, Br_3^-, and Cl_3^- ions are all known, the only one of any importance at this level is the I_3^- ion, the Br_3^- and Cl_3^- ions being very much more unstable and rare.

The solubility of iodine in water is found to be greatly enhanced by the presence of iodide ions which promote the formation of the soluble triiodide anion.

$$I_2(aq) + I^-(aq) \rightleftharpoons I_3^-(aq); \quad K = 725 \text{ at } 25\,°C$$

This complex anion has a linear structure and forms stable, ionic salts (e.g. potassium triiodide, KI_3) with large cations. The above equilibrium finds use in iodine titrations where, because of the low solubility of iodine in water, it is dissolved in aqueous potassium iodide. On titration with a reducing agent, such as sodium thiosulphate, the free iodine reacts with the reducing agent, displacing the equilibrium towards the left until finally no triiodide remains. The iodine is liberated so readily that the solution behaves as if the dissolved iodine were all free iodine.

The ability of molten iodine to conduct electricity is generally accounted for in terms of its tendency to undergo self-ionization into I_3^+ and I_3^- ions.

Mixed trihalides (e.g. $KICl_2$) and more complex polyiodides, such as I_5^-, I_7^-, and I_9^-, are also known.

Electronegativity

The halogens provide the most electronegative element in their respective periods, the electronegativity values decreasing with increasing size of the atom (Table 14.6).

Table 14.6

	F	*Cl*	*Br*	*I*
Electronegativity	4.0	3.0	2.8	2.5

Bond enthalpies of the elemental molecules

As the atomic number of the halogen increases, there is a reduction in the degree of overlapping of the bonding orbitals, giving weaker and less stable bonds. The regular decrease in the bond enthalpy from chlorine to iodine (Table 14.7) serves to substantiate this. The bond enthalpy of fluorine, however, is abnormally low for the position that it occupies in the Periodic Table, a property which is generally attributed to the repulsive forces between the non-bonding electrons in the molecules, and which assumes importance for a number of molecules consisting of small atoms. It is, in fact, these factors, coupled with the high electronegativity of fluorine, which account for the *great reactivity of the fluorine molecule.*

The small covalent radius of fluorine, and its low bond enthalpy, enables it to extend the covalency of other elements to a maximum, e.g. SF_6, IF_7, and BrF_5.

Reactions

With metals and non-metals

Because of the considerable reactivity of the halogens, they all combine directly with most metals and also with many non-metals.

e.g. $$2P + 3X_2 \rightarrow 2PX_3$$

Fluorine is the most reactive element of the group, the reactivity of the others gradually decreasing on descending the group; but, with heating, even iodine will combine directly with a large number of elements. Because of their considerable oxidizing powers (see page 199) both fluorine and chlorine tend to bring out the higher oxidation states of other elements,

e.g. $$2Fe(s) + 3Cl_2(g) \rightarrow 2FeCl_3(s)$$
$$2P(s) + 5F_2(g) \rightarrow 2PF_5(g)$$
$$2P(s) + 5Cl_2(g) \rightarrow 2PCl_5(s)$$

whereas this is not necessarily the case with bromine and iodine.

e.g. $$2P(s) + 3Br_2(l) \rightarrow 2PBr_3(l)$$

With alkalis

The reactions of the halogens with alkalis depend upon the conditions. Chlorine reacts with excess aqueous alkali in the *cold* to form reasonably pure solutions of chloride and chlorate(I) (hypochlorite) ions, ClO^-

$$2OH^-(aq) + Cl_2(g) \rightarrow Cl^-(aq) + ClO^-(aq) + H_2O(l)$$

In hot solution (approximately 75 °C), chlorate(V) ions, ClO_3^-, and further chloride ions are formed,

$$6OH^-(aq) + 3Cl_2(g) \rightarrow 5Cl^-(aq) + ClO_3^-(aq) + 3H_2O(l)$$

as a result of disproportionation of the chlorate(I) ions.

$$3ClO^-(aq) \xrightarrow[\text{approx. 75 °C}]{\text{disproportionate}} 2Cl^-(aq) + ClO_3^-(aq)$$

(The above processes are somewhat over-simplified and are not precisely represented by the stoichiometric equations given.) Bromine and iodine react with alkalis in a similar manner.

Disproportionation of the bromate(I) (hypobromite) ion, BrO^-, occurs quite rapidly, even at room temperature, and this ion can therefore only be retained if prepared and kept at 0 °C. The iodate(I) (hypoiodite) ion, IO^-, disproportionates rapidly at all temperatures and is unknown in solution.

Oxidizing ability

By virtue of the ease with which they gain an electron to form singly charged anions, the halogens are powerful oxidizing agents. Their relative strength as oxidizing agents diminishes on descending the group and is dependent upon the algebraic sum of several energy terms. These may be represented by means of a Born-Haber cycle:

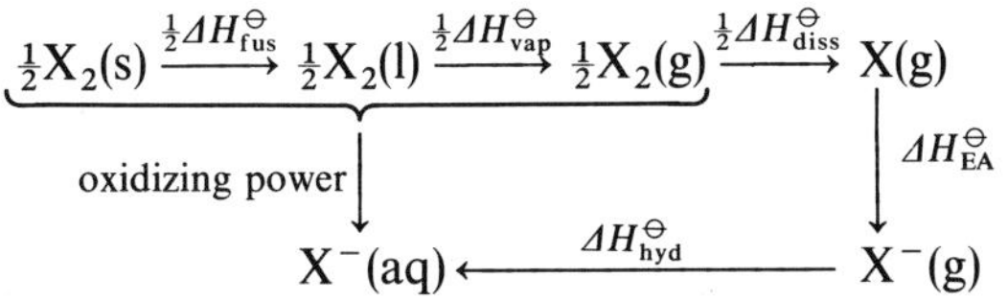

$$\underbrace{\tfrac{1}{2}X_2(s) \xrightarrow{\frac{1}{2}\Delta H^{\ominus}_{fus}} \tfrac{1}{2}X_2(l) \xrightarrow{\frac{1}{2}\Delta H^{\ominus}_{vap}} \tfrac{1}{2}X_2(g)} \xrightarrow{\frac{1}{2}\Delta H^{\ominus}_{diss}} X(g)$$

$$\text{oxidizing power}\downarrow \qquad\qquad \downarrow \Delta H^{\ominus}_{EA}$$

$$X^-(aq) \xleftarrow{\Delta H^{\ominus}_{hyd}} X^-(g)$$

Fig. 14.1

For the gaseous halogens, fluorine and chlorine, the first two terms are omitted and for bromine, which is a liquid, the enthalpy of fusion is not relevant.

Table 14.7 Relative oxidizing powers of the halogens as indicated by the Born-Haber cycle

Enthalpy term, $\Delta H^{\ominus}$/kJ mol^{-1}	*F*	*Cl*	*Br*	*I*
½ Enthalpy of fusion, $\frac{1}{2}\Delta H^{\ominus}_{fus}$	—	—	—	$\frac{+15.74}{2}$
½ Enthalpy of vaporization, $\frac{1}{2}\Delta H^{\ominus}_{vap}$	—	—	$\frac{+30}{2}$	$\frac{+44}{2}$
½ Enthalpy of dissociation, $\frac{1}{2}\Delta H^{\ominus}_{diss}$	$\frac{+158}{2}$	$\frac{+242}{2}$	$\frac{+193}{2}$	$\frac{+157}{2}$
Electron affinity, $\Delta H^{\ominus}_{EA}$	−354	−370	−348	−320
Enthalpy of hydration, $\Delta H^{\ominus}_{hyd}$	−506	−364	−335	−293
Sum	−781	−613	−571.5	−504.73

The relative oxidizing powers indicated by the Born-Haber cycle (Table 14.7) are supported by the relative redox potentials (Table 14.8) of the elements. These values are dependent on the enthalpy of dissociation, the electron affinity, and the hydration energy of the negative ion.

Table 14.8

	F	*Cl*	*Br*	*I*
Redox potential, $E^{\ominus}$/V	+2.87	+1.36	+1.07	+0.54

Comparison of the electron affinities of the halogens (a process which relates solely to the molar energy change, $X(g)+e \rightarrow X^-(g)$), shows a maximum at chlorine. Nevertheless, fluorine is the better oxidizing agent in aqueous solution owing to its low bond enthalpy and high enthalpy of hydration. Even when dry, fluorine will usually replace chlorine from its compounds. Despite the fact that both bromine and iodine have lower bond enthalpies than chlorine, they are weaker oxidizing agents because of the lower exothermic values of their electron affinities and hydration energies.

The relative oxidizing powers of the halogens are well illustrated by their reactions with water, in which many of the hydrated molecular species present in solution disproportionate rapidly to give ions.

$$2F_2(g)+2H_2O(l) \rightarrow 4H^+(aq)+4F^-(aq)+O_2(g); \quad E = +1.64\,V$$
$$2I_2(s)+2H_2O(l) \rightarrow 4H^+(aq)+4I^-(aq)+O_2(g); \quad E = -0.69\,V$$

Fluorine is so powerful in its behaviour as an oxidizing agent that it readily liberates oxygen (together with about 14 per cent trioxygen) from water with a considerable evolution of heat. Iodine, on the other hand, reacts with water endothermically, implying that iodide ions are in fact oxidized by water to form iodine. Chlorine and bromine do not liberate oxygen in water, but instead form a solution of the hydrogen halide and halic(I) (hypohalous) acids.

$$H_2O(l)+Cl_2(g) \rightarrow HCl(aq)+HOCl(aq)$$

The halic(I) acids decompose easily to give oxygen.

Reactions involving the halogens as oxidizing agents are commonplace. Chlorine and iodine are the two halogens which are generally encountered in laboratory procedure. For example, chlorine will oxidize bromide and iodide ions to bromine and iodine respectively. These are often referred to as displacement reactions.

$$2Br^-(aq)+Cl_2(g) \rightarrow 2Cl^-(aq)+Br_2(aq); \quad E = +0.29\,V$$
$$2I^-(aq)+Cl_2(g) \rightarrow 2Cl^-(aq)+I_2(aq); \quad E = +0.82\,V$$

Chlorine will also oxidize Fe(+2) in solution to Fe(+3),

$$2Fe^{2+}(aq)+Cl_2(g) \rightarrow 2Fe^{3+}(aq)+2Cl^-(aq); \quad E = +0.59\,V$$

hydrogen sulphide to sulphur,

$$H_2S(g)+Cl_2(g) \rightarrow 2HCl(g)+S(s)$$

and ammonia (excess) to nitrogen (see page 165).

Iodine can be used to oxidize sulphurous acid to sulphuric acid,

$$SO_3^{2-}(aq)+I_2(aq)+H_2O(l) \rightarrow 2H^+(aq)+2I^-(aq)+SO_4^{2-}(aq); \quad E = +0.37\,V$$

Sn(+2) in solution to Sn(+4),

$$Sn^{2+}(aq)+I_2(aq) \rightarrow 2I^-(aq)+Sn^{4+}(aq); \quad E = +0.39\,V$$

but is most frequently encountered in volumetric analysis where it is used quantitatively to oxidize thiosulphate ions to the tetrathionate ion, $S_4O_6^{2-}$ (see page 193).

Iodine with starch

Iodine combines with starch, $(C_6H_{10}O_5)_n$, to give an intense blue colour due to the formation of a complex clathrate compound† in which the iodine molecules are trapped within the helical structure of the starch molecule. On warming, the blue colour is dispelled. This behaviour of iodine with starch is a useful and sensitive test for the element.

Hydrogen halides

HCl, HBr, and HI are all colourless gases and are essentially covalent in character. Because of the high electronegativity of fluorine, HF is associated by means of intermolecular hydrogen bonds, and exists as a liquid (b.p. 19 °C) at just below normal room temperature.

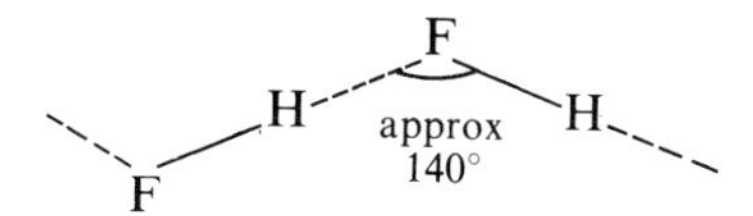

The H—F bond is highly polar, possessing about 55 per cent ionic character.

One general method of preparing hydrogen halides is by direct combination of the constituent elements:

$$H_2 + X_2 \rightarrow 2HX$$

The reactions occur in the gas phase and proceed via a free radical chain reaction, initiated by the homolytic fission of the halogen molecule. The initiation requires a photochemical (photolytic) catalyst. In direct sunlight, the reaction may be explosive, becoming slow in diffuse light, and proceeding to a negligible extent in darkness.

$$X_2 \xrightarrow{h\nu} 2X\cdot \quad \text{Chain initiation}$$

$$\left.\begin{array}{l} H_2 + X\cdot \rightarrow HX + H\cdot \\ H\cdot + X_2 \rightarrow HX + X\cdot \end{array}\right\} \text{Chain propagation}$$

$$\left.\begin{array}{l} H\cdot + X\cdot + S \rightarrow HX + S^* \\ X\cdot + X\cdot + S \rightarrow X_2 + S^* \end{array}\right\} \text{Chain termination}$$

In the chain termination process, the combination of two free radicals occurs with a great evolution of energy. If the radicals are to remain intact after

† A *clathrate compound* is one in which molecules become trapped or caged within the cavities of a more complex crystal lattice or molecular structure.

collision, this energy has to be dissipated to a 'third body', *S*, which can be any of the other molecules present in the system, e.g. HX, X_2, etc., or the walls of the vessel.

Whereas the reaction of fluorine with hydrogen is even more violent than that of chlorine, bromine and iodine react somewhat less readily. Bromine does not combine with hydrogen even in bright sunlight, and a platinum catalyst and a temperature of about 200 °C is required. The reaction between iodine and hydrogen is reversible and does not go to completion, even in the presence of a platinum catalyst and at a temperature of 400 °C.

The observed decline of the affinities of the halogens for hydrogen on descending the group from fluorine to iodine is borne out by the enthalpies of formation of the hydrogen halides (Table 14.9).

Table 14.9

	HF	*HCl*	*HBr*	*HI*
Enthalpy of formation, $\Delta H^{\ominus}/\mathrm{kJ\ mol^{-1}}$	−269	−92.3	−36.2	+25.9

HF and HCl can be prepared in the laboratory by the action of concentrated sulphuric acid on an appropriate salt.

$$CaF_2(s) + H_2SO_4(l) \rightarrow CaSO_4(s) + 2HF(g)$$
$$NaCl(s) + H_2SO_4(l) \rightarrow NaHSO_4(s) + HCl(g)$$

Oxidation of the hydrogen halide to the halogen prohibits the use of this type of reaction for preparing HBr and HI, although HBr is sometimes generated as a reagent during the course of a reaction from sodium bromide and sulphuric acid. It is therefore preferable to prepare these gases by the hydrolysis of a covalent halide. For hydrogen bromide, this is conveniently performed by the addition of bromine to red phosphorus and water.

$$2P(s) + 3Br_2(l) \rightarrow 2PBr_3(l)$$
$$PBr_3(l) + 3H_2O(l) \rightarrow H_3PO_3(aq) + 3HBr(aq)$$

Hydrogen iodide is obtained in a similar way, this time by the action of water on a mixture of iodine and red phosphorus.

The stability of the hydrogen halides, as illustrated by the relative strengths and bond enthalpies of the H—X bond (Table 14.10), decreases with decreasing polarity of the bond, i.e. stability decreases in the order: HF > HCl > HBr > HI.

Table 14.10

	H—F	*H—Cl*	*H—Br*	*H—I*
Bond enthalpy, $\Delta H^{\ominus}/\mathrm{kJ\ mol^{-1}}$	+562	+431	+366	+299
Bond length/nm	0.092	0.128	0.141	0.160

Acid strength

Aqueous solutions of the hydrogen halides behave as strong monobasic acids,

$$HX(aq) \rightarrow H^+(aq) + X^-(aq)$$

although, in dilute solution, HF is only partly dissociated into ions and displays only weak acidic properties.

$$\textit{In dilute solution:}\ HF(aq) \rightleftharpoons H^+(aq) + F^-(aq);\ K_a = 7.2 \times 10^{-4}$$

This contrasts greatly with the strong acidic properties of the solutions of the other hydrogen halides. However, on concentration, a solution of HF adopts the properties of a stronger acid due to the removal of F^- ions from the equilibrium by HF to form the stable complex ion, $[HF_2]^-$, (together with some $[H_2F_3]^-$, $[H_3F_4]^-$, etc.) thus promoting the dissociation of the acid.

$$\textit{In concentrated solution:}\quad F^-(aq) + HF(aq) \rightleftharpoons [HF_2]^-(aq);\ K_a = 5.5$$

It is, in fact, the formation of these complex species which accounts for the strong acidic properties of pure anhydrous liquid HF.

$$2HF \rightleftharpoons H_2F^+ + F^- \xrightarrow{HF} [HF_2]^-$$

(the final step, $F^- + HF \rightleftharpoons [HF_2]^-$, is shown as a vertical equilibrium below F^-)

The considerable acidic strength of *anhydrous* HF is illustrated by the fact that a strong acid like nitric acid behaves as an electron donor (i.e. a base) on dissolving in it.

$$HX(g) + aq \rightarrow H^+(aq) + X^-(aq)$$

The relative strengths of the acids, based upon the overall process,

$$HX(g) + H_2(aq) \rightarrow HO^+(aq) + X^-(aq)$$

can be approximated from a Born-Haber cycle by inserting the approximate enthalpy terms.

$$
\begin{array}{ccccc}
HX(g) & \xrightarrow{\Delta H^{\ominus}_{diss}} & H(g) & + & X(g) \\
\downarrow & & \downarrow \Delta H^{\ominus}_{IE} & & \downarrow \Delta H^{\ominus}_{EA} \\
H^+(aq) & \xleftarrow{\Delta H^{\ominus}_{hyd}} & H^+(g) & & \\
X^-(aq) & \xleftarrow{\Delta H^{\ominus}_{hyd}} & & & X^-(g)
\end{array}
$$

Acid strength decreases in the order: HI > HBr > HCl > HF

The weak acidic properties exhibited by HF in dilute aqueous solution are due mainly to the high bond enthalpy of HF, the irregular electron affinity of fluorine, and intermolecular hydrogen bonding by HF in solution.

Aqueous solutions of the hydrogen halides display a negative deviation from Raoult's law and all form azeotropic mixtures with maximum boiling points (Table 14.11).

Table 14.11

	HF	*HCl*	*HBr*	*HI*
Percentage HX	37	20.24	47.5	57
B.p. of mixture/°C	120	110	126	127

An aqueous solution of HI also behaves as a strong *reducing agent*, reducing concentrated sulphuric acid (S(+6)) to sulphur dioxide (S(+4)) and hydrogen sulphide (S(−2)).

$$2HI(aq)+H_2SO_4(l) \rightarrow I_2(aq)+2H_2O(l)+SO_2(g)$$
$$8HI(aq)+H_2SO_4(l) \rightarrow 4I_2(aq)+4H_2O(l)+H_2S(g)$$

Aqueous HBr is made much weaker in this respect, reducing sulphuric acid only as far as sulphur dioxide.

$$2HBr(aq)+H_2SO_4(l) \rightarrow Br_2(aq)+2H_2O(l)+SO_2(g)$$

Hydrochloric acid, on the other hand, has no tendency to reduce sulphuric acid.

The hydrogen halides have many common reactions, both in the gaseous form and when acting as strong acids (with the exception of HF) in aqueous solution. Most of these are well-known and are not conveniently tabulated in a comparative treatment.

Soluble chlorides, bromides, and iodides can be detected by their ability to form the insoluble silver(I) halide, in a solution made acidic by the addition of dilute nitric acid, with aqueous silver(I) nitrate solution. (Silver(I) fluoride is soluble.)

$$Ag^+(aq)+X^-(aq) \rightarrow AgX(s)$$

The silver(I) chloride precipitate is white, the bromide is cream, and the iodide is pale yellow.

Confirmation is provided by examining the solubility of the silver(I) halide precipitate in ammonia solution. The chloride is soluble (see page 229), the bromide is only sparingly soluble, and the iodide is virtually insoluble. Alternatively, tetrachloromethane (carbon tetrachloride), followed by a few drops of chlorine water, may be added to an acidified sample of the original halide solution. The tetrachloromethane, which is immiscible with the aqueous layer, preferentially dissolves the free halogen and the colour it adopts indicates the halide present. If the tetrachloromethane remains colourless, chloride is indicated and a brown or violet colour indicates bromide or iodide respectively.

Oxides of the halogens

Several halogen oxides are known (Table 14.12), although a number of these are highly unstable and even explosive, with chlorine forming the greatest number of compounds. Owing to the comparatively small difference in electronegativity between the halogens and oxygen, the oxygen compounds of the halogens are all essentially covalent in character. As fluorine is the most electronegative element, its compounds with oxygen are, strictly speaking, fluorides of oxygen rather than oxides of fluorine.

Table 14.12

Fluorides of oxygen	*Oxides of chlorine*	*Oxides of bromine*	*Oxides of iodine*
O_2F_2			
OF_2	Cl_2O	Br_2O	
	ClO_2	BrO_2	I_2O_4
			I_4O_9
			I_2O_5
	Cl_2O_6	BrO_3	
	Cl_2O_7		I_2O_7

With the exception of iodine(V) oxide, all of these compounds are thermally unstable. I_2O_5 decomposes only above 300 °C.

The monoxides: OF_2, Cl_2O, and Br_2O

OF_2 is a colourless gas, dissolving in water to give a neutral solution. It is prepared by bubbling fluorine through very dilute (approximately 2 per cent) sodium hydroxide solution.

$$F_2(g) + 2NaOH(aq) \rightarrow 2NaF(aq) + H_2O(l) + OF_2(g)$$

Cl_2O (yellow gas) and Br_2O (brown liquid) are both acidic oxides, dissolving in water to form the halic(I) (hypohalous) acid,

$$X_2O + H_2O(l) \rightarrow 2HOX(aq)$$

and in aqueous alkali to form halates(I) (hypohalites):

$$X_2O + 2OH^-(aq) \rightarrow 2XO^-(aq) + H_2O(l)$$

Both Cl_2O and Br_2O can be prepared by passing the halogen over freshly precipitated mercury(II) oxide at 300 °C.

$$2X_2 + 2HgO(s) \xrightarrow{300\,°C} HgO{\cdot}HgX_2(s) + X_2O$$

The dioxides: ClO_2, BrO_2, and I_2O_4

The most important of these is ClO_2, which is widely used in the purification of water and as a bleach. It exists as an orange gas above 11 °C and is paramagnetic, containing an odd number of electrons.

BrO_2 is a yellow solid which is only stable below −40 °C.

Both ClO_2 and BrO_2 are powerful oxidizing agents and explosive, especially in the presence of reducing agents. They are mixed acid anhydrides, dissolving in water and aqueous alkalis to form halate(III) and halate(V) ions.

$$XO_2 + 2OH^-(aq) \rightarrow \underset{\text{halate(III) ion}}{XO_2^-(aq)} + \underset{\text{halate(V) ion}}{XO_3^-(aq)} + H_2O(l)$$

The bromate(III) (bromite) ion, however, rapidly disproportionates into bromide and bromate(V) ions.

$$3BrO_2^-(aq) \rightarrow Br^-(aq) + 2BrO_3^-(aq)$$

I_2O_4 is a yellow solid, probably composed of $(IO)_n$ and $(IO_3)_n$ chains.

Iodine(V) oxide, I_2O_5

I_2O_5 is a white powder and is the most stable oxide of the halogens, decomposing on heating only above 300 °C.

$$2I_2O_5(s) \xrightarrow{300\,^\circ C} 2I_2(s) + 5O_2(g)$$

It is the anhydride of iodic(V) acid and can be prepared by the dehydration of the acid at 170 °C,

$$2HIO_3(s) \xrightarrow{170\,^\circ C} I_2O_5(s) + H_2O(l)$$

or by oxidizing iodine with fuming nitric acid at 70 °C.

Being a powerful oxidizing agent, it is used analytically for detecting carbon monoxide, the latter being oxidized quantitatively to CO_2 with the liberation of iodine:

$$I_2O_5(s) + 5CO(g) \rightarrow 5CO_2(g) + I_2(s)$$

The iodine can be estimated by titration with sodium thiosulphate. Unlike I_2O_4 and I_4O_9, I_2O_5 is essentially covalent.

The trioxides: Cl_2O_6 and BrO_3

Dichlorine hexaoxide, Cl_2O_6, is a red liquid (f.p. 3.5 °C) and is prepared by mixing ClO_2 with ozonized oxygen at 0 °C.

$$2ClO_2(g) + 2[O] \rightarrow Cl_2O_6(l)$$

It is a powerful oxidizing agent and acidic oxide, reacting with aqueous alkali to form chlorate(V) and chlorate(VII) (perchlorate) ions, ClO_4^-,

$$Cl_2O_6(l)+2OH^-(aq) \rightarrow ClO_3^-(aq)+ClO_4^-(aq)+2H_2O(l)$$

The dimer contains no unpaired electrons and is diamagnetic. In aqueous solution, however, it displays weak paramagnetic properties, suggesting some dissociation into ClO_3.

BrO_3 is a white solid, only stable below $-70\,°C$. It is prepared by the action of a silent electric discharge on a mixture of the component elements. It is an acidic oxide and dissolves in aqueous alkali to form bromate(V), bromate(III) (hypobromite), and bromide ions.

Bromates(VII) (perbromates) have recently been prepared and are powerful oxidizing agents.

Chlorine heptaoxide, Cl_2O_7

This is a colourless, oily liquid and is prepared by dehydrating chloric(VII) acid, of which it is the anhydride, with phosphorus(V) oxide at $-10\,°C$.

$$4HClO_4(l)+P_4O_{10}(s) \rightarrow 4HPO_3(l)+2Cl_2O_7(l)$$

It is a weaker oxidizing agent than the other oxides of chlorine and explodes on heating or striking.

Oxoacids of the halogens

These are formed by all the halogens with the exception of fluorine.

Halic(I) (hypohalous) acids, HOX

The halic(I) acids are all weak and are known only in aqueous solution.

Table 14.13

	HOCl	*HOBr*	*HOI*
K_a	3×10^{-8}	2×10^{-9}	2.5×10^{-11}
pK_a	7.52	8.70	10.60

All are unstable and unobtainable in the pure state. Their stability decreases in the order: HOCl > HOBr > HOI. As bromine and iodine are both more electropositive than the hydroxyl group, bromine(I) (hypobromous) acid and iodic(I) (hypoiodous) acid may be represented as BrOH and IOH, the latter being sufficiently amphoteric as to be regarded as the hydroxide of iodine. The most common of these acids is chloric(I) (hypochlorous) acid, HOCl. It is very pale yellow in colour and is formed quite rapidly on bubbling chlorine into water:

$$Cl_2(g)+H_2O(l) \rightleftharpoons HCl(aq)+HOCl(aq)$$

A better yield is obtained, however, by the addition of mercury(II) oxide, which removes the hydrochloric acid as it is formed, thus enhancing the formation of the chloric(I) acid:

$$HgO(s)+2H^+(aq)+2Cl^-(aq) \rightarrow H_2O(l)+HgCl_2(s)$$

Chloric(I) acid rapidly decomposes, especially when exposed to sunlight. Certain surfaces, such as that of a glass vessel, also speed up the decomposition.

$$2ClO^-(aq) \rightarrow 2Cl^-(aq)+O_2(g)$$

Iodic(I) acid is very much less stable and rapidly disproportionates to the iodate(V) and iodide ions.

$$3IO^-(aq) \rightarrow IO_3^-(aq)+2I^-(aq)$$

The halic(I) acids are all good oxidizing agents, especially in acidic solution. The only salts of these acids to be isolated in the solid state are the chlorate(I) compounds (hypochlorites), the most important being $NaOCl \cdot 7H_2O$, which, in solution, is used commercially as a bleach. In quantity, it is prepared by the electrolysis of brine. The electrolyte is kept agitated to ensure that the chlorine is formed at the anode,

$$2Cl^-(aq) \xrightarrow{\text{electrolysis}} Cl_2(g)+2e$$

reacts with the OH^- ions in solutions.

$$Cl_2(g)+2OH^-(aq) \rightarrow ClO^-(aq)+Cl^-(aq)+H_2O(l)$$

The concentration of hydroxide ions increases with the liberation of hydrogen at the cathode.

$$2H^+(aq)+2e \xrightarrow{\text{electrolysis}} H_2(g)$$

Bromate(I) compounds, such as $NaOBr \cdot 5H_2O$, can be obtained by the addition of bromine to sodium hydroxide solution below 0 °C, but these are less stable than the chlorate(I) compounds and, like the acid, disproportionate into bromate(V) and bromide ions. Iodate(I) compounds cannot be isolated.

Halic(III) (halous) acids, HXO_2

Of these, chloric(III) (chlorous) acid, $HClO_2$, is the only one known and this exists only in solution. It is a stronger acid than chloric(I) acid.

Halic(V) acids, HXO_3

Iodic(V) acid, HIO_3, is the most stable and exists as a white solid; $HClO_3$ and $HBrO_3$ exist only in solution and are therefore usually employed in the form of their more stable sodium or potassium salts.

Sodium chlorate(V) is obtained by the electrolysis of a hot concentrated solution of sodium chloride.

$$6OH^-(aq)+3Cl_2(g) \rightarrow ClO_3^-(aq)+5Cl^-(aq)+3H_2O(l)$$
$$2HOCl(aq)+ClO^-(aq) \rightarrow ClO_3^-(aq)+2H^+(aq)+2Cl^-(aq)$$

After concentration of the solution by partial evaporation, the precipitated sodium chloride is removed by filtering and the chlorate(V) is then crystallized out.

The acids themselves are strong and are capable of dissolving certain metals.

$$Zn(s) + 2HIO_3(aq) \rightarrow Zn(IO_3)_2(aq) + H_2(g)$$

They are also powerful oxidizing agents.

$$XO_3^-(aq) + 6H^+(aq) + 6e \rightarrow 3H_2O(l) + X^-(aq)$$

Bromate(V) ions and iodate(V) ions, in the form of their potassium salts, react with the corresponding halide ion in acid solution to give a quantitative yield of the halogen.

$$BrO_3^-(aq) + 5Br^-(aq) + 6H^+(aq) \rightarrow 3Br_2(aq) + 3H_2O(l)$$
$$IO_3^-(aq) + 5I^-(aq) + 6H^+(aq) \rightarrow 3I_2(aq) + 3H_2O(l)$$

Both of these reactions are utilized in volumetric analysis. This is particularly true of the iodate(V) ion which is stable in aqueous solution and can be obtained in a high state of purity, enabling it to be used as a primary standard.

The action of heat on potassium chlorate(V), $KClO_3$, to just below 400 °C results in the formation of the chlorate(VII), $KClO_4$. At higher temperatures this decomposes to the chloride and evolves oxygen.

$$4KClO_3(s) \rightarrow KCl(s) + 3KClO_4(s)$$
$$KClO_4(2) \rightarrow KCl(s) + 2O_2(g)$$

This reaction can be catalysed by, e.g. manganese(IV) oxide or copper(II) oxide.

Halic(VII) (perhalic) acids

$HClO_4$, $HBrO_4$, and HIO_4 are known.

Chloric(VII) (perchloric) acid can be obtained by distilling under vacuum a mixture of potassium chlorate(VII) (perchlorate) and concentrated sulphuric acid. It is an unstable oily liquid which decomposes at its boiling point of 90 °C. In cold, aqueous solution, chloric(VII) acid behaves as a strong acid and a weak oxidizing agent.

$$ClO_4^-(aq) + 2H^+(aq) + 2e \rightleftharpoons ClO_3^-(aq) + H_2O(l); \quad E^\ominus = +1.23\,V$$

However, in hot, concentrated solution, it is a powerful oxidizing agent.

Bromate(VII) (perbromate) compounds have been prepared for the first time in comparatively recent years by the oxidation of bromate(V) ions. This can be achieved by fluorine, XeF_2, or electrolytically. Bromic(VII) acid is stable in aqueous solution and a strong acid. It is a more powerful oxidizing agent than ClO_4^-.

Several iodic(VII) (periodic) acids have been obtained. These are weaker acids than the other halic(VII) acids, but are strong oxidizing agents. Potassium iodate(VII) is used in organic chemistry for the cleavage of 1,2-diols.

Halides (other than those of hydrogen)

Halides are formed by all four halogens. For convenience, we will consider the general nature and properties of the chlorides.

The chlorides formed by the more electropositive metals, of which the alkali metal chlorides are typical, exist in the solid state mainly as ionic crystals and dissolve in water to give the stable hydrated Cl^- ion. The less electropositive metals, however, form chlorides which are predominantly covalent in character, Examples of these include $HgCl_2$ with a discrete linear molecule; $CuCl_2$ and $PdCl_2$ with a chain structure; $CdCl_2$, $FeCl_3$, and $AlCl_3$ with layer structures in the crystalline state; and $SbCl_3$ with a pyramidal structure.

Chlorides of the less electropositive elements very often undergo hydrolysis with water in which the chlorine is replaced by an OH group.

e.g.
$$MgCl_2(s)+2H_2O(l) \rightleftharpoons 2HCl(g)+Mg(OH)_2(s)$$
$$BCl_3(g)+3H_2O(l) \rightleftharpoons 3HCl(g)+B(OH)_3(aq)$$

If, on the other hand, the element is significantly electronegative, then its chloride may be hydrolysed to the halic(I) acid.

$$NCl_3(l)+3H_2O(l) \rightleftharpoons NH_3(aq)+3HOCl(aq)$$

Where a high degree of covalency is present, the halide may be quite inert towards water, e.g. SF_6 and CCl_4, even though such reactions are thermodynamically possible.

The halide ion can often be found as a ligand (see page 218) in complex ions, e.g. $[FeCl_4]^-$, $[HgI_4]^-$, $[Co(NH_3)_4Cl_2]^+$, etc.

These generalizations are also widely applicable to the other halides. The main difference is that the metal fluorides tend to be more ionic in character than the corresponding chlorides, while bromides and iodides tend to be more covalent. This is illustrated by the crystalline halides of aluminium, where the fluoride is predominantly ionic, the chloride has a good deal of covalent character and has a layer lattice, and the bromide and iodide are both covalent dimers.

Interhalogen compounds

The halogens form four types of stable, binary interhalogen compounds (Table 14.14). These correspond to the general formulae, XX′, XX'_3, XX'_5 and XX'_7. Such behaviour between elements of the same group is unique.

These compounds can be made by direct combination in a nickel tube, the product being dependent upon the conditions.

e.g.
$$\underset{\text{(equimolar)}}{Cl_2(g)+F_2(g)} \xrightarrow[\text{temperature}]{\text{moderate}} 2ClF(g)$$

$$Cl_2(g)+\underset{\text{(excess)}}{3F_2(g)} \xrightarrow{280\,^\circ C} 2ClF_3(g)$$

$$I_2(s) + 3Cl_2\ (\text{liquid})\ \underset{\text{(excess)}}{} \longrightarrow 2ICl_3(s)$$

$$Br_2(l) + 5F_2(g)\ \underset{\text{(excess)}}{} \xrightarrow{50\,^\circ C} 2BrF_5(l)$$

Table 14.14 Physical states indicated are those in which the compounds exist at a temperature of 20 °C

XX′	*XX′₃*	*XX′₅*	*XX′₇*
ClF(g)	ClF_3(g)		
BrF(l)	BrF_3(l)	BrF_5(l)	
BrCl(g)			
		IF_5(l)	IF_7(g)
ICl(s)	ICl_3(s)		
IBr(s)			

For each type, the boiling points increase as the difference in the electronegativity values of the combined halogens increases.

With the exception of fluorine, the bond enthalpy of the X—X′ bond is less than that of the X—X bond, making the interhalogen compounds more reactive than the halogens themselves. They all behave as oxidizing agents and undergo hydrolysis with water.

$$XX' + H_2O(l) \rightarrow HOX(aq) + X'^-(aq) + H^+(aq)$$

With alkenes they form addition compounds:

$$\gt C{=}C\lt \; + XX' \rightarrow -\underset{X}{\overset{|}{\underset{|}{C}}}-\underset{X'}{\overset{|}{\underset{|}{C}}}-$$

Of the XX'_3 compounds, ClF_3 and BrF_3 are probably the most widely used, being particularly valuable as fluorinating agents.

15 Group 0: The Noble Gases

Table 15.1

	Helium	*Neon*	*Argon*	*Krypton*	*Xenon*	*Radon*
Symbol	He	Ne	Ar	Kr	Xe	Rn
Outer electronic structure	$1s^2$	$2s^22p^6$	$3s^23p^6$	$4s^24p^6$	$5s^25p^6$	$6s^26p^6$

All the noble gas elements are present in the atmosphere, argon being the most plentiful and radon existing only in extremely minute quantities (Table 15.2).

Table 15.2

	He	*Ne*	*Ar*	*Kr*	*Xe*	*Rn*
Percentage in the atmosphere	5.2×10^{-4}	1.8×10^{-3}	0.93	1.1×10^{-4}	9×10^{-6}	6×10^{-18}

Helium is usually found associated with α-emitting materials such as pitchblende (which contains uranium), the α-particles acquiring electrons from the atmospheric gases to form helium gas. In the United States, helium also occurs in some natural gas wells to the extent of about 7 per cent.

Radon is the radioactive decay product of radium and has three isotopic forms, ^{219}Rn, ^{220}Rn, and ^{222}Rn, which are themselves radioactive, decaying into polonium. The most stable isotope of radon is ^{222}Rn with a half-life of 3.825 days.

$$^{226}_{88}Ra \xrightarrow{t_{\frac{1}{2}} = 1622 \text{ years}} {}^{222}_{86}Rn + {}^{4}_{2}He$$

$$^{222}_{86}Rn \xrightarrow{t_{\frac{1}{2}} = 3.825 \text{ days}} {}^{218}_{84}Po + {}^{4}_{2}He$$

General properties

With the exception of helium, which has only the two electrons filling the first principal quantum level, i.e. $1s^2$, the noble gases have the stable outer electronic configuration ns^2np^6. It is because of these filled outer levels of electrons that the noble gases have little tendency to form chemical bonds and possess extremely high ionization energies (Table 15.3).

Table 15.3

	He	*Ne*	*Ar*	*Kr*	*Xe*	*Rn*
First ionization energy, $\Delta H^{\ominus}$/kJ mol^{-1}	+2370	+2080	+1520	+1350	+1170	+1040

The physical properties of these elements in general show an excellent correlation with increasing relative atomic mass (Table 15.5).

The elements are *monatomic* and the *atoms tend to be the largest (by comparison of covalent radii) in their respective periods* (Table 15.4).

Table 15.4

	He	*Ne*	*Ar*	*Kr*	*Xe*
Atomic (covalent) radius/nm	0.12	0.16	0.19	0.20	0.22

Interaction between atoms is minimal and the gases are not easily liquefied, properties which are reflected in the short temperature range of their liquid states and also by the fact that they have much lower boiling points than compounds of comparable relative molecular mass (Table 15.5).

Table 15.5

	He	*Ne*	*Ar*	*Kr*	*Xe*	*Ra*
M.p./°C	—	−249	−189	−157	−112	−71
B.p./°C	−269	−246	−186	−152	−108	−62
Density, $\times 10^{-3}$/g cm^{-3}	0.179	0.900	1.784	3.749	5.897	—

Interatomic forces are, in fact, so small that helium cannot be made to solidify at ordinary atmospheric pressure. It liquefies at −268.97 °C (4.18 K) and behaves as a normal liquid until cooled to −270.97 °C (2.18 K), at which point it becomes transformed into a liquid which possesses the properties of a gas (a *superfluid*) and is referred to as Helium II. This unique liquid has remarkable properties and is of considerable interest in the field of low temperature research work. It has a thermal conductivity which is 600–800 times that of copper, a viscosity about 1/1000th that of gaseous hydrogen, and the peculiar ability to flow up the sides of the containing vessel.

In the solid state, neon, argon, krypton, and xenon all adopt a close-packed cubic arrangement.

Uses

The lighter members of the group are widely used in discharge tubes where their distinctive colours make them particularly useful in coloured signs. They are similarly used in fluorescent tubes for interior and road lighting and, in the case of neon, which has fog-penetration properties, in illuminating airfields.

Helium finds use in gas-cooled nuclear reactors where it acts as an inert body and, furthermore, does not become radioactive. An oxygen-helium mixture is used by deep-sea divers, the helium having the advantage over nitrogen in that it is less easily absorbed into the blood stream. The major use of argon is as an inert atmosphere in arc-welding operations for processes where prevention of oxidation is necessary, e.g. with aluminium.

Preparation of compounds of the noble gases

Until fairly recently, the only known 'compounds' of the noble gases were *clathrate compounds* (see page 201). For example, hydrates of the heavier noble gases, such as $Xe{\cdot}6H_2O$ (approximate empirical formula), have been formed when water soldifies in an atmosphere of the gas. In the case of the lighter gases, this occurs only under conditions of considerable pressure. A similar situation arises on compression (10 to 40 atmospheres) with aqueous benzene-1,4-diol (quinol) in an atmosphere of Ar, Kr, or Xe, resulting in the formation of a stable crystal clathrate corresponding to an empirical formula: 3 benzene-1,4-diol:1 noble gas.

The production of authentic noble gas compounds originates from a discovery by N. Bartlett who, having found that PtF_6 was a sufficiently strong oxidizing agent to convert molecular oxygen into the ionic compound $O_2^+[PtF_6]^-$, reasoned that because the first ionization energies of molecular oxygen and xenon are similar,

$$O_2(g) \rightarrow O_2^+(g) + e; \quad \Delta H^\ominus = +1177\,\text{kJ mol}^{-1}$$
$$Xe(g) \rightarrow Xe^+(g) + e; \quad \Delta H^\ominus = +1169\,\text{kJ mol}^{-1}$$

xenon might behave on a similar way. In 1962, Bartlett reacted the deep red vapour of the PtF_6 with colourless xenon and obtained a yellow powder which he believed to be $Xe^+[PtF_6]^-$, although it was later shown to contain a mixture of complex xenon fluorides.

Since this initial discovery, three fluorides of xenon, XeF_2, XeF_4, and XeF_6, have been obtained by direct synthesis of the elements,

e.g.

$$\underset{(1\text{ vol})}{Xe(g)} + \underset{(5\text{ vol})}{2F_2(g)} \xrightarrow[\text{(ii) rapid cooling}]{\text{(i) 400 °C for 1 hour in Ni vessel}} XeF_4(s)$$

The lower ionization energy of radon compared to that of xenon should, on the basis of this criterion, make radon compounds easier to prepare; but, because of the short radioactive half-life of this element, this is not borne out in practice, and only a fluoride of unknown structure has been prepared.

Because of the higher ionization energies of the lighter gases, compound formation involving these elements is very much more difficult, the only two of significance being KrF_2 and KrF_4.

The chemistry of the noble gases is therefore largely that of xenon, and a sufficient number of compounds have now been isolated to indicate oxidation states of xenon ranging from two to eight. In all its well-known compounds, xenon is always found bonded to either or both of the two most electronegative elements, namely fluorine and oxygen.

Chemistry of xenon compounds

The fluorides, XeF_2, XeF_4, and XeF_6

The three fluorides are the best characterized of the xenon compounds. They all exist as white or colourless solids at ordinary temperatures, but above 42 °C XeF_6 is pale yellow; it melts at 46 °C to form a yellow liquid and vapour.

All three fluorides are reduced by hydrogen, yielding the noble gas and hydrogen fluoride,

e.g. $$XeF_2(s)+H_2(g) \rightarrow Xe(g)+2HF(g)$$

With water rapid hydrolysis occurs, but, whereas the difluoride yields oxygen, xenon and fluoride ions, XF_4 and XF_6, form XeO_3 after evaporation,

e.g. $$XeF_6(s)+3H_2O(l) \rightarrow XeO_3(aq)+6HF(g)$$

Of the three fluorides, XeF_6 is by far the most reactive. With silicon(IV) oxide it forms xenon oxytetrafluoride, $XeOF_4$, a volatile, colourless liquid:

$$2XeF_6(s)+SiO_2(s) \rightarrow 2XeOF_4(l)+SiF_4(g)$$

Xenon trioxide, XeO_3

Xenon trioxide is a colourless, non-volatile solid which is a sensitive and dangerous explosive. An aqueous solution of it, however, is reasonably stable and behaves both as a weak monobasic acid and, like solutions of all other xenon compounds, as an oxidizing agent, liberating chlorine from chlorides, bromine from bromides, etc., together with elemental xenon. If the solution is made strongly alkaline, $[HXeO_4]^-$ ions predominate. The $[HXeO_4]^-$ ion is unstable and slowly disproportionates:

$$\underset{Xe(+6)}{2[HXeO_4]^-(aq)}+2OH^-(aq) \rightarrow \underset{Xe(0)}{Xe(g)}+\underset{Xe(+8)}{[XeO_6]^{4-}(aq)}+O_2(g)+2H_2O(l)$$

Stable perxenates, such as $Na_4XeO_8 \cdot 8H_2O$, have been isolated and are among the strongest oxidizing agents known.

Structure of some xenon compounds

Physical evidence indicates that XeF_2 and XeF_4 (see page 33) have linear and square planar arrangements of atoms respectively. The precise structure of XeF_6 has not yet been fully elucidated, but it is generally considered to have some form of distorted octahedral arrangement, probably involving intra-molecular rearrangements and is sometimes described as a 'floppy' molecule. X-ray analysis shows XeO_3 to possess a trigonal pyramidal structure.

16 Complexes

Complexes and double salts

Compounds formed by the stoichiometric combination of two or more apparently saturated substances which are themselves capable of independent existence are known as *addition* or *molecular compounds.* These fall into two main categories:

Double salts

These exist only in the crystalline form and undergo decomposition into their constituents on dissolving in water, e.g. $KAl(SO_4)_2{\cdot}12H_2O$, aluminium potassium sulphate–12–water (potassium or potash alum) and $Fe(NH_4)_2(SO_4)_2{\cdot}6H_2O$, ammonium iron(II) sulphate–6–water. However, double salts very often contain complex ions which may be present in either part of the double salt. For example, $KAl(SO_4)_2{\cdot}12H_2O$ is more precisely written, $K^+[Al(H_2O)_6]^{3+}(SO_4^{2-})_2{\cdot}6H_2O$, the $[Al(H_2O)_6]^{3+}$ ion existing in both the solid state and in aqueous solution.

Complexes

These retain their identity in solution and may be recovered from it.

On dissolving in water, complexes dissociate into the complex ion and the (usually) simple ion,

e.g. $$K_3[Fe(CN)_6](s) \xrightarrow{\text{water}} 3K^+(aq) + [Fe(CN)_6]^{3-}(aq)$$

Detection of the existence of these separate ions can be achieved by colligative property techniques (e.g. depression of freezing point) which depend upon the number of separate particles present in solution.

In addition, there is also some dissociation of the complex ion itself, although this may be only slight, which gives rise to a *stability constant* for each complex (see page 228). From a study of the values of these stability constants, it is apparent that in certain cases there is no clear dividing line between complexes and double salts, the complex ion in such examples being highly unstable and splitting up into simple ions. If, on the other hand, aluminium potassium sulphate–12–water (potassium alum) is dissolved in water, measurements of its colligative properties indicate that the solution contains separate potassium and aluminium cations and sulphate anions:

$$KAl(SO_4)_2{\cdot}12H_2O(s) + aq \rightarrow K^+(aq) + Al^{3+}(aq) + 2SO_4^{2-}(aq) + 12H_2O(l)$$

Fundamental structure of complexes

A complex consists of a central metal atom or ion surrounded by a number of oppositely charged ions or neutral molecules possessing lone pairs of electrons which are capable of being donated to the vacant orbitals of the metal atom or ion. Such complexing agents are referred to as **ligands**.

The total number of ligands present in a complex corresponds to the co-ordination number of the central metal atom or ion,

e.g. $K_3[Fe(CN)_6]$ Potassium hexacyanoferrate(III)

In this complex, the central iron(III) ion is surrounded by six CN^- ligands. The complex ion is written enclosed within square brackets, i.e. $[Fe(CN)_6]^{3-}$. In the compound potassium hexacyanoferrate(III), the molar ratio of potassium ions to the $[Fe(CN)_6]^{3-}$ ions is 3:1 in order to achieve electrical neutrality.

Naming complexes

The systematic IUPAC (International Union of Pure and Applied Chemistry) nomenclature for complex compounds is summarized as follows:

1. The cation is always named before the anion.
2. The names of ligands precede that of the central ion.
3. The oxidation state of the central ion is denoted by Roman numerals in brackets immediately after its name.
4. Metals forming complex anions end in *-ate*, whereas complex cations and neutral molecules use their usual names.
5. The sequence for specifying ligands is: anions, neutral molecules, and lastly cations. Within each of these categories alphabetical order is adopted when there is more than one type of ligand.
   Anionic ligands end in *-o*, e.g. CN^- *cyano*, Cl^- *chloro*, OH^- *hydroxo*, etc.
   Neutral ligands rarely change, but note H_2O *aqua*, NH_3 *ammine*, CO *carbonyl* (aqua is named first).
   Positive ligands end in *-ium*, e.g. $NH_2—NH_3^+$ *hydrazinium*.
6. When there is a number already stated in the name of the ligand (e.g. ethane-1,2-*di*amine) then, to avoid confusion, the ligand is placed in a bracket and the prefix *bis*, *tris*, *tetrakis*, etc. used instead of di, tri, tetra, etc.

Study the following examples until you can reconcile their names with the principles outlined above:

$[Cr(H_2O)_6]Cl_3$	Hexaaquachromium(III) chloride
$[Ni(CO)_4]$	Tetracarbonylnickel(0)
$K_2[PtCl_4]$	Potassium tetrachloroplatinate(II)
$[Pt(NH_3)_4]Cl_2$	Tetraammineplatinum(II) chloride
$[Co(NH_3)_5NO_2]Cl_2$	Nitropentaamminecobalt(III) chloride
$[Cu(NH_3)_4(H_2O)_2]SO_4$	Diaquatetraamminecopper(II) sulphate
$[Co(NH_2CH_2CH_2NH_2)_3]Cl_3$	Tris(ethane-1,2-diamine)cobalt(III) chloride

Coordination number and the shape of complex ions

Table 16.1

Coordination number	*Shape*	*Examples*
2	Linear	$[Ag(NH_3)_2]^+$ $[Ag(CN)_2]^-$
4	Square planar	$[Ni(CN)_4]^{2-}$ $[Pt(NH_3)_4]^{2+}$ $[Pt(NH_3)_3Cl]^+$ $[Pd(NH_3)_4]^{2+}$
4	Tetrahedral	$[Cd(CN)_4]^{2-}$ $[FeCl_4]^-$ $[Ni(NH_3)_4]^{2+}$ $[NiCl_4]^{2-}$ $[Ni(CO)_4]$
6	Octahedral	$[Cr(H_2O)_6]^{3+}$ $[CrCl_6]^{3-}$ $[Co(NH_3)_6]^{3+}$ $[Fe(CN)_6]^{3-}$ $[Fe(CN)_6]^{4-}$ $[Mn(H_2O)_6]^{2+}$

The concept of hybridization can be extended to account for the bonding encountered in the majority of simple complexes and also their shapes, the the most common shapes being octahedral, tetrahedral, square planar, and linear.

Where the central metal ion is that of a transition metal, the difference in energy between the electrons in the *d* orbitals of the penultimate shell and those in the *s* and *p* orbitals of the outermost shell is very small, explaining the comparative ease with which *d* electrons can be excited for use as valency electrons.

Introduction to crystal field and ligand field theories

The *crystal* and *ligand field theories* have been developed on the basis that the central metal ion and the individual ligands of a complex have their own electrical fields. The field exerted by the central ion influences the ligands, and that exerted by the ligands affects the electronic arrangement of the central ion.

The crystal field theory considers only electrostatic interactions between the *d* orbitals of the central ion and the external electric field of the ligands, whereas in the ligand field theory, these factors are refined by considering the molecular orbitals which can be formed by the orbitals of the ligands and those of the metal ion.

In the isolated or 'free' transition metal ion of the first transition series, all five 3*d* orbitals are energetically identical and are said to be *degenerate*. However, when they are in the presence of a ligand field, because they do not all have the same shape, they split into two groups, each of slightly different energy. One group consists of three orbitals, and is referred to as the t_{2g} group, and the other consists of two orbitals, and is referred to as the e_g group. The three t_{2g} orbitals correspond to *d* orbitals which are concentrated between the *x*, *y*, and *z* axes, and the two e_g orbitals correspond to *d* orbitals concentrated along the *x*, *y*, and *z* axes.

If we now consider an octahedral complex in which the ligands are at six corners, then each of these will exert an electrostatic field in directions which would correspond to *x*, *y*, and *z* axes. Therefore the electrons in the two e_g orbitals of the central metal ion experience a greater repulsive force than those in the three t_{2g} orbitals, causing the e_g orbitals to have a higher energy than the t_{2g}. Consequently, the five *d* orbitals are no longer degenerate. The energy difference between these two levels is assigned the symbol Δ.

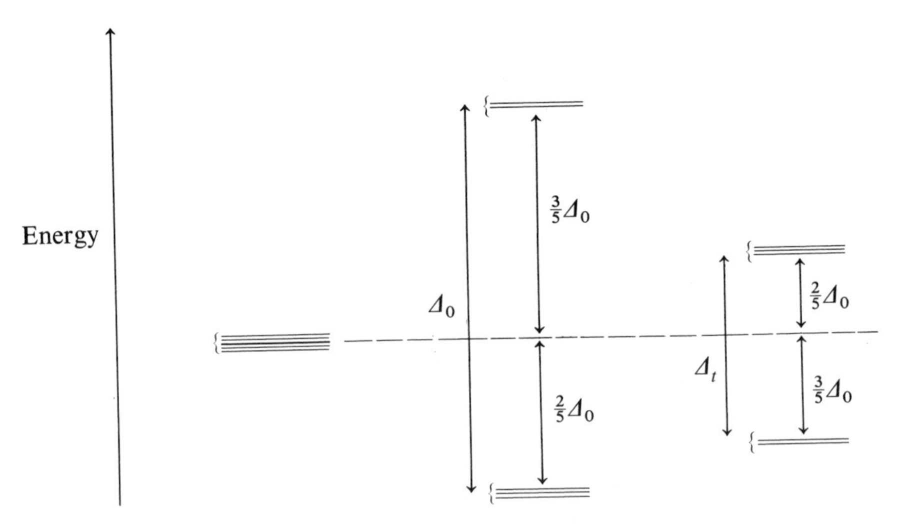

Fig. 16.1 (a) Five degenerate *d* orbitals split (b) by six ligands octahedrally surrounding the central metal ion and (c) by four ligands tetrahedrally surrounding the metal ion

If the mean energy of all five *d* orbitals is considered to be zero, then the two e_g orbitals are raised by an amount equal to $\frac{3}{5}\Delta_o$ and the three t_{2g} orbitals are lowered by an amount equal to $\frac{2}{5}\Delta_o$.

With a tetrahedral arrangement of four ligands around the central ion, the t_{2g} group has the higher energy.

The strength of crystal field effect of ligands

The value of Δ is largely dependent on the strength of the crystal field. For a given central metal ion, the decreasing order of crystal field effect for some of the more common ligands is:

$$CN^- > NO_2^- > NH_3 \text{ (and amines)} > H_2O > F^- > OH^- > Cl^- > Br^- > I^-$$

This series is sometimes referred to as the *spectrochemical series.*

Octahedral complexes

	3d					4s	4p		
Fe atom, ground state:	↑↓	↑	↑	↑	↑	↑↓			
Fe^{2+} ion:	↑↓	↑	↑	↑	↑				
Fe^{2+} ion in the complex ion $[Fe(CN)_6]^{4-}$:	↑↓	↑↓	↑↓	X	X	X	X	X	X

d^2sp^3 hybridized

X represents a lone pair of electrons donated by a CN^- ion.

In the hexacyanoferrate(II) ion, $[Fe(CN)_6]^{4-}$, pairing of electrons in the Fe^{2+} ion is induced by the powerful complexing properties of the CN^- ligands, which have a strong ligand field. The six vacant orbitals available to accept lone pairs of electrons from the CN^- ions are hybridized. As only inner d orbitals have been used in forming this complex ion, it is described as an *inner orbital complex*, the hybridization being written d^2sp^3 to denote this. This satisfactorily accounts for the equivalence of bonds in the complex ion, its octahedral structure, and its diamagnetic properties (see page 240). (The simple Fe^{2+} ion, which contains four unpaired electrons, is strongly paramagnetic.)

Less powerful ligands have a greater tendency to leave the metal ion electrons unpaired, as in the case of the $[CoF_6]^{3-}$ ion.

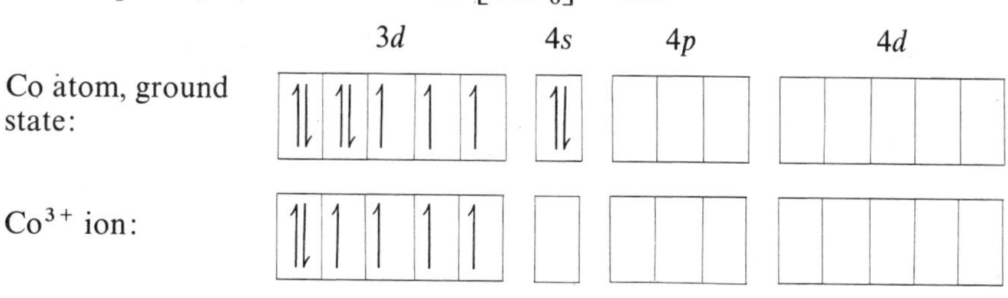

Co^{3+} ion in the complex ion $[CoF_6]^{3-}$: ⇅ ↑ ↑ ↑ ↑ | X | X X X | X X | |

sp^3d^2 hybridized

X represents a lone pair of electrons donated by a F^- ion.

As only outer *d* orbitals have been used in forming this complex ion, it is referred to as an *outer orbital complex*, the hybridization being written sp^3d^2 as opposed to d^2sp^3 for the inner orbital complex. Owing to the presence of unpaired electrons in outer orbital complexes of this type, they exhibit paramagnetic properties. Other examples are shown in Table 16.2:

Table 16.2

Complex ion	*Number of unpaired electrons*
$[Ni(H_2O)_6]^{2+}$	2
$[Fe(CN)_6]^{3-}$	1
$[Fe(H_2O)_6]^{2+}$	4

An extension of this reasoning may be used to account for the ease of conversion of complexes containing the Co^{2+} ion into those containing Co^{3+} ion despite the well-known stability of the Co^{2+} ion.

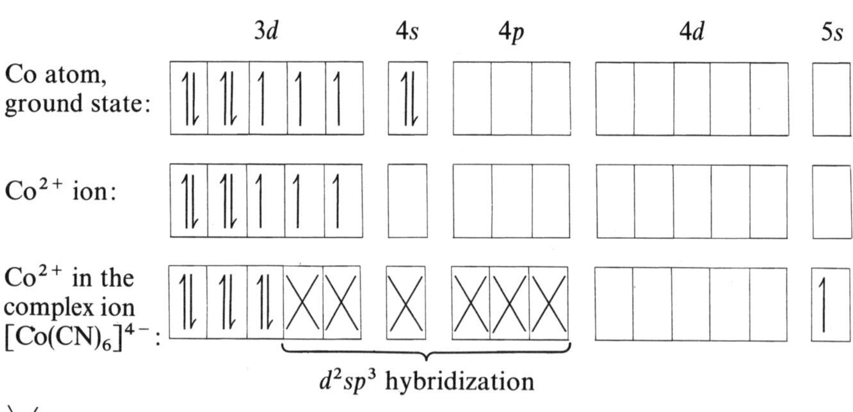

X represents the lone pair of electrons donated by a CN^- ion.

In the presence of a strong ligand field, pairing of the 3*d* electrons occurs with the extra electron probably being promoted to the 5*s* level. This electron is easily lost, enabling the Co(+2) state to be easily oxidized to the Co(+3) state in the complex.

Tetrahedral complexes

Tetrahedral complexes, as exemplified by the complex ion $[Cd(CN)_4]^{2-}$, can be explained in terms of sp^3 hybridization of orbitals of the central metal ion.

	4d					*5s*	*5p*		
Cd atom, ground state:	↑↓	↑↓	↑↓	↑↓	↑↓	↑↓			
Cd^{2+} ion:	↑↓	↑↓	↑↓	↑↓	↑↓				
Cd^{2+} ion in the complex ion $[Cd(CN)_4]^{2-}$:	↑↓	↑↓	↑↓	↑↓	↑↓	X	X	X	X

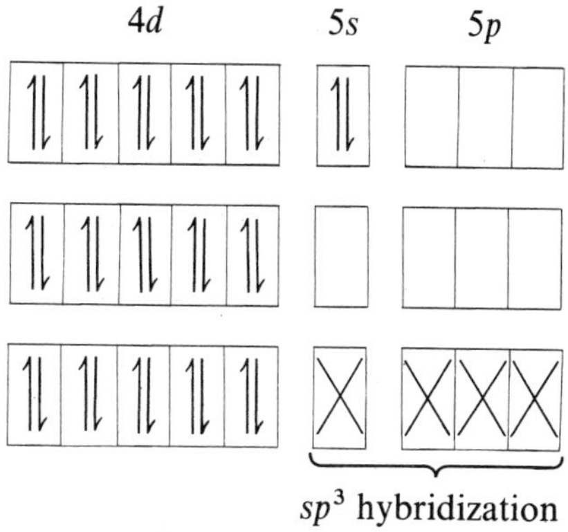

sp^3 hybridization

X represents a lone pair of electrons donated by a CN^- ion.

The formation of diamagnetic tetrahedral copper(I) complexes may be similarly explained, e.g. $[Cu(CN)_4]^{3-}$.

	3d					*4s*	*4p*		
Cu atom, ground state:	↑↓	↑↓	↑↓	↑↓	↑↓	↑			
Cu^+ ion:	↑↓	↑↓	↑↓	↑↓	↑↓				
Cu^+ ion in the complex ion $[Cu(CN)_4]^{3-}$:	↑↓	↑↓	↑↓	↑↓	↑↓	X	X	X	X

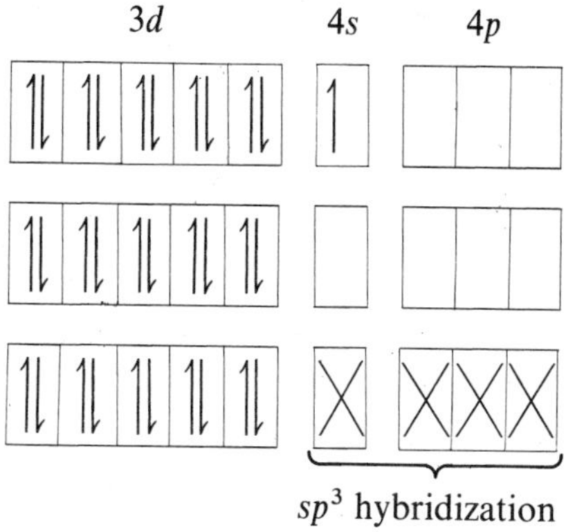

sp^3 hybridization

X represents a lone pair of electrons donated by a CN^- ion.

X-ray analysis of certain nickel complexes shows that weak ligands such as Cl^- give rise to paramagnetic tetrahedral structures.

	3d					*4s*	*4p*		
Ni atom, ground state:	↑↓	↑↓	↑↓	↑	↑	↑↓			
Ni^{2+} ion:	↑↓	↑↓	↑↓	↑	↑				

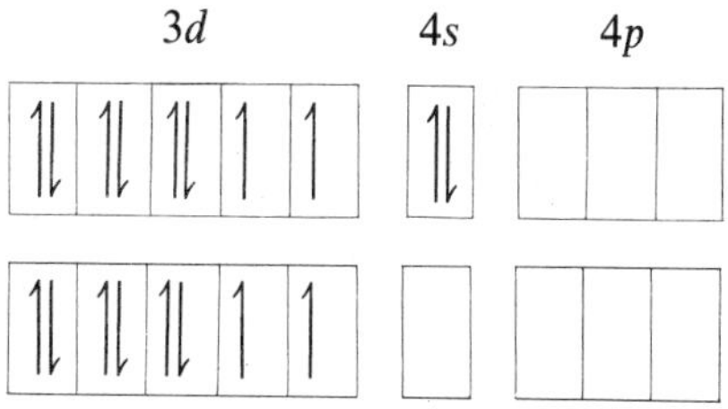

Ni^{2+} ion in the complex ion $[NiCl_4]^{2-}$:

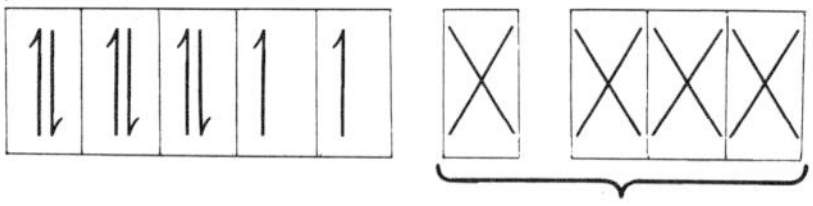

3d					4s	4p		
↑↓	↑↓	↑↓	↑	↑	X	X	X	X

sp^3 hybridization

X represents a lone pair of electrons donated by a Cl^- ion.

Square planar complexes

dsp^2 hybridization of orbitals satisfactorily accounts for the experimentally observed square planar structures of a number of complexes, e.g. $[Ni(CN)_4]^{2-}$, $[Pt(NH_3)_4]^{2+}$, and $[Pd(NH_3)_4]^{2+}$.

	3d					4s	4p		
Ni atom, ground state:	↑↓	↑↓	↑↓	↑	↑	↑↓			
Ni^{2+} ion:	↑↓	↑↓	↑↓	↑	↑				
Ni^{2+} ion in the complex ion $[Ni(CN)_4]^{2-}$:	↑↓	↑↓	↑↓	↑↓	X	X	X	X	

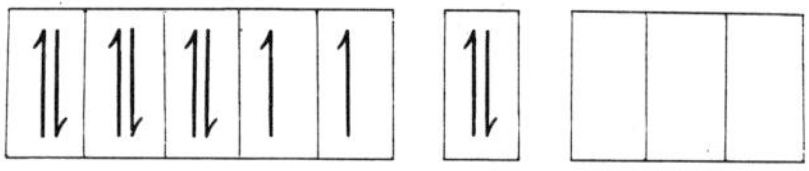

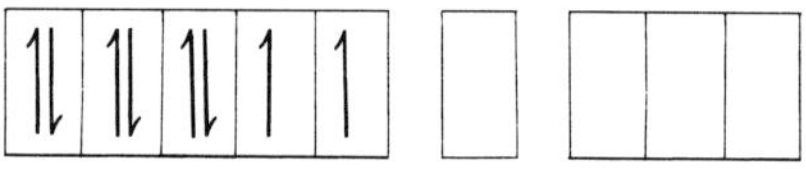

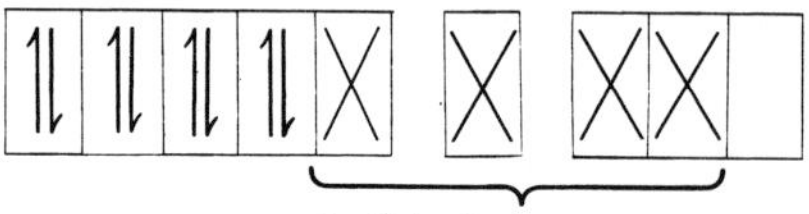

dsp^2 hybrid orbitals

X represent a lone pair of electrons donated by a CN^- ion.

Metal carbonyls

Metal carbonyls are the compounds formed by the bonding of neutral carbon monoxide molecules with a central metal atom. Only mononuclear carbonyls (i.e. those containing one metal atom per molecule) are considered here, in which the metal exhibits a zero oxidation state. The structure of simple carbonyls of this type can be accounted for in terms of simple hybridization theory, as illustrated below for $[Ni(CO)_4]$ and $[Fe(CO)_5]$.

Tetracarbonylnickel(0), $[Ni(CO)_4]$

	3d					4s	4p		
Ni atom, ground state:	↑↓	↑↓	↑↓	↑	↑	↑↓			
Ni atom in the complex molecule $[Ni(CO)_4]$:	↑↓	↑↓	↑↓	↑↓	↑↓	X	X	X	X

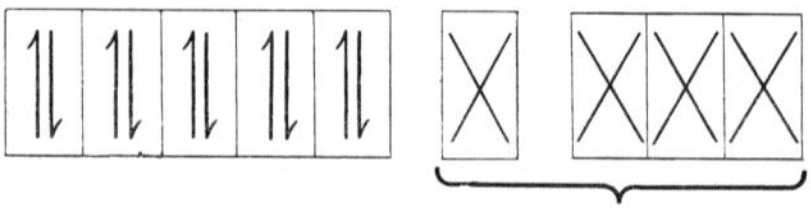

sp^3 hybridization

X represents a lone pair of electrons donated by a CO molecule.

Although this satisfactorily explains the tetrahedral symmetry and the diamagnetic properties of tetracarbonylnickel(0), it represents an over-simplification of the bonding involved. As in all other carbonyls, other types of bond sharing or bond pairing occurs, probably leading to some kind of delocalization of electrons.

Pentacarbonyliron(0), [Fe(CO)$_5$]

Fe atom, ground state:

Fe atom in the complex molecule [Fe(CO)$_5$]:

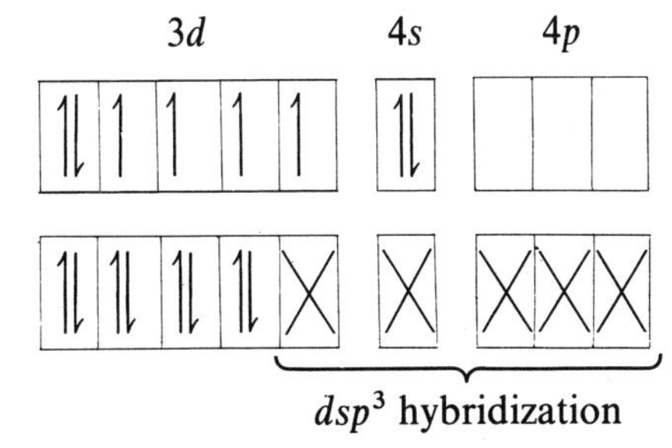

X represents a lone pair of electrons donated by a CO molecule.

The shape of the pentacarbonyliron(0) molecule is trigonal bipyramidal.

Isomerism

Because of the many different possible arrangements in the structural formulae of complex compounds, several types of isomerism are possible.

Geometric isomerism

In disubstituted complexes the same ligands may be adjacent, giving rise to the *cis* isomer, or opposite, giving rise to the *trans* isomer. For example, [Pt(NH$_3$)$_2$Cl$_2$], which is square planar:

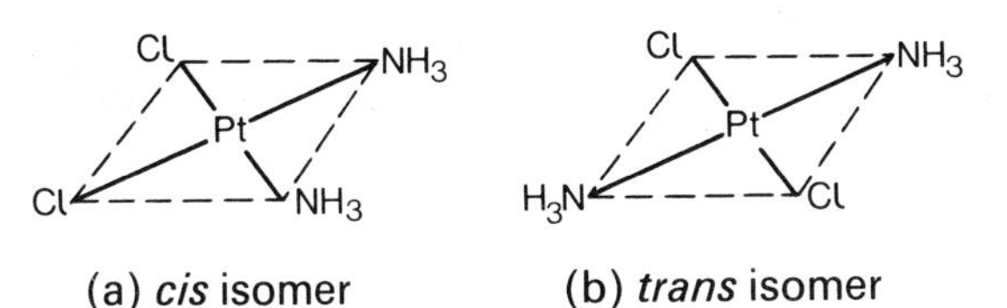

Fig. 16.2 (a) *cis* isomer (b) *trans* isomer

Geometric isomers can be distinguished by means of their dipole moments.

Optical isomerism

Two isomers which are not superimposable on their mirror images are described as optical isomers because of their ability to rotate plane-polarized light in opposite directions. These isomers are referred to as *enantiomers* or *enantiomorphs* of each other and their non-superimposable structures are described as being asymmetric. They are distinguished by the prefixes *dextro* ((+) or *d*) and *laevo* ((−) or *l*) in accordance with the direction in which they rotate plane-polarized light.

Optical isomerism in inorganic chemistry is generally encountered in octahedral structures containing bidentate groups such as ethane-1,2-diamine (ethylenediamine) (abbreviated in the structures to *en* (see page 231).

Consider the complex cation dichlorobis(ethane-1,2-diamine)cobalt(III) ion $[Co(en)_2Cl_2]^+$. While the *trans* form has a plane of symmetry, the *cis* form is asymmetric and displays optical activity.

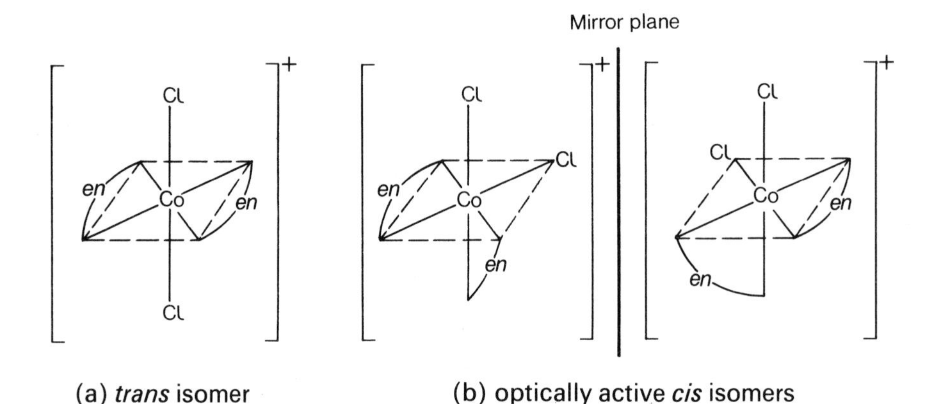

(a) *trans* isomer (b) optically active *cis* isomers

Fig. 16.3

Ionization isomerism

Groups may be directly bonded as ligands to the central metal ion or exist as ions outside the complex. As a result different ions may be produced when the complexes are dissolved in water.

$$[Pt(NH_3)_4Cl_2]Br_2 \xrightarrow[\text{(ii) Ag}^+\text{ ions}]{\text{(i) water}} \text{Cream precipitate } (Br^-)$$

$$[Pt(NH_3)_4Br_2]Cl_2 \xrightarrow[\text{(ii) Ag}^+\text{ ions}]{\text{(i) water}} \text{White precipitate } (Cl^-)$$

Hydrate isomerism

Water molecules possess lone pairs of electrons enabling them to function as ligands and bond with the central metal ion as well as remaining outside the complex. Quantitative precipitation in aqueous solution of the chlorine present in $CrCl_3{\cdot}6H_2O$ shows there to be three hydrates.

$[Cr(H_2O)_6]Cl_3$	Violet
$[Cr(H_2O)_5Cl]Cl_2{\cdot}H_2O$	Green
$[Cr(H_2O)_4Cl_2]Cl{\cdot}2H_2O$	Green

This is, in effect, a special case of ionization isomerism.

Linkage isomerism

Linkage isomerism arises when a ligand contains two different atoms each of which is capable of bonding with the central metal ion. The most commonly encountered case is the NO_2^- ion where either the nitrogen *or* the oxygen can donate electron pairs.

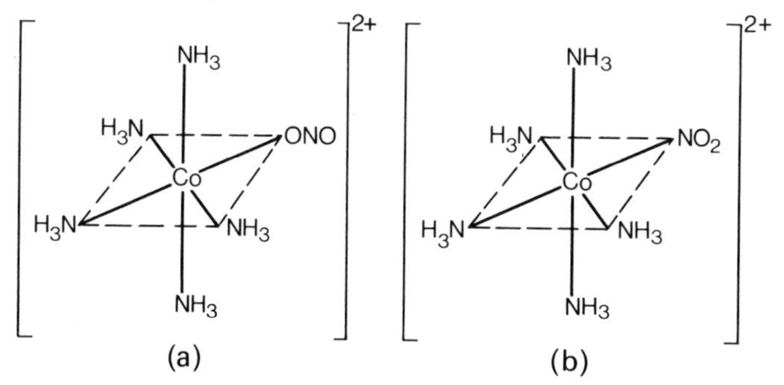

Fig. 16.4 (a) Nitritopentaamminecobalt(III) ion (red)
(b) Nitropentaamminecobalt(III) ion (yellow)

Coordination isomerism

An interchange of ligands between positive and negative ions, both of the ions themselves being complex, gives rise to coordination isomerism:

$$[Cu(NH_3)_4]\,[PtCl_4] \quad \text{and} \quad [Pt(NH_3)_4]\,[CuCl_4]$$

Polymerization isomerism

Polymerization isomerism occurs among complexes which have the same empirical formula but different relative molecular masses

The empirical formula $Pt(NH_3)_2Cl_2$ gives rise to the following polymerization isomers:

$$[Pt(NH_3)_2Cl_2]$$
$$[Pt(NH_3)_4]\,[PtCl_4]$$
$$[Pt(NH_3)_4]\,[Pt(NH_3)Cl_3]_2, \quad \text{etc.}$$

Strictly speaking, although it is generally accepted as such, this is not true isomerism because the isomers do not all have the same molecular composition. Furthermore, they are not really polymers in the generally accepted sense of the word as no polymerization of simple units actually occurs.

Coordination position isomerism

This results from an interchange of ligands between two 'central' metal ions in a polynuclear complex (i.e. one containing two or more 'central' metal ions).

For example,

and

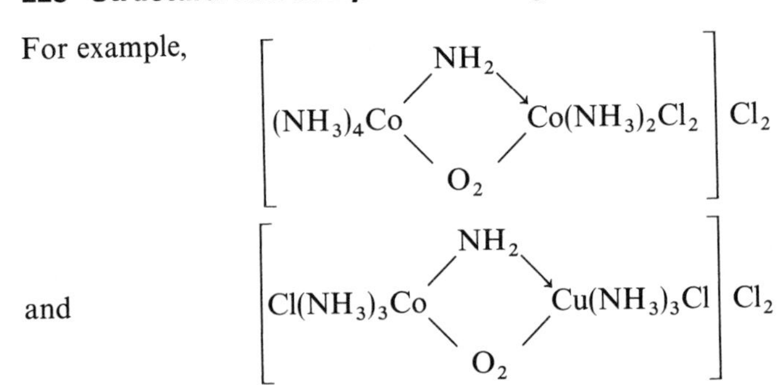

Stability of complex ions

In solution, complex ions establish an equilibrium between themselves and their components. The *stability constant*, K_{stab}, corresponds to the reciprocal of the dissociation constant which, in this context, is sometimes referred to as the instability constant. The equilibrium may be written so as to exclude any specific reference to the extent of hydration of the ions. Thus, for $[Ag(NH_3)_2]^+$ ion, the equilibrium corresponds to:

$$Ag^+(aq) + 2NH_3(aq) \rightleftharpoons [Ag(NH_3)_2]^+(aq)$$

the stability constant being given by:

$$K_{stab} = \frac{[Ag(NH_3)_2^+]}{[Ag^+][NH_3]^2}$$
$$= 1.7 \times 10^7 \text{ mol dm}^{-3}$$

The higher the value of the stability constant, the more stable the complex ion. For greater ease of comparison, the values are often employed in the logarithmic form, lg K_{stab} (Table 16.3).

Table 16.3

Complex ion	*Stability constant, K_{stab} /mol dm^{-3}*	*Lg K_{stab}*
$[Ag(CN)_2]^-$	1.0×10^{21}	21.0
$[Ag(NH_3)_2]^+$	1.7×10^{7}	7.2
$[Cd(CN)_4]^{2-}$	7.1×10^{16}	16.9
$[Co(NH_3)_6]^{2+}$	7.7×10^{4}	4.9
$[Co(NH_3)_6]^{3+}$	4.5×10^{33}	33.7
$[Cu(CN)_4]^{3-}$	2.0×10^{27}	27.3
$[Cu(NH_3)_4]^{2+}$	1.4×10^{13}	13.1
$[FeCl_4]^-$	8.0×10^{-2}	−1.1
$[Ni(NH_3)_6]^{2+}$	4.8×10^{7}	7.7

Application of complex formation to qualitative analysis

Solubility of the diamminesilver(I) ion, $[Ag(NH_3)_2]^+$

The most commonly encountered form of qualitative analysis involves dissolving an unknown substance in a suitable aqueous solvent and then detecting the cations and anions separately. Any dissolved metal ions therefore exist in solution as their aqua complexes.

Although there are a number of different schemes of qualitative analysis of cations, the first group of many of these involves the precipitation of those metal ions with insoluble chlorides. One of these is silver, whose presence can be confirmed by the ready solubility of its chlorides in dilute ammonia.

Solid silver(I) chloride is virtually insoluble in water and a saturated solution contains only 10^{-5} moles of Ag^+ ions per dm^3, the equilibrium between the solid and those ions present in solution being given by:

$$Ag^+Cl^-(s) + aq \rightleftharpoons Ag^+Cl^-(aq) \rightleftharpoons Ag^+(aq) + Cl^-(aq)$$

Addition of dilute ammonia causes the Ag^+ ions to be removed from solution as the comparatively stable $[Ag(NH_3)_2]^+$ ions, enabling the equilibrium to adjust so far towards the right that, provided sufficient ammonia is present, the solid silver(I) chloride totally dissolves, forming a solution of the soluble complex.

$$Ag^+Cl^-(s) \rightleftharpoons Ag^+(aq) + Cl^-(aq)$$

$$\Big\Updownarrow + 2NH_3(aq)$$

$$[Ag(NH_3)_2]^+(aq)$$

Re-precipitation of the silver(I) chloride may be achieved by the subsequent addition of dilute acid, which breaks down the complex ion by forming NH_4^+ ions from the NH_3 molecules.

Some of the many other examples of compounds which dissolve with the formation of complex ions are given in Table 16.4.

Table 16.4

Compound or ion	*Solution*	*Complex formed*
$AgCl$	$KCN(aq)$	$[Ag(CN)_2]^-(aq)$
$AgCl$	$Na_2S_2O_3(aq)$	$[Ag(S_2O_3)_2]^{3-}(aq)$
Au^{3+}	$NaCN(aq)$	$[Au(CN)_4]^-(aq)$
$CuSO_4$	$NH_3(aq)$	$[Cu(NH_3)_4(H_2O)_2]^{2+}(aq)$
I_2	$KI(aq)$	$[I_3]^-(aq)$
KCN (excess)	$FeSO_4(aq)$	$[Fe(CN)_6]^{3-}(aq)$
KI (excess)	$Hg^{2+}(aq)$	$[HgI_4]^{2-}(aq)$
$PbCl_2$	HCl (conc.)	$[PbCl_4]^{2-}$

Many of the complexes formed find application in the laboratory and in industry, e.g. $[Ag(CN)_2]^-$ and $[Ag(S_2O_3)_2]^{3-}$ in photography, $[Cu(NH_3)_4(H_2O)_2]^{2+}$ in dissolving cellulose (Schweitzer's solution), $[I_3]^-$ in iodine titrations, $[Au(CN)_4]^-$ in the extraction of gold, and $[HgI_4]^{2-}$ in Nessler's solution for the detection and estimation of ammonia.

Preferential precipitation using complexes

A complex ion may dissociate sufficiently in solution to enable its simple component ions to undergo precipitation reactions with certain ions but not with others, the process depending upon the degree of solubility of the compound to be precipitated. For example, silver(I) cyanide dissolves in alkali metal cyanide solutions forming the highly stable complex ion, $[Ag(CN)_2]^-$, which is only very slightly dissociated.

$$Ag^+(aq) + 2CN^-(aq) \rightleftharpoons [Ag(CN)_2]^-(aq)$$

The concentration of free Ag^+ ions is therefore so low that the subsequent addition of chloride ions will not bring about the precipitation of silver(I) chloride. If, however, sulphide ions, in the form of an alkali metal sulphide, are used instead of chloride ions, the precipitation is observed owing to the extremely low solubility of silver(I) sulphide.

Detection of Cd^{2+} in the presence of Cu^{2+}

In most qualitative analysis schemes, Cd^{2+} ions are detected in the same group as Cu^{2+} ions as their insoluble sulphides, and may be present in the same solution.

For the analysis of this group, the solution must initially be slightly acidic with hydrochloric acid in order to suppress the concentration of free sulphide ions, so that only those sulphides with low solubilities are precipitated. Unfortunately, the presence of cadmium ions may be missed, their precipitation as the yellow cadmium sulphide being masked by the black copper(II) sulphide precipitate.

Before distinguishing between Cd^{2+} and Cu^{2+} ions, it is necessary that they should be isolated from any bismuth ions which may also be present in this group. This is achieved by making the solution alkaline with dilute ammonia; this causes bismuth ions to be precipitated as the insoluble hydroxide while holding the cadmium and copper ions in solution in the form of their ammine complexes, $[Cd(NH_3)_4]^{2+}$ and $[Cu(NH_3)_4]^{2+}$. Excess potassium cyanide is then added to the solution of the complex ions; cyanide ligands displace the ammonia ligands and reduce the copper(II) ions to copper(I) ions to form the complex ions $[Cd(CN)_4]^{2-}$ and $[Cu(CN)_4]^{3-}$:

$$[Cd(NH_3)_4]^{2+}(aq) \rightleftharpoons Cd^{2+}(aq) + 4NH_3(aq)$$

$$\upharpoonleft\downharpoonright 4CN^-(aq)$$

$$[Cd(CN)_4]^{2-}(aq)$$

$$[Cu(NH_3)_4]^{2+}(aq) \rightleftharpoons Cu^{2+}(aq) + 4NH_3(aq)$$

$$\Big\updownarrow \; 4CN^-(aq) + e$$

$$[Cu(CN)_4]^{3-}(aq)$$

On the addition of sulphide ions to the solution, only cadmium sulphide is precipitated, as the stability constant of $[Cd(CN)_4]^{2-}$ is small by comparison with that of $[Cu(CN)_4]^{3-}$ (see page 228) and there are insufficient copper ions present to enable the copper(II) sulphide to be precipitated.

Chelates

The stability of a complex is enhanced by ligands which are capable of forming more than one bond with the central metal ion. Ligands of this type produce ring structures which are known as *chelates*; the ring-forming groups are described as *polydentate* groups, and ligands which provide two lone pairs for bonding are described as *bidentate* groups.

Ethane-1,2-diamine (ethylenediamine), $H_2NCH_2CH_2NH_2$, which is often abbreviated to *en* for convenience, is a bidentate chelating agent which has been encountered earlier in this chapter (Figure 16.5).

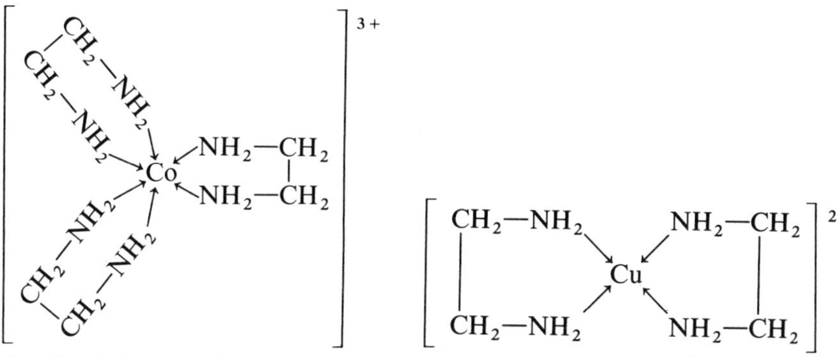

(a) Tris(ethane-1,2-diamine) cobalt(III) ion, $[Co(en)_3]^{3+}$

(b) Bis(ethane-1,2-diamine) copper(II) ion, $[Cu(en)_2]^{2+}$

Fig. 16.5

An important chelating agent, and probably the one most widely used, is edta (ethylenediaminetetraacetic acid, sequestrol or, systematically, bis[di-(carboxylmethyl)amino]ethane), which contains six bonding positions (four via the carbonyl oxygens of the carboxyl groups and two via the nitrogen atoms) and is capable of forming five ring structures with a metal ion:

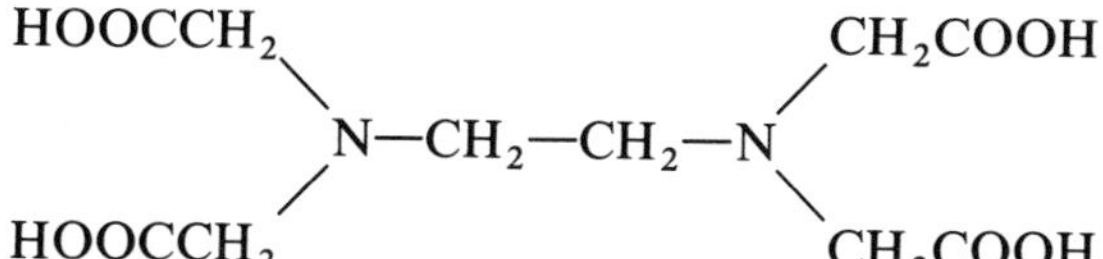

$$(HOOCCH_2)_2N{-}CH_2{-}CH_2{-}N(CH_2COOH)_2$$

edta

Edta is also extensively used in volumetric analysis for the quantitative estimation of Ca^{2+} and Mg^{2+} ions. The removing of ions from solution in this way to form stable complexes is sometimes referred to as *sequestration* and the compound bringing about the change is called a *sequestrating agent*.

Another substance, known by the trade name of Calgon and containing polyphosphate(V) ions, acts in a similar way to edta and is also commonly employed for softening both industrial and domestic water supplies.

Insoluble complexes

Most of the complexes encountered in chemistry are soluble in aqueous media, but occasionally the formation of insoluble complexes is important, especially in the field of qualitative analysis. For example, sodium hexanitrocobaltate(III) (sodium cobaltinitrite) forms a yellow precipitate with K^+ ions, and the organic reagent 2-hydroxy-1,2-diphenylethanone oxime (known as cupron reagent) gives a green precipitate with Cu^{2+} ions. Another particularly useful organic reagent is butanedione dioxime (dimethylglyoxime), which gives a series of different coloured solutions and precipitates with a number of ions, e.g.

Fe^{2+} Red colour
Ni^{2+} Red precipitate
Bi^{3+} Yellow precipitate

Detection of complex formation

How does one know that a complex has been formed in solution? Suspicions are often aroused as a result of the changes produced in the properties of the solution, and also from a knowledge of the nature of the ions used. More precise information may be obtained by means of several techniques, some of which include:

(1) Colligative properties

The formation of a complex is accompanied by a change in the number of particles dissolved in a solvent and this in turn will affect the vapour pressure, freezing point, boiling point, etc. of the solvent; for example,

$$\underbrace{Ag^+ + 2CN^-}_{\text{3 particles}} \qquad \underbrace{[Ag(CN)_2]^-}_{\text{1 particle}}$$

(2) Partition (distribution) coefficient

When ammonia, for example, is distributed between copper(II) sulphate solution and trichloromethane (chloroform), the formation of the copper complex is indicated by the values obtained for the partition coefficient. The experiment can also be designed to determine the value of x in the formula $[Cu(NH_3)_x(H_2O)_2]^{2+}SO_4^{2-}$.

(3) E.m.f. measurements

The e.m.f. of an electrochemical cell depends upon the ionic concentrations of the components present. The formation of a complex ion brings about a change in these ionic concentrations, and is therefore accompanied by an alteration in the e.m.f. of the cell.

(4) Conductance measurements

Conductance and ionic mobility are not included in this text. This method is mentioned, however, because of its importance in work of this kind, and the reader is advised to refer to a physical chemistry text for further information.

17 The d-block Elements

In the *d*-block elements, which are generally referred to as the *transition metals* as their properties are intermediate between those of the *s*- and *p*-blocks, it is the penultimate *d* orbitals which are partially filled with electrons.

The *d*-block is composed of three complete rows, or series, of ten elements and an incomplete fourth row (Table 17.1). Although the simple general patterns that occur in the *s*- and *p*-blocks do not emerge quite so clearly from a study of the transition elements, there are, nonetheless, certain characteristic patterns among their physical and chemical properties.

In the first section of this chapter, the general properties of the series are considered, making reference, for the most part, to the *first series, scandium to zinc*. The second part of the chapter deals with elements from the first series in more detail, with the emphasis on their oxidation states and the behaviour of their commonly encountered ions.

The transition elements are all metals and generally possess the properties which characterize metals, being good conductors of heat and electricity, hard, malleable, ductile, etc. Because of the similarity of their atomic radii, they are able to form alloys between themselves as well as with other metals.

General properties

Density

The densities of the transition metals are significantly higher than those of the *s*-block elements of the same period. This is attributable to the increase in the nuclear masses and to the lower atomic volumes which result from the outer electrons being pulled in by the increased nuclear charge.

Table 17.2

	Sc	*Ti*	*V*	*Cr*	*Mn*	*Fe*	*Co*	*Ni*	*Cu*	*Zn*
Density/ $g\ cm^{-3}$	2.99	4.54	5.96	7.19	7.20	7.86	8.90	8.90	8.92	7.14

Melting and boiling point

In contrast to the *s*-block elements, the melting and boiling points are generally high, the notable exceptions being Zn, Cd, and Hg where the *d* orbitals are all

Table 17.1

First series	21 Scandium Sc	22 Titanium Ti	23 Vanadium V	24 Chromium Cr	25 Manganese Mn	26 Iron Fe	27 Cobalt Co	28 Nickel Ni	29 Copper Cu	30 Zinc Zn
Second series	39 Yttrium Y	40 Zirconium Zr	41 Niobium Nb	42 Molybdenum Mo	43 Technetium Tc	44 Ruthenium Ru	45 Rhodium Rh	46 Palladium Pd	47 Silver Ag	48 Cadmium Cd
Third series	57 Lanthanum La	72 Hafnium Hf	73 Tantalum Ta	74 Tungsten W	75 Rhenium Re	76 Osmium Os	77 Iridium Ir	78 Platinum Pt	79 Gold Au	80 Mercury Hg
Fourth series	89 Actinium Ac	90 Thorium Th	91 Protactinium Pa	92 Uranium U						

completely filled (Table 17.3, and Figure 17.1).

Table 17.3

	Sc	*Ti*	*V*	*Cr*	*Mn*	*Fe*	*Co*	*Ni*	*Cu*	*Zn*
M.p./°C	1540	1675	1900	1890	1240	1535	1492	1453	1083	420
B.p./°C	2730	3260	3000	2482	2100	3000	2900	2730	2595	907

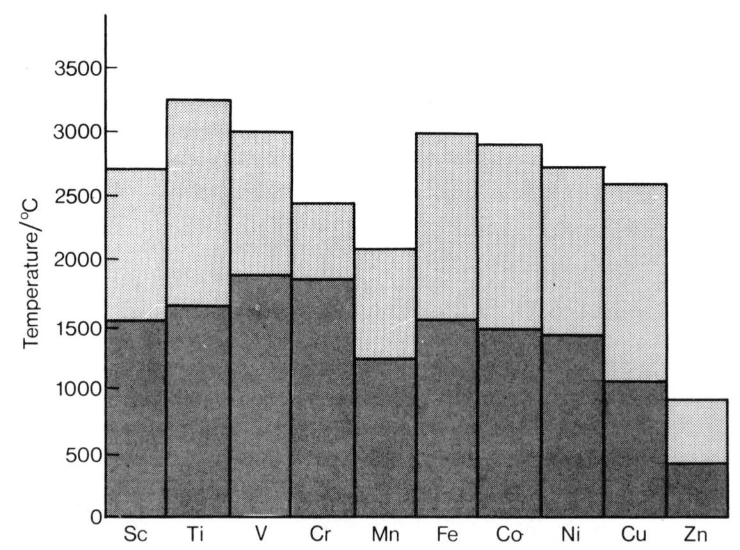

Fig. 17.1 Melting and boiling points of the elements of the first transition series

Ionization energy

Owing to the increase in the nuclear charge, the outer electrons of the *d*-block elements are more firmly held than those in the *s*-block and have predictably higher first ionization energies. The ionization energies of the transition elements are nevertheless relatively small, which is in accordance with their observed metallic properties (Table 17.4 and Figure 17.2).

Table 17.4 Ionization energies, $\Delta H^{\ominus}$/kJ mol^{-1}

	Sc	*Ti*	*V*	*Cr*	*Mn*	*Fe*	*Co*	*Ni*	*Cu*	*Zn*
1st	*637*	*666*	*653*	*658*	*722*	*767*	*762*	*741*	*750*	*913*
2nd	*1240*	*1310*	*1370*	*1590*	*1510*	*1560*	*1640*	*1750*	*1960*	*1730*
3rd	*2390*	*2720*	*2870*	*2990*	*3250*	*2960*	*3230*	*3390*	*3550*	*3830*

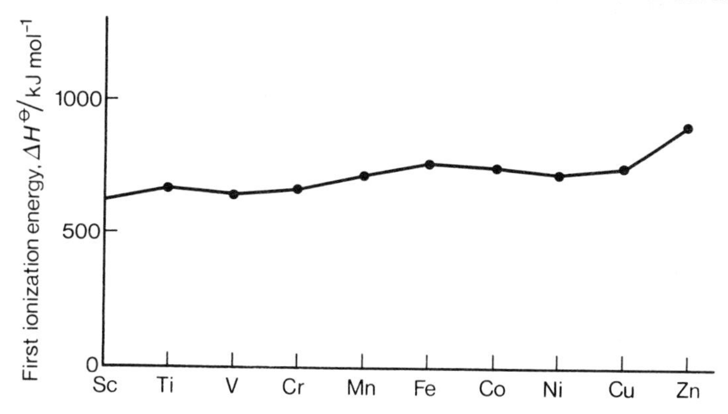

Fig. 17.2 Plot of the first ionization energies of the elements of the first transition series

Electronegativity values

These elements maintain the general pattern of increasing electronegativity on traversing the Periodic Table from left to right (Table 17.5), although the last member in each series proves exceptional and has a significantly lower value than the element immediately preceding it.

Table 17.5

	Sc	*Ti*	*V*	*Cr*	*Mn*	*Fe*	*Co*	*Ni*	*Cu*	*Zn*
Electronegativity	1.3	1.5	1.6	1.6	1.5	1.8	1.8	1.8	1.9	1.6

Because they are more electronegative than the *s*-block elements, it is no surprise that covalency features more in the properties of the transition metals. The tendency to form covalent compounds increases as the atoms get smaller, i.e. from Sc to Cu. The variation in size of the atoms is only very small by comparison with that observed on traversing Periods 2 and 3 of the *s*- and *p*-blocks, the elements Cr to Cu having almost identical atomic radii (Table 17.6).

Variable oxidation state

The transition metals generally portray a much wider range of oxidation states than do the elements of the *s*- and *p*-blocks. (Table 17.7). The existence of these variable states is ascribed to the comparative ease with which the *d* electrons can be promoted.

The electronic structures of chromium and copper do not fit the general pattern. In each case, greater electronic stability of the free atom in the ground state is acquired by the movement of an electron from the *s* level, providing chromium with an exactly half-filled *d* level (five electrons) and copper with a completely filled *d* level (ten electrons).

Table 17.6

Element	Sc	Ti	V	Cr	Mn
Atomic (covalent) radius/nm	0.144	0.132	0.122	0.117	0.117
Element	Fe	Co	Ni	Cu	Zn
Atomic (covalent) radius/nm	0.116	0.116	0.115	0.117	0.125

Table 17.7 Oxidation states exhibited by the transition metals (less common ones are enclosed in brackets)

	Sc	*Ti*	*V*	*Cr*	*Mn*	*Fe*	*Co*	*Ni*	*Cu*	*Zn*
Electronic structure of 3*d* and 4*s* levels	d^1s^2	d^2s^2	d^3s^2	d^5s^1	d^5s^2	d^6s^2	d^7s^2	d^8s^2	$d^{10}s^1$	$d^{10}s^2$
Oxidation states	3	2 3 4	2 3 4 5	(1) 2 3 (4) (5) 6	2 3 4 5 6 7	2 3 (4) (5) 6	2 3 4 (5)	2 (3) 4	1 2 (3)	2

Zero oxidation state is also a feature of the transition elements, notably in their carbonyl compounds (see page 224).

The stability of the highest oxidation states exhibited by each element of the first transition series decreases progressively on passing from left to right across the series (i.e. from Sc to Zn). Compounds in which the elements exhibit these oxidation states tend readily to undergo reduction, i.e. they are usually oxidizing agents.

As is general amongst elements having more than one oxidation state, the lower states have a greater tendency to form basic oxides and compounds which possess a high degree of ionic character, whereas the higher states tend to form acidic oxides and largely covalent compounds.

Colour

Colour arises as a result of the promotion of electrons to unoccupied orbitals of higher energy. This electronic transition corresponds to a precise wavelength of light, the transmitted or reflected light therefore being deficient in the wavelength absorbed and its corresponding colour.

The various colours displayed by a particular transition metal in different complexes may be explained as resulting from electronic transitions between the t_{2g} and the e_g levels. Consider, for example, the $[Cu(CN)_4]^{2-}$ ion, which is colourless and the $[Cu(H_2O)_6]^{2+}$ ion, which is blue. In the strong crystal field produced by the CN^- ions, the value of Δ (see page 221 and Figure 16.1) is too large to allow electronic transitions in the visible region and the complex appears colourless. On the other hand, in the comparatively weak ligand field of the H_2O molecules, Δ is small and the absorption of yellow light from the visible spectrum gives the hydrated copper(II) ion its characteristic blue appearance.

Table 17.8 shows the colour, oxidation state, and number of unpaired electrons of some of the hydrated ions of the first transition series.

Table 17.8 Aqua ions of some of the first transition series and their respective colours

Oxidation state and colour of the hydrated ion of type $[M(H_2O)_6]^{n+}$		*Number of unpaired electrons*
Ti(+4), colourless		0
Ti(+3), violet;	V(+4), blue	1
V(+3), green		2
V(+2), violet;	Cr(+3), violet	3
Cr(+2), blue;	Mn(+3), violet	4
Mn(+2), pink;	Fe(+3), yellow	5
Fe(+2), green		4
Co(+2), pink		3
Ni(+2), green		2
Cu(+2), blue		1
Cu(+1), colourless		0

In contrast to the transition metal ions, the ions of the *s*-block elements are typically white or colourless owing to the electrons being firmly held, so requiring absorption of light from the ultra-violet spectrum for excitation. Therefore, these compounds appear white or colourless unless, of course, the anion is itself coloured, e.g. chromate(VI) ion (yellow), dichromate(VI) ion (orange), or manganate(VII) (permanganate) ion (purple), in which case the colour of the anion predominates.

Magnetic properties

All substances are influenced to some degree by a magnetic field. Substances containing no unpaired electrons, when placed in a non-uniform magnetic field, tend to move to the weakest part of the field, repelling the magnetic lines of force, and are said to be **diamagnetic**. Those materials containing unpaired electrons, however, have the opposite effect, attracting the magnetic lines of force and tending to move from the weakest to the strongest part of the field; such substances are said to be **paramagnetic**.

Most transition metal ions possess unpaired electrons and show paramagnetic properties. Furthermore, the greater the number of unpaired electrons, the greater is the magnetic moment of the ion (Figure 17.3).

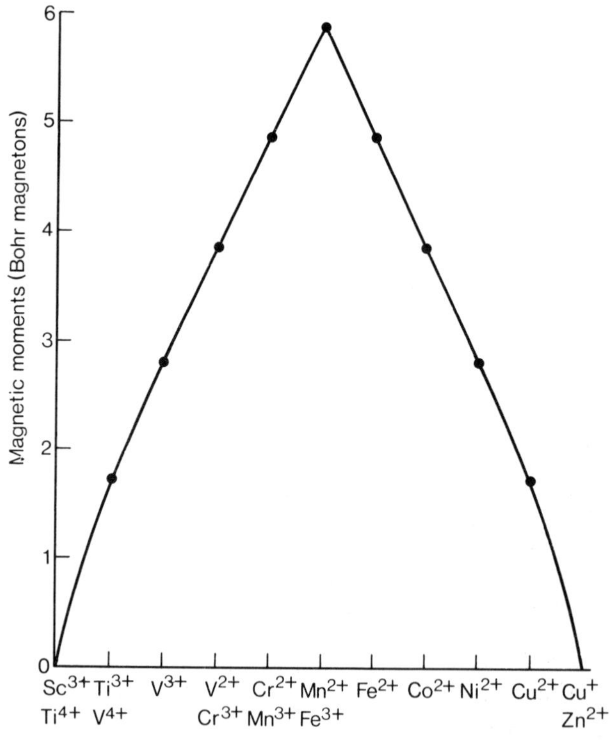

Fig. 17.3 Calculated magnetic moments of some of the ions of the first transition series in aqueous solution

A few crystalline substances are capable of attracting a magnetic field very strongly and may themselves be magnetized, their magnetism being permanent, even in the absence of a magnetic field. This property is referred to as **ferromagnetism** and is exhibited by iron, cobalt, and nickel, being strongest in iron.

Non-stoichiometry

All compounds suffer to some extent from some defects in the crystal state. At normal temperatures these are only small and do not alter the stoichiometric ratio to any appreciable extent. We are thus able to write Na^+Cl^- for sodium chloride even though minor crystal defects are present.

Some of the compounds of the transition metals, however, have their atoms in a far-from stoichiometric ratio, and may even be of indefinite proportions. Non-stoichiometry is particularly significant among the oxides and sulphides where the transition metal is present in more than one oxidation state. For example, 'FeO' can vary in composition from $Fe_{0.94}O$ to $Fe_{0.84}O$, and undoubtedly has some iron(II) replaced by iron(III); 'FeS' is obtained varying from $Fe_{0.89}S$ to $Fe_{0.86}S$, and 'WO_3' from $WO_{2.88}$ to $WO_{2.92}$.

Non-stoichiometric compounds often possess much more intense colour; white zinc(II) oxide, for example, becomes yellow when hot due to the effects of lattice defects.

Interstitial compounds

Elements of very small atomic radii, e.g. H, B, C, and N, sometimes form non-stoichiometric interstitial compounds with the transition metals by occupying the interstices of the fairly close-packed crystal lattice of the metal. This property is favoured by those metals whose cationic radii are large, such as V, Ti, Zn, Nb, Hf, and Ta. Hydrogen always occupies tetrahedral holes in the lattice, whereas carbon and nitrogen tend to occupy the larger octahedral holes (see page 42).

These compounds are of great technical and industrial importance. All steels, for example, are interstitial compounds of iron and carbon, and tungsten carbide is well known for its very hard refractory properties.

Catalytic properties

The reader can hardly be unaware of the catalytic properties of the elements Pt, Ni, Fe, and Pd – to mention just a few – and the extensive use that is made of them in industry. In some cases they provide a suitable surface for the reaction and in others they actually form unstable intermediates; both processes are facilitated by the variable oxidation states of the metals.

The first transition series of elements, scandium to zinc; a more detailed study

Scandium: [Ar]$3d^1 4s^2$

The chemistry of scandium is unimportant in the context of the first transition series. Its chemistry shows a much closer similarity to the lanthanoid elements ('rare earths'), and with the element lanthanum ($[Xe]5d^16s^2$) in particular, than with the first series. Unlike the other elements of the first series, scandium does not form a divalent ion and is virtually always trivalent in solution, utilizing all three of its outer electrons. Therefore, the Sc^{3+} ion has neither *d* nor unpaired electrons, and is predictably diamagnetic and colourless. Although its ability to form complexes is limited, a few, well characterized examples are known, e.g. $K_3[ScF_6]$.

Titanium $[Ar]3d^24s^2$

General

When pure, titanium is a grey, hard metal and a good conductor. Because of its low density, resistance to corrosion, and tensile strength, it finds particular application in the aero-space industries.

The chemistry of titanium is mainly that of its +2, +3, and +4 oxidation states. Other oxidation states occur (Table 17.7), but these are comparatively unimportant. All the lower oxidation states are oxidized by air and water to the +4 state, which is the most stable and common oxidation state.

Complexes are not easily formed by titanium, the stable $[TiF_6]^{2-}$ and $[TiCl_6]^{2-}$ ions being two of the few common examples, and even these are readily hydrolysed.

The metal combines with many non-metals, particularly above 500 °C. Examples include oxygen (forming TiO_2), fluorine and chlorine (forming TiF_4 and $TiCl_4$), nitrogen (forming TiN), carbon (forming TiC), and boron (forming TiB_2). The compounds formed with nitrogen, carbon, and boron are interstitial, refractory substances.

Titanium(+2)

There are relatively few compounds corresponding to this oxidation state of titanium. The most common are the halides $TiCl_2$, $TiBr_2$, and TiI_2 and the oxide TiO, the latter being a non-stoichiometric compound.

The absence of any aqua ion chemistry of the +2 state, because of its immediate oxidation by water, is unique in the first transition series.

Titanium(+3)

Titanium(III) ions are formed when the element reacts with dilute acid in the absence of air, or by the reduction of an aqueous solution of a titanium(IV) salt. The hydrated titanium(III) ion is violet and is composed mainly of $[Ti(H_2O)_6]^{3+}$ ions. On exposure to air, these are rapidly oxidized to the +4 state.

Titanium(III) salts are mild reducing agents and can be used in volumetric analysis. Titanium(III) sulphate is prepared by the action of the metal on moderately concentrated sulphuric acid:

$$2Ti(s) + 6H_2SO_4(aq) \rightarrow Ti_2(SO_4)_3(aq) + 3H_2O(l) + 6SO_2(g)$$

The initial concentration of titanium(III) ions in solution remains almost constant in the moderately concentrated acid for several hours and can be used to determine iron(III), chlorate(V), and chlorate(VII) (perchlorate) ions, reducing them respectively to iron(II) and chlorides. Titanium(III) chloride solution in moderately concentrated sulphuric acid can be used in the same way.

Crystalline, hydrated titanium(III) chloride has a green and a violet form (cf., its chromium counterparts, page 246), the latter being the hexaaqua compound $TiCl_3{\cdot}6H_2O$. Anhydrous titanium(III) chloride is prepared by the reduction of $TiCl_4$ with hydrogen above 500 °C.

Titanium(+4)

The simple titanium(IV) ion does not exist as such; the oxocation, TiO^{2+}, has been suggested but the evidence available indicates that more complex ions are present.

Titanium(IV) oxide, TiO_2, occurs naturally in three crystalline forms, the most common being rutile. This is a very stable compound which, because of its whiteness and opacity, is used in paints and in paper-making. Although the oxide is amphoteric, it shows mainly acidic properties, forming hydrated compounds such as $Na_2TiO_3{\cdot}xH_2O$ with concentrated alkalis. These compounds are referred to as 'titanates', but they are of unknown structure and there is in fact no TiO_3^{2-} ion present. With acids, the oxide forms basic salts, such as $TiOSO_4{\cdot}nH_2O$ (approximate formula).

Titanium (+4) can be reduced to the +3 state by zinc and hydrochloric acid, but not by sulphur dioxide or hydrogen sulphide.

$$TiO^{2+}(aq)+2H^+(aq)+H_2O(l)+e \rightleftharpoons Ti^{3+}(aq); \quad E^{\ominus} = +0.1\text{ V}$$

The halides TiF_4, $TiCl_4$, $TiBr_4$, and TiI_4 are well-known. $TiCl_4$ is a colourless covalent liquid, which fumes in air because of hydrolysis:

$$TiCl_4(l)+2H_2O(l) \rightarrow TiO_2(s)+4HCl(aq) \quad \text{(approximate equation)}$$

In the presence of excess chloride ions, titanium(IV) chloride forms the complex ion, $[TiCl_6]^{2-}$.

In hydrogen peroxide, an acidified solution of a titanium(IV) salt forms a yellow/orange colour. This reaction provides a very sensitive test for low concentrations of titanium(IV) compounds, the colour being due to the formation of ionic peroxo species.

Vanadium: [Ar]$3d^34s^2$

General

Vanadium is a hard, pale grey metal which, like titanium, is resistant to corrosion, and is used mainly as an alloying ingredient in steel.

The principal oxidation states of vanadium are +2 to +5 inclusive, the +4 and +5 states being the most important. Under ordinary conditions the +4 state is the most stable.

Ionic character and basic properties of its compounds predictably increase with decreasing oxidation state of the elements (see page 158). This means that:

vanadium(II) compounds are ionic, and the oxide, VO, is basic;
vanadium(III) compounds are mainly ionic, and the oxide, V_2O_3, is mainly basic;
vanadium(IV) compounds are mainly covalent, and the oxide, VO_2, is amphoteric and mainly basic;
vanadium(V) compounds are covalent, and the oxide, V_2O_5, is amphoteric and mainly acidic.

Vanadium is also similar to titanium in its behaviour on heating with non-metals. It reacts with oxygen (forming mainly V_2O_5), chlorine (forming VCl_4), nitrogen (forming VN), and carbon (forming VC and V_4C_3). As with titanium the nitride and carbides are interstitial.

Vanadium(+2)

This is the least important oxidation state of vanadium. Vanadium(II) ions can be obtained by the reduction of any of the higher oxidation states of vanadium with zinc and acid. In solution, the ion exists as the violet hexaaqua complex, $[V(H_2O)_6]^{2+}$. Solutions containing this ion are readily oxidized by air or water to mainly the V(+3) state, unless preserved in the presence of a strong reducing agent.

Several crystalline salts containing the hydrated ion are known, a typical example being $VSO_4 \cdot 6H_2O$, which exists as violet crystals.

Vanadium(II) oxide, VO, a black solid, is very basic, dissolving in acids to give the violet hexaaqua cation.

Vanadium(+3)

In aqueous solution, the vanadium(III) ion exists mainly as the green hexaaqua cation, $[V(H_2O)_6]^{3+}$. This is a reducing agent and is readily oxidized by air to the V(+4) state.

Vanadium(III) oxide, V_2O_3, is a black solid which is essentially basic in character, and can be obtained by reducing vanadium(V) oxide in a stream of hydrogen:

$$V_2O_5(s) + 2H_2(g) \rightarrow V_2O_3(s) + 2H_2O(l)$$

It forms salts, such as the sulphate, $V_2(SO_4)_3 \cdot 3H_2O$, which, on crystallizing from aqueous solution with an equivalent amount of potassium, forms an alum, $KV(SO_4)_2 \cdot 12H_2O$. Because of the greater stability of the V(+4) state, V_2O_3 undergoes gradual conversion, even in the cold, to vanadium(IV) oxide, VO_2, if exposed to oxygen.

Vanadium(+4)

Because of the ease of oxidation of V(+3) and the reduction of V(+5) to V(+4),

$$VO^{2+}(aq) + 2H^+(aq) + e \rightleftharpoons V^{3+}(aq) + H_2O(l); \quad E^\ominus = +0.36\,V$$
$$VO_2^+(aq) + 2H^+(aq) + e \rightleftharpoons VO^{2+}(aq) + H_2O(l); \quad E^\ominus = +1.0\,V$$

the +4 state is recognized as the most stable oxidation state of vanadium under ordinary conditions.

Vanadium(IV) oxide, VO_2, is a dark blue solid and can be obtained by the reduction of vanadium(V) oxide by sulphur dioxide. This process can be reversed by heating the vanadium(IV) oxide in air. Although it is amphoteric in character, the basic properties of vanadium(IV) oxide are more typical than its acidic properties. With acids, it forms salts, many of which contain the blue oxovanadium(IV) ion, $[VO(H_2O)_5]^{2+}$. With alkalis, it forms a solution which contains a variety of ions and from which vanadate(IV) compounds can be isolated.

Vanadium(+5)

Vanadium(V) oxide, V_2O_5, is an orange solid which can be prepared by the action of heat on ammonium vandate(V):

$$2NH_4VO_3(s) \rightarrow V_2O_5(s) + 2NH_3(g) + H_2O(l)$$

Although the simple formula is used in this equation, ammonium vanadate(V) has a polymer-type structure.

The oxide is amphoteric in nature but its acidic properties are much more pronounced than its basic. In aqueous alkali, it dissolves to give a colourless solution which contains orthovanadate(V) ions, VO_4^{3-}.

$$V_2O_5(s) + 6OH^-(aq) \rightarrow 2VO_4^{3-}(aq) + 3H_2O(l)$$

Hydrated vanadate(V) compounds of general formula, M_3VO_4, can be crystallized from solution. If the solution is made *slightly* acidic (pH about 6.5), then it becomes bright orange in colour due to the presence of more complex vanadate(V) ions. At a higher acid concentration (pH about 2), V_2O_5 precipitates out as a brown solid but redissolves in more acid to form the dioxovanadium(V) ion, VO_2^+. The VO_2^+ ion is the cationic species generally encountered for the V(+5) state, usually in acid solution.

With hydrogen peroxide, an acidified, colourless solution of a vanadium(V) salt forms an orange solution. This behaviour, like that of titanium(IV) salts, is due to the formation of complex peroxo ions.

If a solution of ammonium vanadate(V) is reduced with zinc and dilute acid, the different colours of the aqueous cationic species associated with each oxidation state of vanadium can be observed.

$$\underset{\text{colourless}}{\underset{(+5)}{VO_2^+(aq)}} \rightarrow \underset{\text{blue}}{\underset{(+4)}{VO^{2+}(aq)}} \rightarrow \underset{\text{green}}{\underset{(+3)}{[V(H_2O)_6]^{3+}(aq)}} \rightarrow \underset{\text{violet}}{\underset{(+2)}{[V(H_2O)_6]^{2+}(aq)}}$$

Chromium: **[Ar]*$3d^54s^1$***

General

Chromium is a hard, bluish-white metal which is extremely resistant to corrosion at ordinary temperatures and has the highest melting point (1890 °C) of all of the elements of the first transition series. It is used for plating steel objects and for alloying with steel in manufacturing stainless steel.

When pure, it dissolves very slowly in dilute acids to form chromium(II) salts.

The principal oxidation states of chromium are +2, +3, and +6, the others being rare and sometimes unstable (Table 17.7).

As ionic and basic character is more usually found among the lower oxidation states, it may reasonably be expected that:

chromium(II) compounds are essentially ionic and that CrO is basic;
chromium(III) compounds are partly ionic and that Cr_2O_3 is predominantly basic with some amphoteric character;
chromium(VI) compounds are essentially covalent and that CrO_3 is acidic.

Chromium(+2)

The chromium(II) ion, Cr^{2+}, is formed in solution when a solution containing chromium(VI) (e.g. a chromate(VI) or a dichromate(VI)) or chromium(III) ions is reduced by zinc and hydrochloric acid. Hydrated, Cr^{2+} ions, or more precisely $[Cr(H_2O)_6]^{2+}$ ions, are sky-blue in colour and their formation from dichromate(VI) ions is accompanied by a distinctive colour change:

$$\underset{\text{orange}}{Cr_2O_7^{2-}(aq)} \rightarrow \underset{\text{green/violet}}{Cr^{3+}(aq)} \rightarrow \underset{\text{sky-blue}}{Cr^{2+}(aq)}$$

Chromium(II) compounds are essentially ionic in character and easily oxidized in air to the more stable chromium(III) compounds; in solution they are one of the strongest reducing agents known.

$$Cr^{3+}(aq) + e \rightleftharpoons Cr^{2+}(aq); \quad E^{\ominus} = -0.41\text{ V}$$

Owing to the ease of oxidation of chromium(II) compounds, their general chemistry is not especially important, although this oxidation state may be stabilized by the formation of complexes. The most important complexes of chromium are, however, confined to the oxidation states of +3 and +6.

Chromium(+3)

This corresponds to the most stable oxidation state of chromium and, furthermore, in aqueous solution Cr^{3+} ions are the most stable of all the trivalent transition metal cations, existing as the violet hydrated cation, $[Cr(H_2O)_6]^{3+}$. (Because of replacement of water molecules by other anions present, this solution often appears green in colour.) $[Cr(H_2O)_6]^{3+}$ ions are predictably very stable in acidic solution but with an alkali, the gelatinous, pale green, hydrated sesquioxide, $Cr_2O_3 \cdot xH_2O$, (the simple hydroxide, $Cr(OH)_3$, does not exist as such) is precipitated.

$$[Cr(H_2O)_6]^{3+}(aq) \underset{H^+(aq)}{\overset{OH^-(aq)}{\rightleftharpoons}} \underset{\text{pale green}}{Cr_2O_3 \cdot xH_2O(s)}$$

With excess alkali, the sesquioxide dissolves to form salts known as *chromate(III) salts*, probably $[Cr(OH)_6]^{3-}$.

$$Cr_2O_3 \cdot xH_2O(s) \xrightarrow{\text{excess } OH^-(aq)} [Cr(OH)_6]^{3-}(aq)$$

deep green solution

Despite the stability of Cr(+3), it undergoes oxidation relatively easily to Cr(+6). This may be achieved by fusion with sodium peroxide or by adding sodium peroxide to a solution of the chromium(III) salt. The use of sodium peroxide ensures an alkaline medium. The products are usually a mixture of yellow chromate(VI), CrO_4^{2-}, and orange dichromate(VI), $Cr_2O_7^{2-}$, ions.

$$\underset{Cr(+3)}{[Cr(H_2O)_6]^{3+}(aq)} \xrightarrow[\text{or heat in atmosphere of oxygen}]{Na_2O_2} \underbrace{CrO_4^{2-}(aq) \text{ and } Cr_2O_7^{2-}(aq)}_{Cr(+6)}$$

Stable complexes containing Cr(+3) are extremely common, the chromium being present in both anionic and cationic complexes, e.g.

$$[Cr(H_2O)_6]^{3+}, \quad [Cr(NH_3)_6]^{3+}, \quad [CrCl_6]^{3-}, \quad \text{and} \quad [Cr(CN)_6]^{3-}.$$

Most of these complexes are octahedral and paramagnetic, containing three unpaired electrons.

Various mixed aqua/ammine and aqua/chloro complexes exist as a result of ligand substitution, e.g. $[Cr(H_2O)(NH_3)_5]^{3+}$ and $[Cr(H_2O)_5Cl]^{2+}$.

Chromium(+6)

The acidic nature of chromium(VI) oxide, CrO_3, is shown by its ability to react with an alkali to form the chromate(VI) ion.

$$CrO_3(s) + 2OH^-(aq) \rightarrow CrO_4^{2-}(aq) + H_2O(l)$$

An aqueous solution of chromium(VI) oxide contains the acids H_2CrO_4, $H_2Cr_2O_7$, $H_2Cr_3O_{10}$, and $H_2Cr_4O_{13}$. None of these has actually been isolated although their salts are well known.

As a result of the high oxidation state, chromium(VI) compounds are powerful oxidizing agents, the two most common ions used in this capacity being the yellow chromate(VI), CrO_4^{2-}, and the orange dichromate(VI), $Cr_2O_7^{2-}$, ions.

The CrO_4^{2-} ion has a tetrahedral structure and the $Cr_2O_7^{2-}$ ion adopts the shape of a double tetrahedron joined by a common oxygen atom. In solution, a ready equilibrium exists between these two ions which is dependent on the pH of the solution; a low pH favours the $Cr_2O_7^{2-}$ ion.

$$2CrO_4^{2-}(aq) + 2H^+(aq) \underset{\text{alkali}}{\overset{\text{acid}}{\rightleftharpoons}} Cr_2O_7^{2-}(aq) + H_2O(l)$$

Potassium dichromate(VI), $K_2Cr_2O_7$

The oxidizing properties of the dichromate(VI) ion, in the form of the potassium compound, are utilized in volumetric analysis, where it may be employed as a primary standard.

As an oxidizing agent, the $Cr_2O_7^{2-}$ ion operates in acidic solution (usually dilute sulphuric) in accordance with the equation:

$$Cr_2O_7^{2-}(aq)+14H^+(aq)+6e \rightleftharpoons 2Cr^{3+}(aq)+7H_2O(l); \quad E^\ominus = +1.33\,V$$

The end-point of the process is not particularly distinct, the orange colour of the dichromate(VI) being replaced by the green colour of the hydrated Cr^{3+} ion (see page 246). It is usual, therefore, to use a suitable oxidation-reduction indicator such as diphenylamine which, at the end-point, turns from colourless to blue.

The many oxidation reactions of the dichromate(VI) ion in aqueous solution include:

Iodide to iodine, $2I^-(aq) \rightleftharpoons I_2(aq)+2e$
Hydrogen sulphide to sulphur, $H_2S(aq) \rightleftharpoons S(s)+2H^+(aq)+2e$
Tin (+2) to tin (+4), $Sn^{2+}(aq) \rightleftharpoons Sn^{4+}(aq)+2e$
Iron (+2) to iron (+3), $Fe^{2+}(aq) \rightleftharpoons Fe^{3+}(aq)+e$

Manganese: [Ar]*3d⁵4s²*

General

Manganese is a hard, grey metal, similar in appearance to iron, and exists in three allotropic forms which are stable over various temperature ranges. It is not readily attacked by air but dissolves in dilute acids to form manganese(II) salts.

Manganese exists readily in a greater number of oxidation states than any other transition metal, the principal states being +2, +4, +6, and +7, although all values from +1 to +7 are known. Of these, the oxidation state of +2 is the most stable.

As in the case of chromium, the covalency of the compounds and the acidity of the oxides increases with increasing value of oxidation state:

	MnO	MnO_2	MnO_4^{2-}	Mn_2O_7
Oxidation state	+2	+4	+6	+7

$\xrightarrow{\hspace{6em}}$
increase in covalency and acidity

Manganese(+2)

Compounds of Mn(+2) are quite common and include the hydroxide $Mn(OH)_2$, the chloride, $MnCl_2{\cdot}4H_2O$, the sulphate $MnSO_4{\cdot}5H_2O$ and the carbonate, $MnCO_3$. The Mn(+2) state is more stable than the divalent ions of other transition metals and is therefore more resistant to oxidation than the corresponding Cr^{2+} and Fe^{2+} ions.

The Mn^{2+} ion is formed in solution on reacting the metal with a dilute acid,

$$Mn(s)+2H^+(aq) \rightarrow \underset{\text{pale pink}}{Mn^{2+}(aq)}+H_2(g)$$

although the ion is more correctly represented as the hydrated complex, $[Mn(H_2O)_6]^{2+}$, which is paramagnetic since it possesses five unpaired electrons, and is pale pink in colour.

Manganese(II) oxide is a grey-green solid prepared by heating the carbonate (or any higher oxide of manganese) in hydrogen. It is distinctly basic and is readily oxidized in air, even in the cold, to form a higher oxide of manganese.

Manganese(II) salts are stable in acidic solution, but yield a white precipitate of the hydroxide in alkaline solution.

$$Mn^{2+}(aq) + 2OH^-(aq) \rightarrow \underset{\text{white}}{Mn(OH)_2(s)}$$

On exposure to air, the hydroxide is oxidized and darkens to the oxide of Mn(+4) via the formation of the oxide of Mn(+3), i.e. the sesquioxide, Mn_2O_3.

$$Mn(OH)_2(s) \xrightarrow{\text{air oxidation}} \underset{\text{black}}{Mn_2O_3(s)} \longrightarrow \underset{\text{black}}{MnO_2(s)}$$

Chlorate(V) oxidation of the hydrated Mn^{2+} ion in acidic solution gives the manganese(IV) oxide, MnO_2,

$$Mn^{2+}(aq) + 2[O] + 2e \xrightarrow{\text{chlorate(V)/acid}} MnO_2(s)$$

whereas a more powerful oxidizing agent, such as PbO_2 or $NaBiO_3$, brings out the higher oxidation state, Mn(+7), in the form of the purple manganate(VII) (permanganate) ion, MnO_4^-.

$$Mn^{2+}(aq) + 4[O] + 3e \longrightarrow MnO_4^-(aq)$$

Manganese(+4)

The only readily available compound in which manganese corresponds to an oxidation state of +4 is black manganese(IV) oxide (manganese dioxide), MnO_2, which occurs naturally as *pyrolusite*. This compound is non-stoichiometric and exhibits amphoteric behaviour, forming *manganate(IV)* ions with alkalis (although the exact constitution of these is uncertain), and the complex ion, $[MnCl_6]^{2-}$, with dilute hydrochloric acid.

$$MnO_2(s) + 4H^+(aq) + 6Cl^-(aq) \rightarrow [MnCl_6]^{2-}(aq) + 2H_2O(l)$$

Manganese(IV) chloride, $MnCl_4$, has not been isolated and the unstable complex, $[MnCl_6]^{2-}$, readily breaks down to form manganese(II) chloride and free chlorine:

$$[MnCl_6]^{2-}(aq) \rightarrow Mn^{2+}(aq) + 4Cl^-(aq) + Cl_2(g)$$

This reaction is often employed for the laboratory preparation of chlorine gas.

Manganese(IV) oxide is almost insoluble in water, but may be oxidized to the green manganate(VI) ion, MnO_4^{2-}, by fusion with an alkali in the presence of air or a suitable oxidizing agent, such as potassium chlorate(V).

$$2MnO_2(l)+4OH^-(l)+2[O] \rightarrow 2MnO_4^{2-}(l)+2H_2O(l)$$

The MnO_4^{2-} ion can be easily oxidized to the MnO_4^- ion (see below).

Manganese(IV) salts are unstable in aqueous solution, being readily oxidized to the Mn(+6) state in alkaline solution or reduced to the Mn(+2) state in acidic solution.

Manganese(+6)

The trioxide, MnO_3, is thought not to exist. The corresponding manganate(VI) compounds of sodium and barium are, however, quite stable giving rise to the green, tetrahedral manganate(VI) ion in aqueous solution. Only two salts of this ion have been isolated in the pure form, K_2MnO_4 and Na_2MnO_4. Both of these are dark green in colour.

Manganate(VI) ions are only stable in basic solution, the MnO_4^{2-} ion disproportionating in acidic or only slightly basic solution into Mn(+4) and Mn(+7). This can be achieved by bubbling carbon dioxide through the solution.

$$\underset{Mn(+6)}{3MnO_4^{2-}(aq)}+4H^+(aq) \rightarrow \underset{Mn(+7)}{2MnO_4^-(aq)}+\underset{Mn(+4)}{MnO_2(s)}+2H_2O(l)$$

Industrially, manganate(VII) ions are obtained by the electrolytic oxidation of aqueous manganate(VI) ions.

$$MnO_4^{2-}(aq) \xrightarrow{\text{electrolysis}} MnO_4^-(aq)+e$$

Manganese(+7)

With the exception of its existence in the manganate(VII) ion, MnO_4^-, the Mn(+7) state is rarely encountered. It is, however, the most strongly oxidizing of all the higher oxidation states of the transition metals in aqueous solution.

Potassium manganate(VII) (permanganate), $KMnO_4$

Crystalline potassium manganate(VII) is dark purple in colour. In solution, it provides the principal source of the tetrahedrally shaped manganate(VII) ion, MnO_4^-.

In a *slightly* acidic solution, it slowly decomposes, evolving oxygen and precipitating black manganese(IV) oxide.

$$4MnO_4^-(aq)+4H^+(aq) \rightarrow 4MnO_2(s)+3O_2(g)+2H_2O(l)$$

In neutral and basic solution, decomposition is very much slower. As this process is catalysed by light, solutions of potassium manganate(VII) are preferably stored in dark bottles.

Potassium manganate(VII) is commonly encountered as an oxidizing agent in both acidic and basic solution. In the presence of excess acid, usually dilute sulphuric acid, it is reduced by excess of the reducing agent to Mn^{2+}.

$$MnO_4^-(aq)+8H^+(aq)+5e \rightleftharpoons Mn^{2+}(aq)+4H_2O(l); \quad E^\ominus = +1.51\,V$$

In the presence of excess manganate(VII), however, the Mn^{2+} is oxidized in accordance with the equation:

$$3Mn^{2+}(aq)+2MnO_4^-(aq)+2H_2O(l) \rightleftharpoons 5MnO_2(s)+4H^+(aq); \quad E = +0.46\,V$$

In basic solution, MnO_4^- is reduced to manganese(IV) oxide.

$$MnO_4^-(aq)+2H_2O(l)+3e \rightleftharpoons MnO_2(s)+4OH^-(aq); \quad E^\ominus = +1.23\,V$$

In concentrated alkali, excess manganate(VII) slowly evolves oxygen and forms the green manganate(VI) ion.

$$MnO_4^-(aq)+e \rightleftharpoons MnO_4^{2-}(aq); \quad E^\ominus = +0.56\,V$$

For volumetric analysis purposes, potassium manganate(VII) is usually employed in acidic solution as, under these conditions, it acts as its own indicator, discharging its own colour at the end-point. Manganate(VII) oxidations are accelerated by the presence of the hydrated Mn^{2+} ions, which function as an autocatalyst.

The following partial ionic equations represent some instances where potassium manganate(VII) may be used as a means of gaining quantitative information in volumetric analysis.

Ethanedioate to carbon dioxide	$(COO)_2^{2-}(aq) \rightleftharpoons 2CO_2(g)+2e$
Iron (+2) to iron (+3)	$Fe^{2+}(aq) \rightleftharpoons Fe^{3+}(aq)+e$
Nitrite to nitrate	$NO_2^-(aq)+H_2O(l) \rightleftharpoons NO_3^-(aq)+2H^+(aq)+2e$
Iodide to iodine	$2I^-(aq) \rightleftharpoons I_2(aq)+2e$

Iron: [Ar]$3d^64s^2$

General

Pure iron is whitish grey and comparatively soft. The metal exists in three allotropic forms; the stable form or α-form (*ferrite*) is strongly attracted by a magnetic field, a phenomenon known as *ferromagnetism.* At a temperature of 768 °C this allotrope loses its ferromagnetic properties, becoming simply paramagnetic; at 906 °C a transition to the γ-iron occurs and at 1401 °C to the δ-form.

$$Fe_\alpha \xrightleftharpoons{906\,°C} Fe_\gamma \xrightleftharpoons{1401\,°C} Fe_\delta$$

The α- and δ- forms both have metallic, body-centred cubic structures, whereas the γ-form has a face-centred cubic structure. The ferromagnetic property of α-iron is very much stronger than the comparatively weak paramagnetic properties exhibited by the Fe^{3+} ion (see page 240).

Iron is the first element in the first row of transition elements which does not have an oxidation state corresponding to the total number of valence electrons (eight), the highest state being +6, although this is rarely encountered. The two common oxidation states are +2 and +3. These are much closer in terms of stability than the +2 and +3 states of other transition metals and one of the features of the chemistry of iron is the comparative ease with which each of these oxidation states can be converted into the other state by means of quite mild oxidizing and reducing agents. Other oxidation states of iron are less important.

The general pattern of decrease in the basic properties with increasing oxidation state is again maintained; iron(II) oxide, FeO, is basic while iron(III) oxide possesses certain amphoteric characteristics, although it is predominantly basic. The mixed oxide of iron, Fe_3O_4 ($FeO{\cdot}Fe_2O_3$) displays essentially basic properties.

The extraction of iron; the blast furnace

The iron ore, such as haematite, Fe_2O_3, or siderite, $FeCO_3$, is first roasted to remove water, decompose carbonates, and oxidize sulphides. It is then mixed with coke and limestone and fed into the top of the *blast furnace* (Figure 17.4) by a hopper, which has a double cone design to prevent the loss of gases. The furnace is constructed of steel and lined with firebrick, and is about 30 metres high.

Preheated oxygen-enriched air is blasted into the lower part of the furnace through small nozzles, known as tuyères. The temperature of this part of the furnace reaches 1760 °C to 1900 °C, as compared with about 250 °C in the uppermost region.

Although some reduction of the oxide by the coke occurs in the hottest part of the furnace, there is little doubt that the principal reducing agent is carbon monoxide, formed in the middle and lower zones by the blast of air through the coke:

$$C(s)+O_2(g) \rightarrow CO_2(g)$$
$$C(s)+CO_2(g) \rightarrow 2CO(g)$$
$$FeO(s)+CO(g) \rightarrow Fe(l)+CO_2(g)$$

The reduction reaction is often represented as

$$Fe_2O_3(s)+3CO(g) \rightarrow 2Fe(l)+3CO_2(g)$$

but by the time the main process takes place the iron oxide is mostly in the form of FeO. The carbon dioxide formed in the reduction of the ore is itself reduced in the middle and lower regions of the furnace to the monoxide. Thermodynamics dictates that this reduction of carbon dioxide can only occur at temperatures in excess of 700 °C.

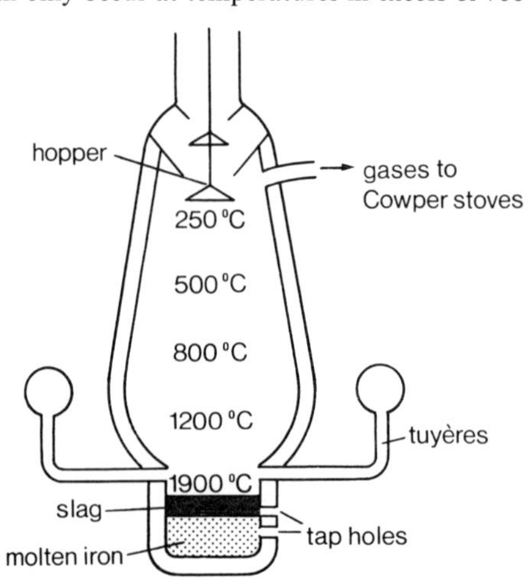

Fig. 17.4
A blast furnace

As the reduction of the ore by carbon monoxide is a reversible process, it is essential to ensure that the conditions are such that the monoxide is in considerable excess. The hot gases which leave the furnace are freed from dust and used in heat exchangers (Cowper stoves) to pre-heat the air before it is blasted into the furnace.

In the upper region of the furnace the charge loses moisture and becomes porous. In the middle region, at about 800–1000 °C, reduction of the ore begins and impurities start to separate out.

The limestone in the charge decomposes at about 800 °C and removes earthy acidic impurities, which are mainly sand, to form a fusible slag consisting largely of calcium silicate.

$$CaCO_3(s) \rightarrow CaO(s) + CO_2(g)$$
$$CaO(s) + SiO_2(s) \rightarrow CaSiO_3(l)$$

The limestone also provides a further source of carbon monoxide, the dioxide evolved on decomposition being reduced by the coke.

At 1100–1200 °C, the crude iron absorbs 3–5 per cent carbon, which lowers the melting point of the iron and causes it to become spongy and then molten. The carbon is present mainly as *cementite*, Fe_3C. Some sulphur, phosphorus, and manganese, formed by the reduction of other oxide impurities present in the ore, may also be absorbed by the molten iron forming an alloy.

The process is continuous, the molten iron and slag being run off periodically through separate tap holes in the well of the furnace. Most of the molten iron (90–95 per cent) is transferred in the liquid state to the steel manufacturing industry. The remaining crude iron, known as *pig iron* or *cast iron*, is transferred to open moulds, referred to as pig-beds, where it solidifies.

Pig iron is extremely hard but, unfortunately, very brittle. It is used for making manhole covers, framework for machinery, and other items which do not require a material of appreciable tensile strength.

Iron is classified as either 'basic' or 'acid' iron, the main difference being that 'basic' iron has a high phosphorus and low silicon content whereas 'acid' iron is low in phosphorus. The type of iron manufactured is governed primarily by the type of ore available and by the purpose for which the iron is to be used. In the U.K., the main ore is haematite, which is more suitable for producing 'basic' iron. This is the more important type of iron for steel-making, although small quantities of 'acid' steel are manufactured.

Wrought iron, which is an almost pure form of iron, is manufactured nowadays only on a very small scale since, for most purposes, *mild steel* is a suitable alternative and considerably cheaper to produce.

Steel making

Two of the most important techniques used for manufacturing steel have been the Bessemer and open-hearth processes. In recent years, however, the industry has been undergoing major changes and the original Bessemer process is gradually being superseded by improved oxygen-injection techniques. Of these, the *basic oxygen steelmaking process*, (B.O.S., Figure 17.5), is rapidly becoming the most popular.

The converter consists of a cylindrical furnace which can be rotated about both horizontal and vertical axes. The tilted furnace is charged with pig iron together with up to 30 per cent scrap iron, and then rotated into the vertical position. A blast of oxygen under great pressure is directed onto the surface of the molten charge by means of a water-cooled lance which is lowered into the furnace. To prevent the nozzles of the lance burning away, the oxygen is diluted with air, steam, or carbon dioxide. During the 'blow', which

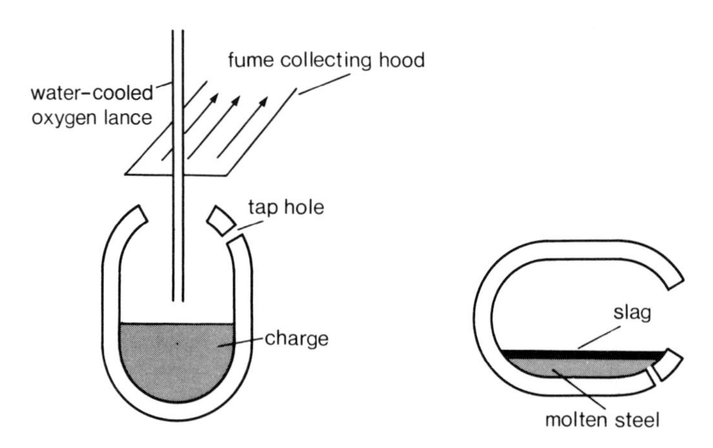

Fig. 17.5 A basic oxygen converter in the vertical and horizontal positions

lasts about fifteen minutes, lime is added as a flux to help remove the oxidized impurities as a slag. The temperature of the furnace is carefully controlled to prevent this rising too high, and samples of the molten metal are taken periodically to check its composition.

The carbon impurity is converted into its oxides and removed from the furnace through a hood. The other oxides of silicon, phosphorus, and manganese form a slag with the lime, most of this slag being run off at a convenient stage.

Once the required concentration of carbon has been achieved, the oxygen blast is turned off and the furnace tipped into the horizontal position. The molten steel is then poured from the tap hole. The residual slag, which contains some unrecovered iron, is left to become part of the next charge.

The properties of steel

The properties of steel depend upon its composition and heat treatment. *Mild steel* is comparatively soft and is an alloy of iron and 0.15 to 0.2 per cent carbon. *Medium steel* contains between 0.2 and 0.6 per cent carbon and *hard steel*, which has a high tensile strength, contains up to 1.5 per cent carbon. For general engineering purposes, mild steel is chiefly used.

In *tempering* or *annealing* steel for making cutting tools, the steel is heated to a temperature of between 230 °C and 315 °C (the precise temperature depending upon the purpose for which it is to be used) and then *quenched* by dropping into cold water or oil.

Other special properties can be afforded to steel by alloying with other elements. Stainless steel is manufactured by adding 12 per cent chromium, and hard steels, which do not soften at high temperatures, may contain Cr, W, V, Ti, and sometimes Mo. Steels required to withstand extremely hard wear often contain up to 10 per cent manganese, and those used for constructional purposes may contain Cr, Ni, and sometimes V and Mn.

Iron(+2) and iron(+3)

Because of the ease with which Fe(+2) and Fe(+3) are interconverted, the two oxidation states are conveniently discussed together.

Fe^{2+} ions are formed when the metal is dissolved in dilute acid. In aqueous solution they always exist as the hydrated complex (the hexaaquairon(II) ion), $[Fe(H_2O)_6]^{2+}$, which is pale green in colour.

$$Fe(s) + 2H^+(aq) + 6H_2O(l) \rightarrow [Fe(H_2O)_6]^{2+}(aq) + 3H_2(g)$$

The hexaaquairon(II) ion is slightly more stable in acidic solution than in basic but, nonetheless, can be oxidized to the Fe(+3) state by ordinary molecular oxygen under both types of conditions with relative ease.

$$\underset{\text{pale green}}{4[Fe(H_2O)_6]^{2+}(aq)} + O_2(g) + 4H^+(aq) \xrightarrow{\text{very low pH}} \underset{\text{pale purple}}{4[Fe(H_2O)_6]^{3+}(aq)} + 2H_2O(l)$$

If the solution is not *strongly* acidic, the pale purple $[Fe(H_2O)_6]^{3+}$ ion rapidly hydrolyses, forming yellow aqua/hydroxo complexes, e.g. $[Fe(H_2O)_5OH]^{2+}$, and, at pH values of about 3 or more, precipitates hydrated iron(III) oxide in the colloidal form. The formula of this compound is represented in various ways: $\frac{1}{2}Fe_2O_3{\cdot}3H_2O$, $Fe_2O_3{\cdot}1.5H_2O$, $Fe_2O_3{\cdot}xH_2O$, sometimes as iron(III) oxyhydroxide $FeO{\cdot}OH$, and occasionally as 'iron(III) hydroxide', $Fe(OH)_3$, although this is used more for convenience than accuracy.

$$[Fe(H_2O)_6]^{3+}(aq) \rightleftharpoons [Fe(H_2O)_5OH]^{2+}(aq) + H^+(aq)$$
$$[Fe(H_2O)_5OH]^{2+}(aq) \rightleftharpoons [Fe(H_2O)_4(OH)_2]^+(aq) + H^+(aq)$$
$$Fe(H_2O)_4(OH)_2(aq) \rightleftharpoons [Fe(H_2O)_3(OH)_3](aq) + H^+(aq)$$
$$\downarrow$$
$$Fe_2O_3{\cdot}xH_2O(s)$$

In *basic solution*, the hydrated Fe^{2+} ions form a dirty, pale green precipitate of iron(II) hydroxide, a typical base,

$$[Fe(H_2O)_6]^{2+}(aq) + 2OH^-(aq) \rightarrow Fe(OH)_2(s) + 6H_2O(l)$$
$$\downarrow \text{air-oxidation}$$
$$Fe_2O_3{\cdot}xH_2O(s)$$

which darkens almost immediately as a result of air-oxidation to the hydrated iron(III) oxide.

Crystalline compounds containing Fe^{2+} ions also undergo slow air-oxidation to Fe^{3+} ions on exposure to the atmosphere.

The amphoteric nature of $Fe(OH)_2$ is indicated by its ability to form iron(II) salts with dilute acids, and to form *ferrate(III)* ions (possibly FeO_2^- or $[Fe(OH)_6]^{3-}$) with concentrated alkalis.

Reducing agents, such as hydrogen sulphide and sulphur dioxide, readily convert Fe(+3) into Fe(+2).

$$2Fe^{3+}(aq) + H_2S(g) \rightarrow 2Fe^{2+}(aq) + 2H^+(aq) + S(s); \quad E = +0.63\ V$$

Complexes

It will already be apparent from the previous chapter that complexes are a feature of the chemistry of iron, the following being among those most commonly formed by the hydrated Fe^{3+} ions:

$[Fe(CN)_6]^{3-}$ with CN^- ions,

$[Fe(SCN)]^{2+}$ with SCN^- ions,

and $\left[Fe\begin{pmatrix}COO\\|\\COO\end{pmatrix}_3\right]^{3-}$ with ethanedioate, $(COO)_2^{2-}$, ions,

the latter finding application in removing rust stains from clothing. Also of interest is the pentacyanonitrosylferrate(II) ion (as sodium nitroprusside) which is used as a sensitive test for sulphur. The deep red colour obtained by adding thiocyanate ions, SCN^-, to a solution containing Fe^{3+} ions provides a discriminating test for ions in this state.

The hexacyanoferrate(II) (ferrocyanide) ion is more stable than the hexacyanoferrate(III) (ferricyanide) ion, a fact which is borne out by the electrode potential for the change:

$$[Fe(CN)_6]^{3-}(aq) + e \rightleftharpoons [Fe(CN)_6]^{4-}(aq); \quad E^\ominus = +0.36\,V$$

The hexacyanoferrate(II) ion is formed by the action of a soluble cyanide (e.g. NaCN, KCN) on an aqueous solution containing Fe^{2+} ions.

$$[Fe(H_2O)_6]^{2+}(aq) + 6CN^-(aq) \rightarrow \underset{\text{yellow}}{[Fe(CN)_6]^{4-}(aq)} + 6H_2O(l)$$

Chlorine and other oxidizing agents convert the hexacyanoferrate(II) ion into the hexacyanoferrate(III) ion, which appears yellow in aqueous solution but red in the crystalline form, e.g. $K_3[Fe(CN)_6]$.

Other examples of Fe(+2) complexes are known, and include haemoglobin (the red pigment in blood), and the complex formed in the brown ring test for detecting nitrates and nitrites (probably $[Fe(H_2O)_5NO]^{2+}$).

Hexacyanoferrates are probably best known as the potassium compounds for producing blue precipitates in detecting and distinguishing Fe^{2+} and Fe^{3+} ions (Table 17.9).

Table 17.9

Reagent	*with* Fe^{2+}	*with* Fe^{3+}
Hexacyanoferrate(III) ion, $[Fe(CN)_6]^{3-}$	Turnbull's blue precipitate $[Fe(Fe(CN)_6)]^-$	Brown solution $[Fe(Fe(CN)_6)]$
Hexacyanoferrate(II) ion, $[Fe(CN)_6]^{4-}$	White precipitate $[Fe(Fe(CN)_6)]^{2-}$	Prussian blue precipitate $[Fe(Fe(CN)_6)]^-$

The Prussian blue and Turnbull's blue precipitates are recognized as being the same compound and, in fact, all the products obtained in this scheme are structurally related. The difference in intensity of the two blue precipitates is probably due to the presence in the Turnbull's blue of the white compound $K_2[Fe(Fe(CN)_6)]$.

Cobalt: [Ar]$3d^74s^2$

General

Cobalt is a hard, bluish-white metal possessing ferromagnetic properties. It is slightly less reactive than iron and is resistant to attack by air and water at ordinary temperatures, but reacts slowly with dilute acids, although the reaction with dilute nitric acid is somewhat more rapid than with dilute hydrochloric and sulphuric acids.

The principal oxidation states of cobalt are +2 and +3. The Co(+1) state also exists but is less common, and the Co(+4) state is extremely rare. The Co(+3) state is encountered mainly in the form of its complexes, as simple cobalt(III) salts are not easily obtained owing to the strong oxidizing powers of the Co^{3+} ion which is capable of liberating oxygen from water.

Cobalt(+2)

The Co(+2) state exists in solution as the pink hydrated complex $[Co(H_2O)_6]^{2+}$, and is fairly resistant to oxidation.

$$[Co(H_2O)_6]^{3+}(aq) + e \rightleftharpoons [Co(H_2O)_6]^{2+}(aq); \quad E^{\ominus} = +1.82\,V$$

The freshly formed $[Co(H_2O)_6]^{2+}$ ion is pink in colour but on warming rapidly becomes blue.

In the presence of strong complexing agents, such as CN^- and NH_3, the stability of the Co^{3+} ion is greatly enhanced and oxidation of the hydrated Co^{2+} ion occurs much more easily, forming a cobalt(III) complex (see page 222). These conditions favour the oxidation reaction to such an extent that the presence of atmospheric oxygen is often sufficient to bring about the change.

$$\underset{\text{pink}}{[Co(H_2O)_6]^{2+}(aq)} + 6CN^-(aq) \xrightarrow{\text{air or } H_2O_2} \underset{\text{yellow}}{[Co(CN)_6]^{3-}(aq)} + 6H_2O(l) + e$$

$$[Co(H_2O)_6]^{2+}(aq) + 6NH_3(aq) \xrightarrow{\text{air or } H_2O_2} \underset{\text{yellow-orange}}{[Co(NH_3)_6]^{3+}(aq)} + 6H_2O(l) + e$$

Most of the complexes of Co(+2) are octahedral in shape, e.g. $[Co(H_2O)_6]^{2+}$, $[Co(NH_3)_6]^{2+}$, and $[Co(CN)_6]^{4-}$, although a few are tetrahedral, e.g. $[CoCl_4]^{2-}$ and $[CoBr_4]^{2-}$. The $[CoCl_4]^{2-}$ ion is easily prepared by the action of Cl^- ions on the hydrated Co^{2+} ion.

$$\underset{\text{pink}}{[Co(H_2O)_6]^{2+}(aq)} + 4Cl^-(aq) \xrightarrow{\text{conc. HCl or NaCl}} \underset{\text{blue}}{[CoCl_4]^{2-}(aq)} + 6H_2O(l)$$

Cobalt(+3)

As has already been implied, the chemistry of the Co(+3) state is essentially that of its stable complexes, of which it forms a greater number than any other element. They are all quite easily prepared and nearly all octahedral in shape. Examples include:

$[Co(NH_3)_6]^{3+}$	Yellow-orange
$[Co(CN)_6]^{3-}$	Yellow
$[Co(NO_2)_6]^{3-}$	Yellow
$[Co(NH_3)_3(NO_2)_3]$	Yellow
$[Co(NH_3)_5Cl]^{2+}$	Purple

The hexanitrocobaltate(III) ion, $[Co(NO_2)_6]^{3-}$, is used in the form of its soluble sodium salt as a test for potassium, whose salt is yellow and insoluble. A cobalt(+3) complex is also present in vitamin B_{12}.

Trivalent cobalt is significantly more stable in basic solution than in acidic.

Nickel: [Ar]$3d^84s^2$

Nickel is a hard, silver-grey metal possessing most of the physical characteristics usually associated with metals. Because of its inertness to corrosion, it is used for electroplating and finds extensive use in industry as a catalyst and as a component of alloys, e.g. nichrome (80 per cent Ni, 20 per cent Cr); constantan (40 per cent Ni, 60 per cent Cu); coinage alloys, 'silver' (25 per cent Ni, 75 per cent Cu), and invar steel (35 per cent Ni). Like iron, nickel is ferromagnetic although much less strongly so.

Owing to the fact that it contains three electron pairs in the 3*d* level, nickel, unlike many of the other transition metals, does not display many stable oxidation states. The +2 state and, to a lesser extent, the +4 state are by far the most important. At 60 °C it forms, with carbon monoxide, tetracarbonylnickel(0) in which it exhibits zero oxidation state (see page 224).

Nickel(+2)

Although oxidation and reduction between two or more oxidation states has featured prominently in the chemistry of the transition metals so far discussed, it does not arise in the chemistry of nickel to any appreciable extent because of the comparative stability of the +2 state.

$$Ni^{2+}(aq) + 2e \rightleftharpoons Ni(s); \quad E^\ominus = -0.24\,V$$

Furthermore, it is only in the formation of complexes that it shows any strong resemblance to the other transition metals.

In aqueous solution, the nickel(II) ion exists as the green hydrated cation, $[Ni(H_2O)_6]^{2+}$, accounting for the green colour observed when many nickel(II) salts are dissolved in water. The most common of the simple compounds of nickel include the nitrate $Ni(NO_3)_2$, sulphate $NiSO_4$, hydroxide $Ni(OH)_2$, and carbonate $NiCO_3$.

The many nickel(II) complexes each belong to one of three geometrical structures. There are the octahedral complexes, e.g. the blue $[Ni(NH_3)_6]^{2+}$, the green $[Ni(H_2O)_6]^{2+}$, and the mixed aqua/ammines such as $[Ni(NH_3)_4(H_2O)_2]^{2+}$, the latter providing an explanation as to why many insoluble aqua complexes dissolve in ammonia. Also quite common are tetrahedral complexes, e.g. $[NiCl_4]^{2-}$. The square planar arrangement provides the third category and is quite rare, the most prominent example being $[Ni(CN)_4]^{2-}$ (see page 224). In each case, the geometry of the resultant complex ion is strongly influenced by the strength and nature of the ligands involved.

Nickel(II) complexes can be reduced to the +1 state, but nickel(I) complexes are, in the main, rarely stable, the $[Ni(CN)_3]^{2-}$ ion being one of the few exceptions.

Copper: [Ar]$3d^{10}4s^1$

Copper is a very familiar element and was one of the first metals used by man, notably in the form of bronze, which is its alloy with tin; bronze objects were made over 5000 years ago. The metal is reddish-brown, malleable and ductile, and an extremely good conductor of heat and electricity. In dry air it is resistant to attack, and in the absence of air it does not react with dilute hydrochloric or sulphuric acids. It is attacked by dilute and concentrated nitric acid, giving mainly nitrogen oxide (nitric oxide) as the gaseous product with dilute acid and mainly dinitrogen tetraoxide with concentrated acid.

Most of the common chemistry of copper is concerned with the Cu(+2) state.

Because of the completed *d* shell, copper(I) compounds are diamagnetic and often colourless (one common exception is Cu_2O, which is red), a phenomenon which is rare among the compounds of the transition elements.

Disproportionation of the Cu^+ ion

The existence of the Cu^+ ion in aqueous solution is very rare owing to its disproportionation:

$$2Cu^+(aq) \rightleftharpoons Cu^{2+}(aq) + Cu(s); \quad E = +0.37\,V$$

$$K = \frac{[Cu^{2+}]}{[Cu^+]^2} \approx 10^6$$

and is generally encountered only in solutions of soluble complexes. The high value of the equilibrium constant, *K*, indicates the low tendency of the Cu^+ ion to exist under these conditions.

Other copper(+1) chemistry

The covalent copper(I) halides, of which the fluoride is unknown, are significantly more stable than Cu(+2) halides, from which they may be formed by thermal decomposition.

$$2CuCl_2(s) \xrightarrow[\text{heat}]{\text{strong}} 2CuCl(s) + Cl_2(g)$$

$$2CuBr_2(s) \xrightarrow[\text{heat}]{\text{gentle}} 2CuBr(s) + Br_2(l)$$

Copper(I) chloride is formed as a white insoluble precipitate on warming either copper(I) oxide, or a mixture of copper(II) chloride and copper turnings, with concentrated hydrochloric acid:

$$\underset{\text{reddish brown}}{Cu_2O(s)} + 2HCl(l) \longrightarrow \underset{\text{white}}{2CuCl(s)} + H_2O(l)$$

$$\underset{\text{blue}}{CuCl_2{\cdot}2H_2O(s)} + Cu(s) \xrightarrow{\text{conc. HCl}} 2CuCl(s) + 2H_2O(l)$$

Both reactions involve the intermediate formation of the complex ion, $[CuCl_2]^-$.

Copper(I) chloride dissolves in aqueous ammonia forming $Cu(NH_3)_2$, which is colourless in the absence of oxygen but dark blue in its presence.

$$CuCl(s) + 2NH_3(aq) \rightarrow \underset{\text{colourless or dark blue}}{[Cu(NH_3)_2](aq)}$$

The reduction of Cu^{2+} ions to the Cu(+1) state is used in the detection of aldehydes and reducing sugars by means of Fehling's reagent. This consists of a solution of copper(II) sulphate added to a solution of Rochelle salt (sodium potassium 2,3-dihydroxybutanedioate) in excess sodium hydroxide. A deep blue colour is obtained on mixing the solutions owing to the formation of a complex copper(II) ion. The two solutions must be kept separate before they are needed, as the deep blue complex is unstable. On warming, aliphatic aldehydes and certain sugars reduce the Cu^{2+} ions to reddish-brown copper(I) oxide.

$$2Cu^{2+}(aq) + 2OH^-(aq) \rightarrow \underset{\text{reddish brown}}{Cu_2O(s)} + H_2O(l)$$

Copper(I) iodide is obtained as a white precipitate by the addition of potassium iodide solution to a soluble copper(II) salt. This reaction is used in the volumetric estimation of copper.

$$2Cu^{2+}(aq) + 4I^-(aq) \rightarrow \underset{\text{white}}{2CuI(s)} + I_2(s)$$

The iodine liberated is then titrated against sodium thiosulphate.

Copper(+2)

This state is attained quite easily by the action of an oxidizing agent, e.g. nitric acid, on the metal, the resulting Cu^{2+} ion having one unpaired electron.

$$Cu(s) \xrightarrow[\text{e.g. nitric(V) acid}]{\text{oxidizing agent,}} Cu^{2+}(aq) + 2e$$

In common with the ions of the other transition metals, the Cu^{2+} ion exists in aqueous solution as a hydrated complex. The formula is $[Cu(H_2O)_6]^{2+}$, and the structure is a distorted octahedron with two of the water molecules much farther away from the central metal ion than the other four.

The hydrated Cu^{2+} ion behaves in the usual way with alkali metal hydroxides, precipitating the gelatinous pale blue insoluble hydroxide:

$$Cu^{2+}(aq) + 2OH^-(aq) \rightarrow \underset{\text{pale blue}}{Cu(OH)_2(s)}$$

Although the hydroxide generally behaves as a typical base, it possesses weak amphoteric characteristics, forming cuprate(II) ions with concentrated alkali:

$$Cu(OH)_2(s) + 2OH^-(aq) \rightarrow \underset{\substack{\text{tetrahydroxocuprate(II) ion} \\ \text{blue}}}{[Cu(OH)_4]^{2-}(aq)}$$

On heating, the hydroxide decomposes to black copper(II) oxide:

$$Cu(OH)_2(s) \longrightarrow CuO(s) + H_2O(l)$$

With aqueous ammonia, it dissolves giving a solution of the deep blue diaquatetraamminecopper(II) complex:

$$Cu(OH)_2(s) + 4NH_3(aq) + 2H_2O(l) \rightarrow \underset{\text{royal blue solution}}{[Cu(NH_3)_4(H_2O)_2]^{2+}(aq)} + 2OH^-(aq)$$

This complex has a distorted octahedral structure similar to that of the hexaaqua ion above in which four ammonia ligands adopt a square planar arrangement and are significantly closer to the central copper ion than the two water molecules (Figure 17.6).

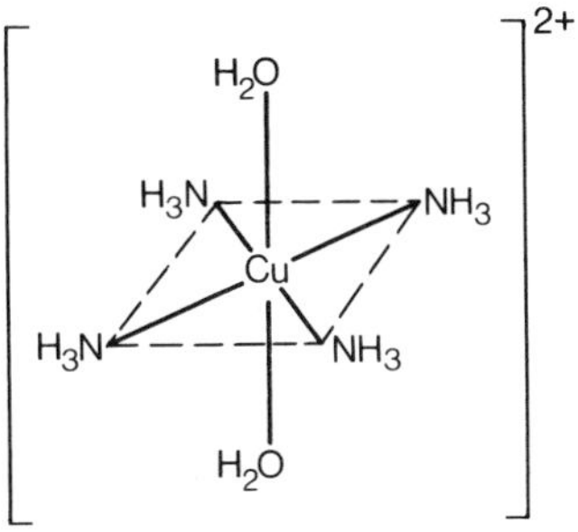

Fig. 17.6 Diaquatetraamminecopper(II) ion: a distorted octahedron

The formation of the diaquatetraamminecopper(II) ion may be interpreted as resulting from the successive displacement of water molecules in the $[Cu(H_2O)_6]^{2+}$ ion with the consequential intermediate formation of $[Cu(NH_3)(H_2O)_5]^{2+}$, $[Cu(NH_3)_2(H_2O)_4]^{2+}$, and $[Cu(NH_3)_3(H_2O)_3]^{2+}$ ions. Replacement of the final two water molecules by ammonia ligands, however, is not possible in aqueous solution although the $[Cu(NH_3)_6]^{2+}$ ion can be obtained in liquid ammonia.

A number of copper(II) complexes exist in which the central copper ion exhibits a coordination number of four. Typical of this is the $[CuX_4]^{2-}$ ion, where X is a halide ion. Where the cation is large, e.g. $Cs_2[CuCl_4]$, the shape of the complex ion is a distorted tetrahedron, and where the cation is small, e.g. $(NH_4)_2\ [CuCl_4]$, the complex has a square planar arrangement.

The yellow-brown $[CuCl_4]^{2-}$ ion, formed by dissolving copper(II) chloride in concentrated hydrochloric acid is, on dilution, converted into the blue

$[Cu(H_2O)_6]^{2+}$ ion. The intermediate solution appears green owing to the presence of both ions, the precise shade depending on the relative amounts of each ion.

$$\underset{\text{yellow-brown}}{[CuCl_4]^{2-}(aq)} + 6H_2O(l) \rightarrow \underset{\text{blue}}{[Cu(H_2O)_6]^{2+}(aq)} + 4Cl^-(aq)$$

Zinc: [Ar]$3d^{10}4s^2$

Zinc is a grey-white metal which is softer and more electropositive than most transition metals. The *d* orbitals are all filled with paired electrons and are not utilized in chemical bond formation. Its chemistry is therefore mainly that of a divalent metal, and displays very few of the properties generally associated with the variable valencies of the transition metals. This lack of *d* orbital bonding explains the absence of coloured salts and of any carbonyl compound, although similarity with the transition metals is observed in the formation of complexes.

The metal itself is employed in galvanizing and in certain alloys, of which brass (40 per cent Zn, 60 per cent Cu) is by far the most common. It is not attacked by dry air at ordinary temperatures, but with moist air it tarnishes forming a basic carbonate which serves as a protective coating and prevents further corrosion. It finds general use in the laboratory with its ready action on mineral acids, giving hydrogen with dilute sulphuric and hydrochloric acids. With nitric acid it forms oxides of nitrogen; the precise composition of the gas evolved depends upon the concentration of the acid. Because of its presence in certain enzymes, zinc is very important biologically.

The chemistry of zinc is often compared to that of magnesium, but it is less basic, readily forms complexes (unlike magnesium), and its compounds generally possess more covalent character.

Zinc(+2)

As a divalent metal, zinc forms a wide range of stable compounds. The majority of zinc compounds in this class comprise giant molecular structures with a varying range of ionic character. Typical of these are zinc oxide, which is amphoteric in nature and, because of lattice defects, becomes yellow when hot yet is white when cold; and zinc hydroxide, which is also amphoteric, and is insoluble in water although it dissolves in aqueous ammonia to form the complex cation, $[Zn(NH_3)_4]^{2+}$. Compounds which are essentially ionic in nature include zinc nitrate, zinc carbonate, zinc sulphate, and zinc fluoride. Zinc chloride is a white, deliquescent solid and only partly ionic. Zinc hydride, ZnH_2, is an unusual compound of unknown structure and is electron deficient.

The complex cation, $[Zn(NH_3)_6]^{2+}$, exists only in the solid state. The anions, $[Zn(CN)_4]^{2-}$ and $[Zn(OH)_4]^{2-}$, are both tetrahedral. The latter, the tetrahydroxozincate(II) ion, although usually referred to simply as the zincate ion, is formed when excess alkali is added to the metal or the metal salt.

$$Zn(s) + 2OH^-(aq) \rightarrow Zn(OH)_2(s) + 2e$$

or

$$Zn^{2+}(aq) + 2OH^-(aq) \rightarrow Zn(OH)_2(s)$$

$$Zn(OH)_2(s) + \underset{\text{(excess)}}{2OH^-(aq)} \rightarrow [Zn(OH)_4]^{2-}(aq)$$

It is probable that at lower hydroxide ion concentrations the zincate ion may be of the mixed aqua/hydroxo type, e.g. $[ZnH_2O(OH)_3]^-$.

Questions: Multiple-Choice

1. The wavelengths, λ, of the spectral lines of monohydrogen are related by the expression:

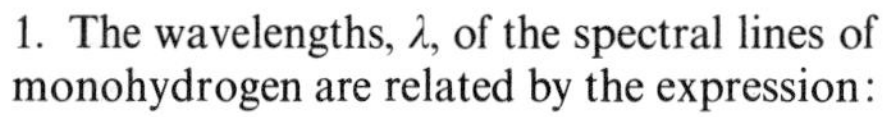

$$\frac{1}{\lambda} = R_H\left(\frac{1}{n_1^2} - \frac{1}{n_2^2}\right)$$

where R_H is the Rydberg constant and n_1 and n_2 are integers.

What is the value of n_1 for observations made in the visible region?

A 1
B 2
C 3
D 4
E 5

2. Which of the following statements relating to the spectrum of monohydrogen is false?

A The lines can be defined by quantum numbers.
B Electronic transitions in the ultra-violet region emanate from the ground state of the atom.
C The line of longest wavelength in the Balmer series corresponds to a transition between the $n = 2$ and $n = 3$ levels.
D The spectral lines are closer together at longer wavelengths.
E A continuum occurs at $n = \infty$.

3. Which one of the following species will give a series of spectral lines similar to that of Mg^{2+}?

A Al^{3+}
B Na
C Mg^{+}
D F
E Cl

4. The maximum number of electrons defined by principal quantum number 3 is:

A 6
B 8
C 10
D 18
E 32

5. Which one of the following electronic structures does *not* obey Hund's rule?

A $1s^2$
B $1s^2 2s^1$
C $1s^2 2s^2 2p_x^1$
D $1s^2 2s^2 2p_x^1 2p_y^1$
E $1s^2 2s^2 2p_x^1 2p_y^2$

6. What would be the most likely molecular formula of a compound formed between the element, X, of atomic number 14 and the element, Y, of atomic number 8?

A XY
B X_2Y
C XY_2
D X_2Y_3
E X_3Y_2

7. Which one of the following species is *not* isoelectronic with the others?

A CO
B N_2
C MgH_2
D O_2
E NO^{+}

8. Which one of the following is *not* isoelectronic with a sodium ion?

A Mg^{2+}
B Al^{3+}
C Ne
D F^{-}
E Cl^{-}

9. Which one of the following does *not* have a planar structure?

A BF_3
B C_2H_4
C PH_3
D H_2O
E XeF_4

10. Which one of the following correctly describes the arrangement of atoms in the xenon tetrafluoride molecule?

A linear.

B square planar.
C tetrahedral.
D octahedral.
E trigonal bipyramid.

11. Which one of the following species does *not* have a tetrahedral structure?

A $[FeCl_4]^-$
B $[AlCl_4]^-$
C $[Ni(CN)_4]^{2-}$
D $SnCl_4$
E $[Ni(CO)_4]$

12. The shape of the PF_5 molecule is best described as:

A tetrahedral.
B planar trigonal.
C trigonal pyramidal.
D octahedral.
E trigonal bipyramidal.

13. The values of the F—S—F bond angles in the SF_6 molecule are:

A 60° only
B 90° only
C 60° and 90°
D 72° and 90°
E 90° and 120°

14. The bond angles in the ammonium ion are equal to:

A 90°
B 104°30′
C 109°28′
D 112°15′
E 120°

15. In which of the following molecules is the bond angle greater than 109°28′?

A H_2O
B CO_2
C CCl_4
D H_2S
E PCl_3

16. Which of the following equations represents Bragg's law, where λ is the wavelength of the X-rays incident on a crystal, θ is the angle of incidence, d is the distance apart of the planes of atoms in the crystal, and n is an integer?

A $\lambda = 2n \sin \theta$
B $n\lambda = 2d \sin \theta$
C $2n\lambda = d \sin \theta$
D $\lambda = 2n \sin 2\theta$
E $n\lambda = 2d \sin 2\theta$

17. X-ray diffraction by crystals is used to provide information on all of the following with the exception of:

A the electronic arrangement of the atom.
B the arrangement of atoms in the crystal.
C the distance between planes of atoms.
D the distance between nuclei.
E angles at which planes of atoms intersect.

18. ABAB . . . . close-packing of identical spheres gives rise to:

A a triclinic structure.
B a face-centred cubic structure.
C a body-centred cubic structure.
D a hexagonal close-packed structure.
E a cubic close-packed structure.

19. Which geometrical arrangement of atoms about a central atom is predicted by a co-ordination number of 4?

A primitive cubic.
B octahedral.
C tetragonal.
D plane triangular.
E linear.

20. The caesium chloride unit cell is best described as:

A primitive cubic.
B body-centred cubic.
C face-centred cubic.
D tetragonal.
E orthorhombic.

21. Which one of the following most appropriately describes the binding forces between molecules of solid naphthalene?

A covalent bonding forces.
B dipole-dipole interactions.
C van der Waals forces.
D ionic bonding forces.
E the donation of electron-pairs.

22. The iodine crystal structure is best described as:

A simple molecular
B macromolecular.
C giant molecular.
D polymeric.
E ionic.

23. Which compound does *not* exist as a 'giant' structure or ionic lattice?

A CO_2
B SiO_2
C GeO_2
D SnO_2
E PbO_2

*

Questions 24 to 27

Given the following standard enthalpies of combustion:

hydrogen, $\Delta H = -286\,kJ\,mol^{-1}$;
carbon, $\Delta H = -394\,kJ\,mol^{-1}$;
methane, $\Delta H = -890\,kJ\,mol^{-1}$;
ethene, $\Delta H = -1390\,kJ\,mol^{-1}$;
ethanol, $\Delta H = -1370\,kJ\,mol^{-1}$.

24. The enthalpy of formation of methane, in $kJ\,mol^{-1}$, is:

A +76
B −76
C −115
D +230
E −230

25. The enthalpy of formation of ethene, in $kJ\,mol^{-1}$, is:

A −30
B +30
C −542
D −710
E +710

26. The enthalpy change for the reaction,

$$CH_2{=}CH_2 + H_2O \rightarrow CH_3CH_2OH$$

in $kJ\,mol^{-1}$, is:

A +20
B −20
C −266
D +306
E −306

27. The enthalpy of combustion of $4.48\,dm^3$ of ethene, in $kJ\,mol^{-1}$, is:

A −1390
B +278
C −278
D −2780
E −(1390 × 4.48)

28. The mean bond enthalpy of the C—H bond in methane refers to the process:

A $C(s) + 2H_2(g) \rightarrow CH_4(g);\ \frac{\Delta H}{2}$

B $CH_4(g) \rightarrow CH_3(g) + H(g);\ \Delta H$

C $CH_4(g) \rightarrow C(s) + 2H_2(g);\ \frac{\Delta H}{4}$

D $CH_4(g) \rightarrow C(s) + 4H(g);\ \Delta H$

E $CH_4(g) \rightarrow C(g) + 4H(g);\ \frac{\Delta H}{4}$

29. The lattice enthalpy for a solid Y^+Z^- refers to the change:

A $Y^+Z^-(s) \rightarrow Y^+(aq) + Z^-(aq)$
B $Y^+(aq) + Z^-(aq) \rightarrow Y^+Z^-(s)$
C $Y^+Z^-(g) \rightarrow Y^+(g) + Z^-(g)$
D $Y^+Z^-(s) \rightarrow Y^+(g) + Z^-(g)$
E $Y^+(g) + Z^-(g) \rightarrow Y^+Z^-(s)$

30. In which species does nitrogen exhibit the highest oxidation state?

A NO_2
B N_2O_3
C NH_2OH
D NH_4^+
E NO_3

31. In which one of the following complexes does the metal *not* have an oxidation state of +2?

A $[Cu(NH_3)_4(H_2O)_2]SO_4$
B $[Zn(OH)_4(H_2O)_2]^{2-}$
C $[Ag(NH_3)_2]^+$
D $[Mn(H_2O)_6]^{2+}$
E $[CoCl_4]^{2-}$

32. The $NO_2^-(aq)$ ion in acid solution may be oxidized to the $NO_3^-(aq)$ ion. In the half equation for this reaction, what is the number of electrons transferred from each $NO_2^-(aq)$ ion oxidized?

A 1
B 2
C 3
D 4
E 5

33. Which one of the following reactions in aqueous solution does *not* involve disproportionation?

A $Cl_2 + 2NaOH \rightarrow NaClO + NaCl + H_2O$
B $2Cu^+ \rightarrow Cu^{2+} + Cu$
C $3MnO_4^{2-} + 4H^+ \rightarrow 2MnO_4^- + MnO_2 + 2H_2O$
D $3HNO_2 \rightarrow HNO_3 + 2NO + 2H_2O$
E $Cr_2O_7^{2-} + 2OH^- \rightarrow 2CrO_4^{2-} + H_2O$

34. Which one of the following can be oxidized?

A Mn_2O_7
B PbO_2
C SO_2
D NO_3^-
E ClO_4^-

35 Select a compound which will not be reduced by an aqueous solution of iron(III) ions.

A $SnCl_2$
B $KMnO_4$
C $AgNO_3$
D $K_2Cr_2O_7$
E Cl_2

Questions 36 to 38

Use the following half reactions to answer questions 36 to 38.

$$Fe^{2+} \rightarrow Fe^{3+} + e$$
$$C_2O_4^{2-} \rightarrow 2CO_2 + 2e$$
$$MnO_4^- + 8H^+ + 5e \rightarrow Mn^{2+} + 4H_2O$$

36. How many moles of iron(II) ions are oxidized by one mole of manganate(VII) ions?

A $\frac{1}{5}$
B $\frac{2}{5}$
C 1
D $\frac{5}{2}$
E 5

37. How many moles of ethanedioate (oxalate) ions are oxidized by one mole of manganate(VII) ions?

A $\frac{1}{5}$
B $\frac{2}{5}$
C 1
D 2
E $\frac{5}{2}$

38. How many moles of iron(II) ethanedioate (oxalate) are oxidized by one mole of manganate(VII) ions?

A $\frac{1}{5}$
B $\frac{2}{5}$
C $\frac{3}{5}$
D $\frac{5}{3}$
E $\frac{5}{2}$

39. In which one of the following reactions does hydrogen behave as an oxidizing agent?

A $H_2 + S \rightarrow H_2S$
B $2K + H_2 \rightarrow 2KH$
C $C_2H_2 + H_2 \rightarrow C_2H_4$
D $2H_2 + O_2 \rightarrow 2H_2O$
E $N_2 + 3H_2 \rightarrow 2NH_3$

40. Which statement is *not* correct?

A Water oxidizes sodium.
B Hydrogen oxidizes lithium.
C Sulphur(VI) oxide reduces hydrogen sulphide.
D Hydrogen peroxide reduces silver(I) oxide.
E Dilute sulphuric acid oxidizes zinc.

41. Two metals are fastened together and placed in dilute hydrochloric acid. In which case would you expect the first named to corrode faster than the second?

A Fe and Sn
B Pb and Zn
C Fe and Mg
D Fe and Zn
E Ag and Zn

42. A piece of pure zinc is placed in dilute sulphuric acid and the reaction is very slow. When a piece of silver is put in the acid and allowed to touch the zinc, then:

A the zinc dissolves rapidly.
B the zinc forms the cathode of the cell.
C the zinc metal is reduced.
D the silver metal is reduced.
E bubbles of hydrogen come from the zinc.

*

43. What is the atomic number of the element which is in the same group of the Periodic Table as the element whose atomic number is 15?

A 5
B 7
C 11
D 17
E 25

44. Which one of the following elements will have an outer electronic configuration in the ground state of $3d^6 4s^2$?

A strontium
B barium
C bismuth
D iron
E zinc

45. The covalent radius of the potassium atom is 0.203 nm. The radius of the potassium ion in nanometres will be:

A 0.013
B 0.133
C 0.231
D 0.234
E 0.258

46. In which of the following are the different iodine species placed in correct order of decreasing size?

A $I > I^+ > I^-$
B $I > I^- > I^+$
C $I^+ > I > I^-$
D $I^- > I > I^+$
E $I^+ > I^- > I$

47. Which will be the order of increasing size of the species Na^+, Mg^{2+}, Cl^-, Ar?

A Cl^-, Ar, Na^+, Mg^{2+}
B Na^+, Mg^{2+}, Cl^-, Ar
C Ar, Cl^-, Mg^{2+}, Na^+
D Mg^{2+}, Na^+, Ar, Cl^-
E Ar, Cl^-, Na^+, Mg^{2+}

48. The first six ionization energies of an element are 1100, 4400, 4600, 6200, 37 800, 47 000 kJ mol^{-1}. On the basis of this evidence, into which one of the following groups of the Periodic Table should the element be placed?

A II
B III
C IV
D V
E VI

49. Which one of the following atoms has the lowest first ionization energy?

A lithium
B rubidium
C caesium
D fluorine
E iodine

50. Which of the following processes requires the greatest amount of energy?

A $Na(g) \rightarrow Na^+(g) + e$
B $Mg^+(g) \rightarrow Mg^{2+}(g) + e$
C $Mg^{2+}(g) \rightarrow Mg^{3+}(g) + e$
D $Al^{2+}(g) \rightarrow Al^{3+}(g) + e$
E $Al^{3+}(g) \rightarrow Al^{4+}(g) + e$

51. Which one of the following changes correctly represents the second ionization energy of calcium?

A $Ca(s) \rightarrow Ca^{2+}(s) + 2e$
B $Ca(s) \rightarrow Ca^{2+}(g) + 2e$
C $Ca(g) \rightarrow Ca^{2+}(g) + 2e$
D $Ca^+(s) \rightarrow Ca^{2+}(g) + e$
E $Ca^+(g) \rightarrow Ca^{2+}(g) + e$

52. Which one of the following statements concerning the electronegativity of elements is incorrect?

A That fluorine has the highest value.
B That carbon is less than that of hydrogen.
C That aluminium is less than that of silicon.
D That bromine is greater than that of iodine.
E That nitrogen is greater than bromine.

53. Which one of the following is the correct order of increasing electronegativity of the elements boron, chlorine, bromine, and aluminium?

A Al < B < Br < Cl
B Cl < Br < B < Al
C Br < Cl < Al < B
D B < Al < Br < Cl
E Al < Br < Cl < B

54. Which one of the following would be accompanied by a negative enthalpy change?

A $O(g) + e \rightarrow O^-(g)$
B $O^-(g) + e \rightarrow O^{2-}(g)$
C $Na(g) \rightarrow Na^+(g) + e$
D $Br_2(g) \rightarrow 2Br(g)$
E $I_2(s) \rightarrow 2I(g)$

55. In which one of the following compounds does ionic character predominate?

A tetrachloromethane
B sodium hydride
C hydrogen chloride
D silicon(IV) oxide
E carbon dioxide

56. Which one of the following compounds possesses the greater degree of covalent character in its bonding?

A BeF_2
B MgF_2
C CaF_2
D NaF
E RbF

57. Which of the following oxides would you expect to have the greatest acidic character?

A CO_2
B P_4O_6
C P_4O_{10}
D SO_2
E SeO_2

Questions 58 to 62 relate to the *s*- and *p*- blocks of the Periodic Table illustrated below:

	Group							
Period	*I*	*II*	*III*	*IV*	*V*	*VI*	*VII*	*O*
1								
2	*a*	*b*				*c*	*d*	
3	*e*	*f*	*g*	*h*	*i*	*j*	*k*	*l*
4	*m*	*n*	*o*			*p*	*q*	*r*
5		*s*	*t*					
6	*u*		*v*				*w*	*x*

8. The most electropositive element is:

a **D** *v*
d **E** *w*
u

). The element with the lowest first ionization energy is:

a **D** *u*
d **E** *x*
l

0. The element in Period 3 which forms the ost basic oxide is:

f **D** *j*
g **E** *k*
i

1. Which pair of elements shows the closest imilarity in physical and chemical properies?

a and *f* **D** *j* and *k*
a and *m* **E** *o* and *l*
h and *i*

2. Which pair of elements will form a comound with the greatest covalent character?

c and *h* **D** *c* and *n*
c and *j* **E** *d* and *w*
j and *s*

*

63. Which one of the following is an incorrect statement concerning the element hydrogen?

A It is not satisfactorily accommodated in any of the groups in the Periodic Table.
B There are three isotopic forms, ^1_1H, ^2_1H, and ^3_1H.
C It forms predominantly ionic hydrides with the highly electropositive elements of Groups I and II.
D It forms predominantly covalent hydrides with the elements of Groups III to VII.
E It is a more powerful reducing agent than sodium.

64. Which of the following statements concerning the elements sodium, potassium, magnesium, and calcium is not wholly correct?

A They are all extracted by electrolytic techniques.
B They all tarnish on exposure to air.
C They all form solid carbonates.
D They all form solid hydrogencarbonates.
E They are all strong reducing agents.

65. The physical property of the Group I metals which increases numerically as the Group is descended is:

A first ionization energy.
B ionic radius.
C melting point.
D electronegativity.
E hydration enthalpy of M^+ ions.

66. Which one of the following reactions is *not* given by an aqueous solution of sodium hydroxide?

A Forms carbonate ions with carbon dioxide.
B Forms sulphate ions with sulphur dioxide.
C Liberates phosphine with phosphorus.
D Liberates ammonia with ammonium salts.
E Forms zincate ions (when in excess) with Zn^{2+}(aq) ions.

67. Which one of the following does *not* produce an alkaline solution in water?

A $NaHCO_3$
B $NaHSO_3$
C Na_2CO_3
D NaOH
E NaH

68. Which one of the following statements concerning the compounds of lithium is false?

A The hydroxide decomposes to the oxide on heating.
B The carbonate decomposes to the oxide on heating.
C It is the most electronegative element in Group I.
D It forms a peroxide but not a superoxide.
E The hydrogencarbonate cannot be isolated as a stable solid.

69. The element possessing the least firmly held outer electron is:

A lithium.
B potassium.
C rubidium.
D calcium.
E barium.

70. A 0.1 M solution of a reagent gives a precipitate when carbon dioxide is bubbled into it. The reagent could be:

A KOH
B CsOH
C Na_2CO_3
D $KHSO_4$
E $Ba(OH)_2$

71. Which of the following decreases on descending Group II (from Be to Ba)?

A solubility of hydroxides in water.
B basic strengths of oxides.
C thermal stabilities of the carbonates.
D sum of the first two ionization energies.
E electropositive character of the element

72. Which one of the following statements is correct? The chlorides of the Group I elements:

A are all hygroscopic with the exception of $BeCl_2$, which hydrolyses in moist air.
B increase in lattice enthalpy from $BeCl_2$ t $BaCl_2$.
C decrease in melting point from $BeCl_2$ t $BaCl_2$.
D are all insoluble except $BaCl_2$.
E cannot be prepared by direct combinatio of the two elements.

*

73. Which of the following statements is in correct for beryllium?

A most of its compounds are largely cova lent.
B it forms a covalent chloride, $BeCl_2$, whic exists in the gaseous state as linear mole cules.
C it has distinctive group properties, attr butable to its comparatively smaller siz and greater electronegativity.
D it forms Be^{2+} ions because of the lo value of the sum of the first two ionizatio energies.
E it is capable of forming tetrahedral com plexes.

74. Which one of the following statement is incorrect? The Group III elements, B to T

A all exhibit an oxidation state of +3.
B all form oxides of formula M_2O_3.
C all display appreciable metallic characte except boron, which is mainly nor metallic in its properties.
D all form trihalides, MX_3.
E all form amphoteric hydroxides, M(OH)

75. Which of the following properties o aluminium are *not* typical of most othe metals?

A considerable temperature difference be tween melting point and boiling point.

B negative standard electrode potential.
C oxide is predominantly basic.
D electrical resistance increases with increase in temperature.
E forms salts with alkalis.

76. Which aluminium compound is most likely to contain the Al^{3+} ion?

A oxide.
B fluoride.
C chloride.
D selenide.
E phosphide.

77. Which one of the following will not rapidly attack and destroy a pure aluminium pan?

A caustic soda solution.
B washing soda solution.
C concentrated nitric acid.
D concentrated hydrochloric acid.
E potassium hydroxide solution.

78. Which of the following statements concerning aluminium oxide, is *not* wholly correct?

A It is formed together with some nitride when aluminium is heated in air.
B It is amphoteric, but its acidic properties are more prominent.
C It has a high exothermic enthalpy of formation.
D A coating of the oxide on the surface of aluminium helps to prevent the metal from further corrosion.
E It forms salts with aqueous sodium hydroxide.

79. Which one of the following will produce an acidic solution in water?

A aluminium sulphate.
B potassium sulphate.
C sodium hydrogencarbonate.
D sodium carbonate.
E magnesium sulphate.

80. Catenation is the ability of:

A atoms to form strong bonds with similar atoms.
B elements to form giant molecules.
C an element to form multiple bonds.
D elements to extend their coordination number by having available *d* orbitals.
E an element to form long chains of identical atoms.

81. The inert pair effect is most prevalent in the compounds of which Group IV element?

A carbon.
B silicon.
C germanium.
D tin.
E lead.

82. The Group IV element oxide exhibiting the strongest basic properties is:

A silicon(IV) oxide.
B tin(II) oxide.
C tin(IV) oxide.
D lead(II) oxide.
E lead(IV) oxide.

83. Which tetrachloride is not readily hydrolysed by water?

A CCl_4
B $SiCl_4$
C $GeCl_4$
D $SnCl_4$
E $PbCl_4$

84. Which are the best materials to use in order to prepare a pure sample of tin(IV) chloride?

A SnO and concentrated hydrochloric acid.
B SnO_2 and dilute hydrochloric acid.
C Sn and concentrated hydrochloric acid.
D Sn and dry hydrogen chloride.
E Sn and dry chlorine.

85. Lead(II) chloride dissolves most appreciably in:

A water.
B dilute alkali.
C concentrated alkali.
D dilute hydrochloric acid.
E concentrated hydrochloric acid.

86. Which one of the following is capable of acting as a Lewis base?

A $[Al(H_2O)_6)]^{3+}$
B CH_4
C BCl_3
D H_2S
E NH_2NH_2

87. Which of the following diatomic molecules has the highest bond enthalpy?

A H_2
B N_2
C O_2
D Cl_2
E Br_2

88. The most volatile trihydride of the Group V elements is:

A NH_3
B PH_3
C AsH_3
D SbH_3
E BiH_3

89. The following equation outlines the industrial production of ammonia by the Haber process:

$$N_2 + 3H_2 \rightleftharpoons 2NH_3; \quad \Delta H = -92\,\text{kJ mol}^{-1}$$

Which of the following does not improve the efficiency of the technique?

A a catalyst.
B catalyst promoters.
C a high pressure.
D the removal of ammonia from the system.
E a temperature of 900 °C.

90. Which one of the following statements concerning ammonia is wrong?

A It is basic in its properties.
B When in excess, it reacts with chlorine and liberates nitrogen.
C It cannot be used to reduce copper(II) oxide.
D It precipitates silver from an aqueous solution of its salt.
E It burns in pure oxygen forming water and nitrogen.

91. Which one of the following gives *no* reaction with ammonia at room temperature?

A BF_3
B Cl_2
C CH_3Cl
D F_2
E H_2S

92. Which one of the following species contains an odd number of electrons?

A N_2O_2
B NO_3^-
C NO
D N_2O_4
E NH_3

93. Odd electron molecules are characteris tically:

A diamagnetic.
B colourless.
C highly reactive.
D unable to dimerize.
E incapable of forming positive ions.

94. With which one of the following metal does nitric acid render to the so-calle 'passive' state?

A aluminium.
B magnesium.
C lead.
D copper.
E mercury.

95. Which chloride is not appreciably hy drolysed by water?

A $AlCl_3$
B $SnCl_2$
C $PbCl_2$
D BCl_3
E PCl_3

96. Which of the following structures bes represents phosphoric(V) acid?

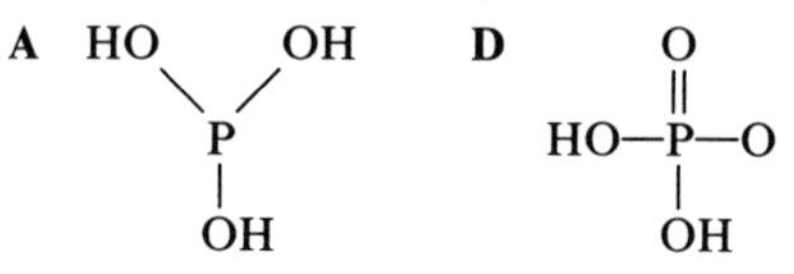

B
```
      O
      ‖
HO—P—OH
      |
      OH
```

C
```
      OH
      |
HO—P—OH
      |
      OH
```

E
```
      O
      ‖
HO→P=O
      |
      OH
```

97. Which of the following statements cor cerning elemental oxygen is false?

A It is limited to a valency of two.
B It exists naturally only as a diatomi molecule.
C It is the second most electronegativ element.

D It is capable of a small degree of catenation.
E It forms divalent anions.

98. In which of the following molecules and ions does sulphur not exhibit its maximum oxidation state?

A SF_6
B SO_3^{2-}
C SO_3
D SO_2Cl_2
E SO_4^{2-}

99. The molecule possessing the smallest angle is:

A H_2O
B H_2S
C H_2Se
D H_2Te
E NH_3

100. In which of the following reactions is hydrogen peroxide not acting as an oxidizing agent?

A $2Fe^{2+} + H_2O_2 + 2H^+ \rightarrow 2Fe^{3+} + 2H_2O$
B $PbS + 4H_2O_2 \rightarrow PbSO_4 + 4H_2O$
C $2I^- + 2H_2O_2 + 2H^+ \rightarrow I_2 + 2H_2O$
D $Mn^{2+} + H_2O_2 + 2OH^- \rightarrow MnO_2 + 2H_2O$
E $Ag_2O + 2H_2O_2 \rightarrow 2Ag + O_2 + H_2O$

101. The compound displaying the greatest acidic strength in aqueous solution is:

A CH_3OH
B NH_2OH
C H_2O
D H_2S
E H_2Se

102. Which one of the following ions in acidic solution is *not* precipitated by hydrogen sulphide?

A Cu^{2+}
B Ni^{2+}
C Pb^{2+}
D Cd^{2+}
E Bi^{3+}

103. Which of the following elements is least likely to combine directly with fluorine?

A bromine.
B hydrogen.
C argon.
D xenon.
E radon.

104. Which one of the following is an incorrect statement regarding the reaction between hydrogen and chlorine in direct sunlight?

A Sunlight causes homolysis of hydrogen molecules.
B Sunlight causes homolysis of chlorine molecules.
C The enthalpy change for the reaction is negative.
D The reaction is a chain reaction.
E The reaction does not require a catalyst.

105. Fluorine is the most reactive halogen because:

A the molecule has the lowest bond enthalpy.
B it is the smallest atom in the group.
C it is the most electronegative element.
D it is the most volatile halogen.
E it has the highest electron affinity.

106. Which one of the following reactions does not occur?

A Fluorine liberates chlorine from an aqueous solution of hydrogen chloride.
B Chlorine liberates bromine from an aqueous solution of hydrogen bromide.
C BrO^- ions disproportionate at room temperature to give Br^- and BrO_3^- ions.
D Fluorine reacts with phosphorus to form PF_5.
E Iodine reacts with phosphorus to form PI_3.

107. When chlorine is bubbled through *hot* alkali, the main products are:

A Cl^- and H_2O.
B ClO^- and H_2O.
C Cl^-, ClO^-, and H_2O.
D Cl^-, ClO_3^-, and H_2O.
E ClO^-, ClO_4^-, and H_2O.

108. What is the minimum number of moles of sodium hydroxide which would be needed to absorb one mole of chlorine gas?

A 1
B 2
C 3
D 4
E 5

109. Which one of the following species would you not expect chlorine to oxidize?

A Mn^{2+}
B Cr^{3+}
C F^-
D Br^-
E I^-

110. Which one of the following species is most likely to be capable of further reaction with fluorine?

A CaF_2
B IF_5
C SF_6
D $[PF_6]^-$
E KF

111. Which reagent will not liberate iodine from acidified potassium iodide solution?

A hydrogen peroxide.
B sodium nitrite.
C concentrated nitric acid.
D sulphur dioxide.
E bromine water.

112. Which statement is true for the halogen hydrides HF, HCl, HBr, and HI?

A All form intermolecular hydrogen bonds with water in aqueous solution.
B Bond length is greatest in HF.
C HF cannot be oxidized to F_2 by any of the other halogens.
D An aqueous solution of HF is the strongest acid.
E Boiling points increase progressively from HF to HI.

113. Which one of the following statements is correct with regard to the reaction:

$$HNO_3 + HF \rightleftharpoons H_2NO_3^+ + F^-$$

A HNO_3^+ acts as a Lewis base.
B HNO_3 acts as a Lewis acid.
C HNO_3 acts as an oxidizing agent.
D HF acts as a Lewis acid.
E HF acts as a Lewis base.

*

114. Which one of the following is the correct value for the coordination number of Co in the complex ion $[Co(NH_3)_5NO_2]Cl_2$?

A 2
B 4
C 5
D 6
E 8

115. Choose a species which cannot act as an electron donor (ligand) to a transition metal ion.

A NH_3
B H_2O
C CO
D CO_2
E CN^-

116. Which one of the following species does not represent a cationic species of vanadium formed in aqueous solution?

A VO_2^+
B VO_2^{2+}
C VO^{2+}
D $[V(H_2O)_6]^{3+}$
E $[V(H_2O)_6]^{2+}$

117. The principal oxidation states exhibited by chromium are:

A +2, +3, +4
B +2, +3, +5
C +2, +3, +6
D +3, +4, +6
E +3, +5, +7

118. Manganese(IV) oxide is amphoteric and the ions that it forms with aqueous alkali and aqueous hydrochloric acid respectively are:

A MnO_4^{2-} and $[MnCl_4]^{2-}$.
B MnO_4^{2-} and $[MnCl_6]^{2-}$.
C MnO_4^- and $[MnCl_6]^{2-}$.
D MnO_4^- and $[MnCl_4]^{2-}$.
E MnO_4^{2-} and MnO_4^-.

119. The ultimate product formed by hydrated iron(III) ions in an aqueous solution of pH 5 is:

A $Fe_2O_3 \cdot xH_2O$
B $[Fe(H_2O)_5OH]^+$
C $[Fe(H_2O)_4(OH)_2]$
D $[Fe(H_2O)_5OH]^{2+}$
E $[Fe(H_2O)_3(OH)_3]$

120. In which one of the following does manganese exhibit its highest oxidation state?

A MnO_2
B MnO_4^{2-}
C MnO_4^-
D MnO
E Mn^{2+}

21. Black manganese(IV) oxide stains are est removed from glassware by placing it in n acidified solution of:

A Na_2S
B $NaHS$
C Na_2SO_4
D $NaHSO_4$
E Na_2SO_3

22. Which one of the following statements oncerning the $[Co(H_2O)_6]^{2+}$ ion is false?

A It has an octahedral structure.
B It is pink when freshly prepared.
C It turns blue on warming.
D In the presence of air, it forms the yellow $[Co(CN)_6]^{3-}$ ion with sodium cyanide.
E It forms the blue $[CoCl_6]^{4-}$ ion with concentrated hydrochloric acid.

23. Which one of the following best describes the complex ion formed when cobalt(II) hloride is added to water?

A $[CoCl_4]^{2-}$
B $[CoCl_6]^{4-}$
C $[CoCl_6]^{3-}$
D $[Co(H_2O)_6]^{2+}$
E $[Co(H_2O)_6]^{3+}$

24. Which one of the following is an incorect representation of a complex transition etal ion?

A $[CuCl_4]^{2-}$
B $[Cu(CN)_4]^{3-}$
C $[Cr(NH_3)_6]^{3+}$
D $[Co(NH_3)_6]^{3+}$
E $[CoCl_6]^{2-}$

25. Which one of the following is formed vhen a soluble copper(I) salt is added to vater?

A $[Cu(H_2O)_6]^{+}$ only.
B $[Cu(H_2O)_6]^{2+}$ only.
C Cu and $[Cu(H_2O)_4]^{+}$.
D Cu and $[Cu(H_2O)_6]^{+}$.
E Cu and $[Cu(H_2O)_6]^{2+}$.

126. Which one of the following statements concerning the copper(+2) state is incorrect?

A It is more stable in aqueous solution than the copper(+1) state.
B Hydrated copper(II) ions are blue and have an octahedral structure.
C Hydrated copper(II) ions form a deep blue, octahedral complex, $[Cu(NH_3)_6]^{2+}$, with aqueous ammonia.
D $CuCl_2$ forms yellow-brown $[CuCl_4]^{2-}$ ions in concentrated hydrochloric acid.
E Hydrated copper(II) ions form a gelatinous, pale blue hydroxide, with aqueous alkali.

127. Which of the following aqueous ions gives no observable change when aqueous hydroxide ions are added drop-wise?

A Al^{3+}
B Fe^{2+}
C CrO_4^{2-}
D Cu^{2+}
E $Cr_2O_7^{2-}$

128. Which one of the following ions in aqueous solution gives a precipitate with excess aqueous alkali?

A Pb^{2+}
B Al^{3+}
C Mg^{2+}
D Zn^{2+}
E Sn^{2+}

129. Which one of the following salts does *not* produce a complex ion $[M(H_2O)_6]^{2+}$ when dissolved in an excess of water?

A $CuSO_4$
B $FeSO_4$
C $Al_2(SO_4)_3$
D $ZnSO_4$
E $Fe_2(SO_4)_3$

130. In which one of the following ions is the transition metal *not* exhibiting its highest common oxidation state?

A CrO_4^{2-}
B $Cr_2O_7^{2-}$
C MnO_4^{2-}
D $[Fe(CN)_6]^{3-}$
E $[Zn(NH_3)_4]^{2+}$

Questions: Essay-Type

1. The lines in the visible atomic spectrum of hydrogen (Balmer series) are represented, by the expression:

$$\frac{1}{\lambda} = R_H\left(\frac{1}{2^2} - \frac{1}{n^2}\right)$$

(a) What do the symbols λ, R_H, and n represent? Indicate the units of these quantities where appropriate.

(b) By means of a series of horizontal lines, draw an energy level diagram for the visible atomic spectrum of hydrogen, and indicate:

(i) the values of n for each line;

(ii) by means of an arrow, the transition which corresponds to the fourth line of the emission spectrum;

(iii) by means of an arrow, the transition which corresponds to the first ionization energy of the hydrogen atom.

(iv) What do you understand by the term *convergence limit* and what is its significance?

(v) Why are discrete lines observed rather than a continuous spectrum?

2. Describe the principal features of the atomic spectrum of hydrogen, and indicate how it may be interpreted in terms of the concept of the principal energy levels in the atom.

3. Outline the spectral evidence, and its interpretation, which led to the adoption of subsidiary quantum numbers.

4. What are the most limiting aspects of the Rutherford-Bohr theory of electronic structure? Briefly explain the reasoning which forms the basis of the modern concept of electronic theory.

5. What do you understand by the term *atomic orbital*? How are *s* and *p* orbitals defined in terms of principal and azimuthal quantum numbers?

6. State what is meant by the *ground state* of an atom? What is observed spectrally when an atom in an excited state returns to the ground state?

7. Account for the shapes of the following molecules and ions:

(a) BCl_3
(b) CH_4
(c) NH_3
(d) NH_4^+
(e) $[PF_6]^-$
(f) SF_6.

8. Briefly outline the factors influencing the symmetry and bond angles of the following molecules and ions:

(a) the NH_3 molecule
(b) the NH_4^+ ion
(c) the H_2O molecule
(d) the H_3O^+ ion
(e) the PCl_5 molecule
(f) the $[PCl_6]^-$ ion.

9. Compare and contrast the valence bond and molecular orbital approaches to covalent bond formation. Illustrate your answer by reference to simple examples.

10. Describe how the concept of hybridization may be interpreted to predict stereochemical shape. By referring to specific examples, indicate the advantages that the theory of hybridization has over the Sidgwick Powell theory.

11. Explain on the basis of electronic theory why the carbonate and nitrate ions are planar while the chlorate(VII) ion is tetrahedral. Draw sketches to supplement your answer.

*

12. Draw a diagram showing the arrangement of metal atoms in a horizontal layer of a

ose-packed metal crystal and indicate, by riting crosses in the appropriate positions, e locations of the centres of those atoms hich lie in a second horizontal plane.

3. Using two coins of suitable sizes (say a and a 2p) draw one face of a sodium loride structure in which the chloride ions e at the corners.

4. At room temperature most metals are od conductors of electricity, but ionic twork-solids are not. Comment on this atement.

5. Discuss the difference in the arrangement ions in solid sodium chloride and in solid esium chloride. Explain why this difference ises.

6. Compare and contrast the macromole- lar structures of diamond and graphite. Use e difference in the type of bonding to count for the differences in their physical operties.

*

7. Illustrate the Born-Haber cycle which presents the enthalpy change for the forma- on of an alkali metal chloride, MCl, from e elements in their standard states.

Name the halogen for which the bond en- alpy value is largest and the alkali metal for hich the ionization energy value is smallest.

8. From the following list of data, calculate e standard enthalpy of formation of potas- um chloride.

$$K(s) \rightarrow K(g);\ \Delta H^{\ominus} = +90\,\text{kJ mol}^{-1}$$
$$K(g) \rightarrow K^{+}(g)+e;\ \Delta H^{\ominus} = +408\,\text{kJ mol}^{-1}$$
$$\tfrac{1}{2}Cl_2(g) \rightarrow Cl(g);\ \Delta H^{\ominus} = +121\,\text{kJ mol}^{-1}$$
$$Cl(g)+e \rightarrow Cl^{-}(g);\ \Delta H^{\ominus} = -370\,\text{kJ mol}^{-1}$$
$$K^{+}(g)+Cl^{-}(g) \rightarrow KCl(s);\ \Delta H^{\ominus} = -701\,\text{kJ mol}^{-1}$$

9. Distinguish between the term *mean bond nthalpy* and *bond dissociation enthalpy*. Il- strate your answer with suitable examples.

*

0. The standard redox potentials for a num- er of half reactions are given in the following ble:

Half reaction	$E^{\ominus}/V$
$Ca^{2+}(aq)+2e \rightleftharpoons Ca(s)$;	−2.87
$Al^{3+}(aq)+3e \rightleftharpoons Al(s)$;	−1.67
$Fe^{2+}(aq)+2e \rightleftharpoons Fe(s)$;	−0.44
$Cu^{2+}(aq)+2e \rightleftharpoons Cu(s)$;	+0.34
$Br_2(l)+2e \rightleftharpoons 2Br^{-}(aq)$;	+1.07
$Cl_2(g)+2e \rightleftharpoons 2Cl^{-}(aq)$;	+1.36
$MnO_4^{-}(aq)+8H^{+}(aq)+5e \rightleftharpoons Mn^{2+}(aq)+4H_2O$;	+1.52

(a) What is meant by the terms (i) *oxidizing agent*, and (ii) *reducing agent*?
(b) Which of the species listed above is (i) the strongest oxidizing agent, and (ii) the strongest reducing agent?
(c) Determine the *standard e.m.f.* of the cell $Cu|Cu^{2+} \vdots Fe^{2+}|Fe$.

21. Illustrate what is meant by the terms *oxidation* and *reduction* by reference to the electrolysis of an aqueous solution of copper (II) sulphate using copper electrodes.

Iron coated with tin rusts rapidly if scratched, whereas iron coated with zinc is much more resistant to rusting. Explain this statement.

22. The table below shows the standard redox potentials of the Group III metal ion/metal systems. Discuss the relative oxidizing and reducing powers of the various species represented.

	$E^{\ominus}/V$
$Al^{3+}(aq)+3e \rightleftharpoons Al(s)$;	−1.67
$Ga^{3+}(aq)+3e \rightleftharpoons Ga(s)$;	−0.53
$In^{3+}(aq)+3e \rightleftharpoons In(s)$;	−0.34
$Tl^{3+}(aq)+3e \rightleftharpoons Tl(s)$;	+0.72
$Tl^{+}(aq)+e \rightleftharpoons Tl(s)$;	−0.34

23. Explain the meaning of the terms (a) electronegativity and (b) ionization energy. How are the values of these properties related to the position of the elements in the Periodic Table?

24. Outline how the electronic configuration of an element is related to (a) its atomic size, (b) its ionic size, (c) its ionization energy, (d) its electronegativity, and (e) its electron affinity.

25. Sketch an approximate graph of the first ionization energies of the first ten elements in the Periodic Table (H to Ne) against their atomic number. Comment briefly upon the important features of the diagram and indicate how it may be interpreted.

26. Discuss the basic and acidic properties of the oxides of the elements in the third period (sodium to chlorine) with regard to their position in the Periodic Table.

27. Discuss the trends observed in the Period 3 elements (Na to Cl) by reference to (a) the physical properties of the chlorides, and (b) the chemical bonding present in the chlorides.

28. Discuss briefly the crude correlation that can be made between the position of metals in the Periodic Table and the chemical nature of their ores.

Outline the principles involved in the extraction of the following metals from their purified ores:

(a) sodium and aluminium from their halides;
(b) iron and zinc from their oxides;
(c) titanium from its chloride.

29. Why cannot the technique of reducing their oxides by carbon be usefully applied to the extraction of aluminium and titanium?

30. Discuss the extraction of (a) sodium and (b) calcium in terms of the idea of electron transfer.

*

31. Compare the properties of hydrogen (protium) and its oxide with those of deuterium and deuterium oxide.

32. Discuss the evidence which appears to classify hydrogen in Group I and that which appears to classify it in Group VII. What conclusions can you make on the basis of this evidence?

33. Describe the structure of and bonding in (a) a water molecule, (b) liquid water, and (c) ice.

34. Discuss the various ways in which hydrc gen is bonded in inorganic compounds, illus trating your answer by means of one exampl in each case.

35. Describe the type of bonding whic occurs in the hydrogen compounds of th Period 3 elements (Na to Cl). In what way i the nature of the bonding reflected in th physical properties of these compounds?

*

36. State which of the elements of Group and Group II has:

(a) the smallest atomic (covalent) radius.
(b) the largest ionic radius.
(c) the highest value for the first ionizatio energy.
(d) the least soluble hydroxide.
(e) the lowest electronegativity value.

37. Discuss the structure and bonding of th different types of oxides formed by the alka metals. Compare the behaviour of each these three types of oxide with water.

38. Show how the properties of the Group elements and their compounds typify tl general characteristics of a Group in th periodic classification.

39. How do the relative stabilities of the ca bonates and hydrogencarbonates reflect th electropositive character of the elements Groups I and II?

40. In many ways the properties of lithiu are not typical of those of the alkali meta but do resemble those of magnesium. Discu this statement, paying particular attention the position in the Periodic Table of the tw elements concerned, the size of their aton and their ions, and their respective electro negativity values.

41. How are the basic properties of the Grou I metal hydroxides illustrated by their r actions with the oxides of aluminium, zin and lead?

:. 'The first element in each group of the
eriodic Table has certain properties which
stinguish it from the other elements in the
oup.' Show how this statement is illustrated
the elements lithium and beryllium and
eir compounds with regard to Groups I and
respectively.

*

3. What do you understand by the term *ectronegativity of an element*? Explain how
e electronegativity varies according to the
osition of the element in the Periodic Table.
riefly account for the so-called *diagonal
lationships*, using the properties of the
ements beryllium and aluminium and of
eir compounds to illustrate your answer.

*

4. Compare the chemistry of boron and
uminium, with particular reference to the
ydrolysis of their chlorides.

5. Aluminium is generally regarded as a
etal. How does this statement conflict with
e properties displayed by the element, its
xide, and its chloride?

6. Account for the exceptionally high en-
alpy of hydration of the trivalent aluminium
on.

7. Why do the halides of boron and alu-
inium function as Lewis acids? Illustrate
our answer by referring to specific examples.

*

8. Briefly discuss the inert pair effect by
onsideration of the chemistry of (a) gallium,
dium, and thallium and (b) germanium, tin,
nd lead.

*

9. In what ways are the oxides and chlorides
f carbon not typical of those of the other
roup IV elements?

0. Carbon dioxide and silicon(IV) oxide are
oth covalent oxides but, whereas carbon
dioxide is a gas, silicon(IV) oxide is a solid
with a high melting point. Explain this statement.

51. Briefly explain why tetrachloromethane is not readily hydrolysed by water whereas the tetrahalides of the other Group IV elements are.

52. By referring to the properties and bond types of their oxides and chlorides, compare and contrast the chemistry of the elements germanium, silicon, tin, and lead.

53. Tin(IV) chloride is more stable than tin(II) chloride, whereas lead(IV) chloride is less stable than lead(II) chloride. Explain this observation.

54. Lead(IV) oxide is a strong oxidizing agent. On the other hand, tin(II) oxide is readily converted to the more stable tin(IV) oxide. How can this be explained?

55. By referring to the elements of Group IV show how metallic character increases as the Group is descended. Illustrate your answer by considering:

(a) the physical properties of the elements.
(b) the relative stabilities of the +4 oxidation state.
(c) the thermal stabilities of the hydrides.
(d) the reaction of the oxides with aqueous alkali.
(e) the hydrolysis of the chlorides.

56. Account for the great solubility of ammonia in water.

*

57. Outline the differences which exist between the hydrides and chlorides of nitrogen and phosphorus.

58. The basic strength of the trihydrides of the Group V elements decreases as the group is descended. Account for this observation.

59. How and under what conditions does ammonia react with (a) potassium, (b) oxygen, (c) chlorine, (d) copper(II) oxide, and (e) sodium chlorate(I) (hypochlorite)?

60. Write equations to illustrate the behaviour of ammonia as (a) a base and (b) a reducing agent.

61. With reference to selected oxides of nitrogen, outline the properties and the nature of the bonding in odd electron molecules.

62. (a) What do you understand by the term *an odd-electron molecule*?
 (b) Give two properties which are characteristic of odd-electron molecules.
 (c) What explanation can you provide for the fact that NO_2 is brown, but pure undissociated N_2O_4 is colourless.

63. Explain why phosphorus forms a compound phosphorus pentachloride whereas nitrogen forms no corresponding compound.

64. Show how the change from non-metallic to metallic character is illustrated by the properties of the Group V elements (N to Bi) and by their compounds.

*

65. In many respects, water and hydrogen sulphide differ quite markedly in both physical and chemical properties. Comment on this statement, illustrating your answers with suitable examples.

66. How is hydrogen peroxide manufactured on an industrial scale? Give a brief account of the reactions of hydrogen peroxide as both an oxidizing agent and as a reducing agent.

67. Discuss the chemical behaviour of sulphur dioxide in aqueous solution.

68. Give the principal oxidation states of sulphur, and select examples of compounds in which it displays these states.

69. Inspect the Periodic Table and state the names of two elements which you would expect to resemble selenium most closely. Explain the reasons for your choices and indicate whether the properties of selenium are primarily those of a metal, a non-meta or both.

*

70. Explain the difference in behaviour whe hydrogen chloride and ammonia are dissolve separately in water.

*

71. Certain elements show certain 'diagona relationships'. Show how this applies to th following pairs of elements: (a) lithium an magnesium, (b) beryllium and aluminium and (c) oxygen and chiorine.

72. The first element in each group in th Periodic Table tends to have propertie markedly different from those of the othe elements in the group. Discuss this statemen with particular reference to carbon, oxygen and fluorine.

*

73. In what ways does the chemistry c fluorine and its compounds differ markedl from that of the other halogens?

74. Explain why fluorine is the most powerfu oxidizing agent among the halogens, bearin in mind that chlorine has a higher electro affinity value.

75. Compare and contrast the chemistry o fluorine and chlorine, with particular re ference to:

(a) their reactions with water.
(b) the reactions of chlorine with aqueou solutions containing (i) fluoride ions an (ii) bromide ions.
(c) the electrolysis of aqueous solutions cor taining fluoride and chloride ions.

76. Comment on the implications of th relative values of the boiling points of th hydrogen halides.

	HF	HCl	HBr	HI
B.p./°C	+19	−85	−67	−36

77. Using your knowledge of the chemistry of the first four halogens, predict the chemical and physical properties of astatine.

78. Indicate the experimental conditions necessary for hydrogen and chlorine to combine directly. Write the mechanism for the reaction and name the active species participating.

79. Give a brief account of the nature and structure of interhalogen compounds and compare their reactivity with the halogens themselves.

80. Discuss the difference in chemical behaviour of the oxides and oxoacids of chlorine with those of iodine.

81. When dry hydrogen chloride gas is dissolved in dry methylbenzene (toluene) the solution is found to be a non-conductor of electricity. It also fumes in air. Account for these observations.

82. Write a comparative account of the acidic strengths of aqueous solutions of the hydrogen halides.

83. Explain what is meant by the term *disproportionation*, and illustrate your answer by reference to the oxoions of the halogens.

*

84. Despite the fact that the water molecule and the neon atom are isoelectronic, they have vastly different boiling points (water 100 °C; neon −246 °C). How can you account for this disparity in the values?

*

85. Outline the importance of the noble gases in the historical development of chemical thinking.

86. Discuss briefly the historical development of the preparation of compounds of the noble gases. What was the significance of the first ionization energy of the oxygen molecule in obtaining the first synthetic compound of xenon?

87. Give a short account of the chemistry of the compounds of xenon.

*

88. Distinguish between what is meant by a double salt and what is meant by a complex compound. Give examples.

89. Describe four of the various type of isomerism possible in complexes and illustrate your answer by means of appropriate examples.

90. Show how the concept of hybridization can be applied to explain the bonding and stereochemistry of complexes.

Give an example of (a) a linear, (b) a tetrahedral, (c) a square planar, and (d) an octahedral complex.

91. Outline the role of complex formation in inorganic qualitative analysis.

92. Discuss briefly the reactions of the ions of the first row of transition metals in aqueous solution with (a) ammonia and (b) chloride ions.

93. On dissolving cobalt(II) chloride in water, a pink solution A is obtained. The addition of concentrated hydrochloric acid to A gives a blue solution B, which in turn gives a yellow-orange solution C when treated with ammonia solution followed by the bubbling of air through the solution.

Write down the formulae of the species A, B, and C and, in each case, illustrate how the ligands are spatially arranged about the central metal ion.

94. What is the stereochemistry and systematic name of the $[Co(NH_3)_6]^{3+}$ ion? Sketch its structure. Indicate how the $[Co(NH_3)_6]^{3+}$ ion can be obtained in aqueous solution from cobalt(II) chloride. State any necessary reagents and conditions.

95. Give the name, formula, and colour of the species formed when the following reagents are added separately to an aqueous solution containing chromium(III) ions:

(a) dilute alkali.
(b) excess alkali followed by sodium peroxide.
(c) as for (b), followed by boiling to remove excess peroxide and then addition of excess dilute sulphuric acid.

96. State *two* ways in which the chemistry of manganese(II) differs from that of iron(III).

97. Survey the stereochemical features of the complexes of the first row of transition metals, paying particular attention to those complexes formed with the following ligands water, ammonia, and chloride ions.

98. By inspection of the Periodic Table, predict the principal oxidation states displayed by chromium, manganese, and iron. Explain your answer in terms of electronic structure.

99. Compare and contrast the chemistry of chromium and manganese, with special reference to (a) the electronic structures of the elements, (b) the stabilities of their principal oxidation states, and (c) the chemistry of their oxides.

100. Discuss paramagnetic and diamagnetic behaviour, selecting your examples from the first row of transition elements.

Answers to Multiple-Choice Questions

Question	Answer
1	B
2	D
3	A
4	D
5	E
6	C
7	D
8	E
9	C
10	B
11	C
12	E
13	B
14	C
15	B
16	B
17	A
18	D
19	C
20	A
21	C
22	A
23	A
24	B
25	B
26	B
27	C
28	E
29	D
30	E
31	C
32	B
33	E
34	C
35	A
36	E
37	E
38	D
39	B
40	C
41	A
42	A
43	B
44	D
45	B
46	D
47	D
48	C
49	C
50	E
51	E
52	B
53	A
54	A
55	B
56	A
57	D
58	C
59	D
60	A
61	B
62	B
63	E
64	D
65	B
66	B
67	B
68	D
69	C
70	E
71	D
72	A
73	D
74	E
75	E
76	B
77	C
78	B
79	A
80	E
81	E
82	D
83	A
84	E
85	E
86	E
87	B
88	A
89	E
90	E
91	C
92	C
93	C
94	A
95	C
96	B
97	B
98	B
99	D
100	E
101	E
102	B
103	C
104	A
105	A
106	A
107	D
108	B
109	C
110	B
111	D
112	C
113	D
114	D
115	D
116	B
117	C
118	B
119	A
120	C
121	E
122	E
123	D
124	E
125	E
126	C
127	C
128	C
129	D
130	C

Table of Useful Data

Physical constant	*Symbol*	*Value*
Speed of light in vacuum	c, c_0	$2.997\,925 \times 10^{8}$ m s^{-1}
Unified atomic mass constant	$m_u = \frac{m_a(^{12}C)}{12}$	$1.660\,531 \times 10^{-27}$ kg
Mass of proton	m_p	$1.672\,614 \times 10^{-27}$ kg
Mass of neutron	m_n	$1.674\,920 \times 10^{-27}$ kg
Mass of electron	m_e	$9.109\,558 \times 10^{-31}$ kg
Charge of proton	e	$1.602\,192 \times 10^{-19}$ C
Faraday constant	F	$9.648\,670 \times 10^{4}$ C mol^{-1}
Gas constant	R	8.314 34 J K^{-1} mol^{-1}
Avogadro constant	L, N_A	$6.022\,169 \times 10^{23}$ mol^{-1}
Planck constant	h	$6.626\,196 \times 10^{-34}$ J s
Rydberg constant for H atom	R_H	$1.096\,776 \times 10^{7}$ m^{-1}
Boltzmann constant	k	$1.380\,622 \times 10^{-23}$ J K^{-1}
Bohr magneton	$m_B = \frac{eh}{4\pi m_e}$	$9.274\,096 \times 10^{-24}$ A m^{2}

Index